PLANT WEALTH OF THE LOWER GANGA DELTA

(An Eco-Taxonomical Approach)

Kumudranjan Naskar

Volume II

1993
DAYA PUBLISHING HOUSE
DELHI - 110035

ISBN 81-7035-117-0 (Set)
ISBN 81-7035-119-7 (Vol. II)

Published by:
Anil Mittal for **Daya Publishing House,**
1123/74, Deva Ram Park, Tri Nagar,
Delhi - 110035. Ph: 7231826.

Laser Typeset by: ShiKi Associates, Ph: 5505295.

Printed at: **Raj Offset Printers,** Delhi.

Bound at: Book Wrap, Delhi-35.

PRINTED IN INDIA

CONTENTS

Vol. I

Vol. II

LXXIV. ASTERACEAE

(*nom. alt.* Compositae)

Key to the Genera

1a	An recently introduced exotic herb; profusely branched and grow gregariously; poisonous to the cattle; heads small, much branched, in terminal or lateral branches; flowers small, white	32. *PARTHENIUM* L.
1b	Generally indigenous or if exotic that introduced very earlier period; generally not much poisonous; heads generally larger than the above and flowers mostly coloured	(2)
2a (1)	Plants armed with spines; heads rosy or white with spiny bracts	10. *CIRSIUM* Adans.
2b (1)	Plants generally free from any spines; heads variable coloured and without spiny bracts	(3)
3a (2)	Herbs with usually milky juice; all flowers ligulate	(4)
3b (2)	Herbs without milky juice; inner flower heads not ligulate	(6)
4a (3)	Stout fistular erect herbs with more or less spinulose serrate leaves and sub-corymbose heads	37. *SONCHUS* L.
4b (3)	Stem slender, often spreading or procumbent; leaves all radical	(5)
5a (4)	Achenes compressed or flattened, ovoid, oblong or narrow	27. *LACTUCA* L.
5b (4)	Achenes truncate at the base as well as at the apex	30. *LAUNAEA* Cass.
6a (3)	Style-arms long, distinct, or if very short or the style sub-entire, these are only in the sterile flowers or heads with dissimilar flowers	(7)

6b (3)	Style arms very short, hairy or thickened towards the base or the style subentire in all the florets, which are similar and tubular to the deeply 5-fid mouth	(46)
7a (6)	Flowers red, purple or white but never yellow; homogamous and tubular, or rarely cleft laterally, involucre of bracts always more than 1-seriate; pappus present, usually setaceous, raceptacle naked or rarely paleaceous	(8)
7b (6)	Flowers yellow, florets homogamous or heterogamous	(14)
8a (7)	Leaves alternate, radical; heads pink or purple	(9)
8b (7)	Leaves opposite; heads blue, pink or white	(11)
9a (8)	Stem leafy; heads many flowered	43. *VERNONIA* Schrab.
9b (8)	Leaves mostly radical; heads few flowered	(10)
10a (9)	Pappus straight	17. *ELEPHANTOPUS* L.
10b (9)	Pappus of 2 twice curved bristles	34. *PSEUDO-ELEPHANTOPUS* Rohr.
11a (8)	Anthers truncate at apex	1. *ADENOSTEMMA* Forsst.
11b (8)	Anthers appendages at apex	(12)
12a (11)	Pappus paleaceous	2. *AGERATUM* L.
12b (11)	Pappus of slender hairs	(13)
13a (12)	Erect shrubs; bracts numerous	21. *EUPATORIUM* L.
13b (12)	Exotic climber; bracts-4	31. *MIKANIA* Willd.
14a (7)	Anthers appendaged at the apex	(15)
14b (7)	Anthers not appendaged at the apex; raceptacle not paleaceous; pappus absent or reduced to a raised rim, rarely scaly and short	(44)

15a (14)	Receptacle naked, smooth or foveolate	(16)
15b (14)	Receptacle paleaceous; anthers sub-entire at the base	(31)
16a (15)	Leaves alternate; involcure bracts many-seriate	(17)
16b (15)	Leaves opposite or alternate; involcure bracts 1-seriate	(28)
17a (16)	Flower-heads heterogamous; style-arms flattened or plano-convex, all or at least those of the disk-florets, tipped by a cons; anther sub entire at the base	(18)
17b (16)	Flower heads both heterogamous or homogamous; style-arms filiform, linear or obtuse, or those of the sterile florets undivided; anthers cleft at the base or rarely with bases, sub entire of the hermophrodite florets filiform	(21)
18a (17)	Flower-heads without a proper ray; pappus generally absent	(19)
18b (17)	Flower-heads with distinct ray-florets	(20)
19a (18)	Achenes minute, oblong, smooth; pappus absent	14. *CYATHOCLINE* Cass.
19b (18)	Achenes flattened or sub-terete, with a terminal toothed or bristly pappus ring	24. *GRANGEA* Adans.
20a (18)	Ray-florets ligulate, ligules 2-3 seriate, neither yellow nor orange	20. *ERIGERON* L.
20b (18)	Ray-florets slender, tubular or with very short ligules; yellow or pinkish	11. *CONYZA* Less.
21a (17)	Female florets if present, filiform	(22)
21b (17)	Female florets if present, ligulate; heads heterogamous, but florets usually all fertile, with linear, style arms rounded or dilated at their tips, receptacle naked	(27)

22a (21)	Style-arms of hermaphrodite florets filiform; flower-heads androgynous	(23)
22b (21)	Style arms of hermaphrodite florets truncate; bracts of the involcure hyaline	(26)
23a (22)	Flower-heads medium, separated, solitary, in corymbs or panicles, not in globose clustered, then the achenes with a copious soft pappus	(24)
23b (22)	Flower-heads small, in dense globose or ovoid masses; herbs with winged stems; pappus absent	38. *SPHAERANTHUS* L.
24a (23)	Herbs; bracts of the involucre narrow; flowers not corymbose	(25)
24b (23)	Shrubs; bracts of the involucre broad; flowers corymbose	33. *PLUCHERA* Cass.
25a (24)	Anther-cells tailed at the base, the tails of adjacent anthers confluent	4. *BLUMEA* DC.
25b (24)	Anther-cells sub-entire at the base or if tailed the tails short and not united	29. *LAGGERA* Sch. ex Oliv.
26a (22)	Flower-heads many flowered; heterogamous. disciform	23. *GNAPHALIUM* L.
26b (22)	Flower-heads 1-flowered, heads homogamous	5. *CAESULIA* Roxb.
27a (21)	Achenes faintly ribbed; flowers usually rayed; pappus hairs all slender, those or ray florets few or none	44. *VICOA* Cass.
27b (21)	Achenes distinctly ribber; flowers either rayed or discoid; pappus of outer florets hair like	35. *PULICARIA* Gaertn.
28a (16)	Leaves alternate; pappus of fine hairs usually soft and generally copious	(29)
28b (16)	Leaves opposite; pappus paleaceous; heterogamous	(30)

29a (28)	Heads all homogamous; florets all purple	18. *EMILIA* Cass.
29b (28)	Heads rayed; heterogamous florets all yellow	36. *SENECIO* L.
30a (28)	Pappus 5-6 scales	41. *TAGETES* L.
30b (28)	Pappus absent, heads simple large, orange	6. *CALENDULA* L.
31a(15)	Anthers free or nearly so; female florets all apetalous; heads 1-sexual	46. *XANTHIUM* L.
31b (15)	Anthers united in a tube	(32)
32a (31)	Heads 1-flowered, in globose clusters; bracts of the involucre forming a 5-fid tube	28. *LAGASCEA* Cav.
32b (31)	Heads many flowered	(33)
33a (32)	Pappus consisting of only 1-4 bristly awns or cup-like or absent	(34)
33b (32)	Pappus of numerous scales, feathery; heads radical, medium	42. *TRIDAX* L.
34a (33)	Corollas of the fertile florets persistent on the achenes; pappus of 1-3 awans	47. *ZINNIA* L.
34b (33)	Corollas of all the florets deciduous	(35)
35a (34)	Achenes all thick, or those of the rayflorets 3-cornered, and those of the disk laterally compressed, pappus cup like or composed	(36)
35b (34)	Achenes more or less depressed from the top; pappus 2 or rarely 3-4, bristles or absent	(41)
36a (35)	Inner-bracts of the involucre embracing and enclosing the achenes of the fertile ray-florets	19. *ENHYDRA* Lour.
36b (35)	Inner-bracts of the involucre all flat	(37)
37a (36)	Scales of the receptacle all flat, very narrow, usually few, disc-florets 4-toothed	16. *ECLIPTA* L.

37b (36)	Scales of the raceptacle concave or less enclosing and embracing the disc-florets	(38)
38a (37)	Achenes wingless, compressed or 4-5 cornered	(39)
38b (37)	Achenes of the disc ciliate or winged on the margins, laterally compressed; heads absent	39. *SPILANTHES* L.
39a (38)	Pappus united at the base in a ring or cup; flower-heads small or medium; ray-florets fertile	(40)
39b (38)	Pappus scales or awans free from the base; flower-heads large; ray-florets sterile	26. *HELIANTHUS* L.
40a (39)	Ray-florets white with small ligules	3. *BLAINVILLEA* Cass.
40b (39)	Ray-florets yellow, ligules large; disc achenes with 1-2 short, deciduous awans	45. *WEDELIA* Jacq.
41a (35)	Outer bracts of the involucre almost equal, inner bracts separate, almost resembling the scales of the receptacle; ray florets fertile	(42)
41b (35)	Outer bracts of the involucre few, small inner bracts connate below, membranous	(43)
42a (41)	Achenes almost 4-cornered without a pappus	25. *GUIZOTICA* Cass.
42b (41)	Achenes flat, margins lacerate and winged, pappus present	40. *SYNEDRELLA* Gaertn.
43a (41)	Style-arms truncate, penicillate or crowned by a short appendage, ray-florets sterile	12. *COSMOS* Cav.
43b (41)	Style arms ending in a long, shortly hairy appendages; ray-florets fertile; achenes long, crowned with 2-3 stiff	22. *GLOSSOGYNE* Cass.
44a (14)	Flower heads radiate	9. *CHRYSANTHEMUM* Rich.

44b (14)	Flower heads discoid, heteogamous, pappus absent	(45)
45a (44)	Heads peduncled; bracts of the involucre 1-2 seriate	13. *COTULA* L.
45b (44)	Heads sub-sessile; bracts of the involucre 2-seriate	8. *CENTIPEDA* Lour.
46a (6)	Flower-heads 1-flowered, crowded into small spherical balls	15. *ECHINOPS* L.
46b (6)	Flower-heads many flowered, separate; achenes glabrous; inserted in the very oblique or quite lateral areoles of the raceptacle	7. *CARTHAMUS* L.

1. *ADENOSTEMMA* Forsst.

1.1. ***Adenostemma lavenia*** (L.) Kuntze, Rev. Gen. Pl. 304.1891; Naskar, Aq. & Semi-Aq. Plants 149.1990. *(Vern. Name: Buro-Keshuti).*

Erect annual herb. Leaves linear or oblong-lanceolate to broadly ovate, cordate, sessile or petioled, crenate or serrate.

Naskar (1.c.) described this plant in respect of its taxonomy, ecology and economic importances.

2. *AGERATUM* L.

2.1. ***Ageratum conyzoides*** L. Sp. Pl. 839. 1753; Hook.f. Fl. Brit. India 3: 243. 1881; Prain Beng. Plants 1: 433. 1903 (Rep. ed. 1963). A. *cordifolium* Roxb. Fl. Ind. 3: 415.1832. *(Vern. Name: Oochunti).*

Erect, terrestrial, annual herb; leaves opposite, cordate-ovate. Head small in dense terminal corymbose, ray-florets pale-blue or white, anther appendaged at apex. Pappus of 5 basally connate paleaceous scales. Achenes sharply angled, sometime glandular, black. *Fls. & Frts.:* Throughout the year, but mostly in March-June.

Very common as a weed in the cultivated fields, along the bank of canals and crevices of old wall.

Field Collections: Canning, *K.N. - 125;* Surjyapur, *K.N. - 299.*

Distributions: Well naturalized throughout India. Native of South America, but at present pantropic; all the tropical countries.

Economic Importances: Leaves used in the wounds and sores; it is also considered as anti-dysenteric agent and also used in the skin diseases, particularly in leprosy.

3. *BLAINVILLEA* Cass.

3.1. ***Blainvillea acmella*** (L.) Philipson, Blumea 6: 350. 1950; Naskar, Aq. & Semi-Aq. Plants 150.1990.

Annual 0.3-0.6 m high. Naskar (1.c.) described this taxon in respect of its taxonomy, ecology and economic importances after collecting from the lower Ganga Delta regions.

4. *BLUMEA* DC.

Key to the Species

1a	Heads numerous; pappus white	(2)
1b	Heads solitary and peduncled on the ends of the branches; flowers all solitary yellow	(6)
2a (1)	Head more or less clustered to form dense, oblong spikes or contracted penicles at the top of the stem	(3)
2b (1)	Heads in open corymb or panicles with divaricate branches; corrolla always yellow	(5)
3a (2)	Corolla purplish; stem erect, sub-simple, very leafy	7. *B. mollis* (D.Don) Merr.
3b (2)	Corolla yellow	(4)
4a (3)	Heads with purplish bracts to the involucre, in small sessile, axillary clusters along the slender branches of the much branched stem	3. *B. fistulosa* (Roxb.) DC.
4b (3)	Heads with greyish bracts to the involucre, collected in dense terminal panicles, stem branched	4. *B. lacera* (Burm.f.) DC.
5a (2)	Leaves on upper part of the stem, few, large, laciniate	5. *B. liciniata* (Roxb.) DC.
5b (2)	Leaves on upper part of stem, many, small, toothed	6. *B. membranacea* DC.

6a (1)	Lobes of the hermophrodite corolla hairy; flowers all solitary	(7)
6b (1)	Lobes of the hermophrodite corolla glabrous, flowers solitary peduncled of few corymbose, pappus white	8. *B. oxyodonata* DC.
7a (6)	Pappus reddish; leaves dentate, all alternate	1. *B. amplectens* DC.
7b (6)	Pappus white; leaves serrate uppermost opposite	2. *B. bifoliata* (L.) DC.

4.1. ***Blumea amplectens*** DC., Wight Contrb. 13. 1834; Hook.f. Fl. Brit. India 3: 250.1881; Prain Beng. Plants 1: 438.1903 (Rep. ed. 1963). *B. obliqua* (L.) Druce.; Datta & Majumder, Bull. bot. Soc. Bengal 20(2): 116. 1966.

Small bushy herb. Stem terete, divaricately branched, more or less hairy. Leaves alternate, dentate. Flower-heads terminal, solitary, yellow, peduncles hairy. Fruits suberect achene; achenes minute, oblong, compressed, minute hairy. Pappus reddish or pink. *Fls. & Frts.:* November-January.

Small bushy herb on the river banks; village thickets.

Field Collection: Duttapukur, *K.N. - 1707.*

Distributions: More or less throughout India; cosmopolitan.

4.2. ***Blumea bifoliata*** (L.) DC., Wight. Contrib. 14. 1838; Hook.f. Fl. Brit. India 3: 261.1881; Prain Beng. Plants 1: 438.1903 (Rep. ed. 1963).

Prostrate, bushy herb. Stem terete, branched from the base. Leaves sub-opposite, serrate, elliptic oblong, acute sessile. Heads 1 cm broad or more, solitary, long peduncles yellow. Fruit angled, achene, pappus white. *Fls. & Frts.:* November-April.

Common on the waste lands, road-sides and railway tracks.

Field Collection: Malancha, *K.N. - 2154.*

Distributions: Almost throughout India; cosmopolitan

4.3. ***Blumea fistulosa*** DC. Prodr. 5: 442.1836. *Conyza fistulosa* Roxb. Fl. Ind. 3: 429.1832. *B. glomerata* DC., Wight. Contrib. bot. Ind.15. 1834; Hook.f. Fl. Brit. India 3: 262.1881; Prain Beng. Plants 1: 438.1903 (Rep. ed. 1963).

Branched, pubescent herbs, young shoots villous. Basal leaves oblanceolate or obovate or elliptic-oblong, minutely incised and dentate. Heads sessile 6 mm diam., in compact clusters, forming interrupted spikes, involucral bracts, bright purple, outer short and linear-lanceolate. Achenes brown, oblong, puberulous, 4-angled, pappus upto 8 mm long, white. *Fls. & Frts.:* December-April.

Common in the waste places, roadsides and railway tracks.

Field Collection: Budge Budge, *K.N. - 752.*

Distributions: Eastern parts of India. Hotter parts of the Old World.

4.4. ***Blumea lacera*** (Burm.f.) DC., Wight Contrib.bot. Ind. 14. 1834; Hook.f. Fl. Brit. India 3: 263. 1881; Watt Dict. Econ. Prod. 6(1): 459.1883; Prain Beng. Plants 1: 438.1903 (Rep. ed. 1963). *Conyza lacera* Burm.f. Fl. Ind. 180. t. 59. f. 1.1768. *B. leptocladea* Dalz. & Gibs, 126 (not of DC.). *(Vern. Name: Kukur-sunga* or *Kukur-mita).*

Erect annual, aromatic, branched herb. Leaves alternate with glandular hairs, toothed serrate. Flower head yellow, numerous, in short axillary cymes and terminal speciform panicles. Achenes not ribbed, small oblong, glabrate, sub-tetragonous. Pappus white. *Fls. & Frts.:* November-May.

Common in the road-sides and railway tracks.

Field Collection: Arbelia/Dhanyakuria, *K.N. - 554.*

Distributions: Almost throughout India. Malaya, China and Tropical Africa.

Economic Importances: The root mixed with black pepper is given in cholera. Juice of the leaves are used in blood disease and in piles. Commonly used as hedge plant in the village gardens.

4.5. ***Blumea laciniata*** (Roxb.) DC. Prodr. 5: 436.1834: Hook.f. Fl. Brit. India 3: 264.1881; Naskar Aq. Semi-Aq. Plants 150. 1990.

Naskar (1.c.) described this taxon after colleting from the Lower Ganga Delta (Frazergunj, *K.N. - 1910).*

4.6. ***Blumea membranacea*** DC. Prodr. 5: 440.1836; Hook.f. Fl. Brit. India 3: 265.1881; Prain Beng. Plants 1: 438.1903 (Rep. ed. 1963). *Conyza diffusa* Roxb. Fl. Ind. 3: 427.1832.

Slender, much-branched, annual herb. Leaves thinly membranous. Heads numerous, yellow small or slender glandular-hairy, peduncles. Pappus white. Achenes 8-10 ribbed, hairy. *Fls. & Frts.:* January-May.

Commonly found as weed in the waste lands and garden edges.

Field Collection: Namkhana, *K.N. - 1917.*

Distributions: Throughout India. Sri Lanka and Malay Islands.

4.7. ***Blumea mollis*** (D.Don.) Merr., Philip. Journ. Sci. (Bot)5: 395. 1910; Randeria, Blumea 10: 261.1960. *Erigeron molle* D. Don., Prodr. Fl. Nepal. 172.1825. *Blumea wightiana* DC., Wight Contrib. bot. Ind. 14. Ind. 14.1834; Hook.f. Fl. Brit. India 3: 261.1881; Prain Beng. Plants 1: 438.1903 (Rep. ed. 1963). *B. neilgherriensis* Hook. f. Fl. Brit. India 3: 261.1881.

Erect, villous, leafy herb. Leaves petioled, 3-6 cm long, ovate, toothed or serrate. Heads many, in terminal or axillary cymes or panicles. Fruits 4-5 angled; achenes sparsely hairy: pappus white. *Fls. & Frts.:* November-April.

Frequently found in waste and unused grounds.

Field Collection: Basanti. *K.N. - 2276.*

Distributions: Almost throughout India. Tropical countries.

4.8. ***Blumea oxyodonata*** DC., Wight, Contrib. bot. Ind. 15. 1884; Hook.f. Fl. Brit. India 3: 266.1881; Prain Beng. Plants 1: 438.1903 (Rep. ed. 1963).

Erect, much-branched, tomentose or shortly villous annual. Leaves often crowded at base, elliptic-obovate or obovate-oblong, rounded or shortly narrowed at base, acute to rounded, apiculate at apex, distantly toothed. Heads terminal, sub-solitary or in lax panicles, involucral bracts slightly longer than the florets, 3-5 cm long, linear, acute. Achenes oblong, brown pubescent, not ribbed of central florets mostly abortive. Pappus 3-4mm long, white, deciduous. *Fls. & Frts.:* February-May.

Common in the road-sides and waste places.

Field Collection: Titagarh, *K.N.- 2214.*

Distributions: Common throughout India; China and Philippines

5. *CAESULIA* Roxb.

5.1. ***Caesulia axillaris*** Roxb. Pl. Cor. 1: 64 .t. 93.1798; Naskar, Aq. & Semi-Aq. Plants 151. 1990.

Naskar (1.c.) described this taxon after collecting from the lower Ganga Delta (Santoshpur/near Budge Budge, *K.N. - 1955).*

6. *CALENDULA* L.

6.1. ***Calendula officinalis*** L. Sp. Pl. 921.1753; Watt Dict. Econ. Prod. 2: 24. 1889. *(Vern. Name: Genda).*

Exotic horticultural herb. Heads heterogamous. Ray florets female, some disc-florets sterile; anther tips with lanceolate flap. Fruits polymorphic. *Fls. & Frts.:* January-April.

Commonly cultivated in the ornamental gardens for its showy flowers.

Field Collection: Santoshpur, *K.N. - 1963.*

Distributions: Throughout India. Native of South Europe.

Economic Importances: It has an excellent property for the remedy on ordinary cuts and bruises, also largely used in surgical dressing. Flowers also very showy.

7. *CARTHAMUS* L.

7.1. ***Carthamus tinctorius*** L. Sp. Pl. 830.1753; Hook.f. Fl. Brit. India 3: 386.1881; Watt Dict. Econ. Prod. 2: 184.1883; Prain Beng. Plants 1: 460.1903 (Rep. ed. 1963). *(Vern. Name: Kusum-Bech).*

Cultivated thistle-like spinous herb. Leaves alternate, rigid, spinescent. Flower-heads large, sub-corymbose, homogamous, fertile. Seeds white; directly edible in fried conditon. *Fls. & Frts.:* November-March.

A cold weather field crops, in general.

Field Collection: Bahuru, *K.N. - 1622.*

Distributions: Cultivated throughout the larger parts of India. Native of the Old World Tropics; cultivated in Spain, Germany, Italy, Persia, China, Egypt, America and S.Russia.

Economic Importances: An essential oil is extracted from the seeds. The seed oil is used for paints, soap, varnishes, linoleum and an illuminant properties. Oil is also edible; oil cake edible for cattle. Flowers used as dye for colouring food and clothes.

8. *CENTIPEDA* Lour.

8.1. ***Centipeda minima*** (L.) A. Br. & Aschers., Ind. Sm. Hort. Berol. App. 6.1867. *Artemisia minima* L. Sp. Pl. 849.1753. *Centipeda orbicularis* Lour. Fl. Cochinch. 2: 493.1790; Hook.f. Fl. Brit. India 3: 317.1881; Prain Beng. Plants 1: 456.1903 (Rep. ed. 1963). *(Vern. Name: Mechilta).*

Small, diffuse perennial weed. Leaves, alternate obovate, toothed. Flower heads, white, solitary, heterogamous, sessile; bracts of the involucre 2-seriate. Pappus absent. Achenes 4-angled minute. *Fls. & Frts.:* January-August.

Frequently found in waste places.

Field Collection: Throughout India, in moist places. Eastern Asia to China, Malaya. Tropical Australia and Polynesia.

Economic Importances: Seeds are curative. Leaves and seed powder used as colds snuff.

9. *CHRYSANTHEMUM* Rich.

9.1. ***Chrysanthemum coronarium*** L. Sp. Pl. 889.1753; Hook.f. Fl. Brit. India 3: 314. 1881; Prain Beng. Plants 1: 455.1903 (Rep. ed. 1963). *(Vern. Name: Chandra Mallika).*

Perennial herbs with sub-aerial stem. Leaves alternate pinnatifid. Flower heads large, terminal, heterogamous, radiate; bracts broad, receptacle not paleaceous; anthers not appendaged at the apex. Pappus very short, cupular. *Fls. & Frts.:* December-April.

Common cultivated as a beautiful ornamental garden flowers.

Field Collections: Calcutta, *Debbarman-1* (CAL).

Distributions: Cultivated throughout India. Native of Mediterranean region.

Economic Importance: Common ornamental garden plants.

10. *CIRSIUM* Adans.

10.1. ***Cirsium arvense*** (L.) Scop. Fl. Garn. 2: 126.1772.*Serratula arvensis* L. Sp. Pl. 820.1753. *Cnicus arvensis* Hoffm., Deutschi. Fl.1(2): 130; 1804; Hook.f. Fl. Brit. India 3: 362.1881; Prain Beng. Plants 1: 458.1903 (Rep. ed. 1963). *(Vern. Name:. Silkanta).* (Fig.30).

Erect leafy herb. Leaves sessile alternate, serrate or pinnately lobed. Flower-heads homogamous, peduncled or sub-sessile: anthers with sagittate base. Furit 4-angled achene; achenes smooth, shinning. Pappus white or light brown, feathery. *Fls. & Frts.:* October -December.

Commonly found as a weed of cultivated fields and waste lands.

Field Collection: Rahara, *K.N. - 1379.*

Distributions: Throughout India. North temperate regions.

11. *CONYZA* Less. *nom. cons.*

11.1. ***Conyza semipinnatifida*** Wall. Cat. 3058. 1829-1832: Majumder, Ind. Agr. 6 (1&2): 142.1962; Naskar & Guha Bakshi, Mangr. Swamps Sundarbans 91.1987; Naskar, Aq. & Semi. Aq. Plants 151.1990.

Naskar & Guha Bakshi (l.c.) and Naskar (l.c.) described this taxon from the lower Ganga delta after collecting from Sarisha/near Diamond Harbour, *K.N. - 1946.*

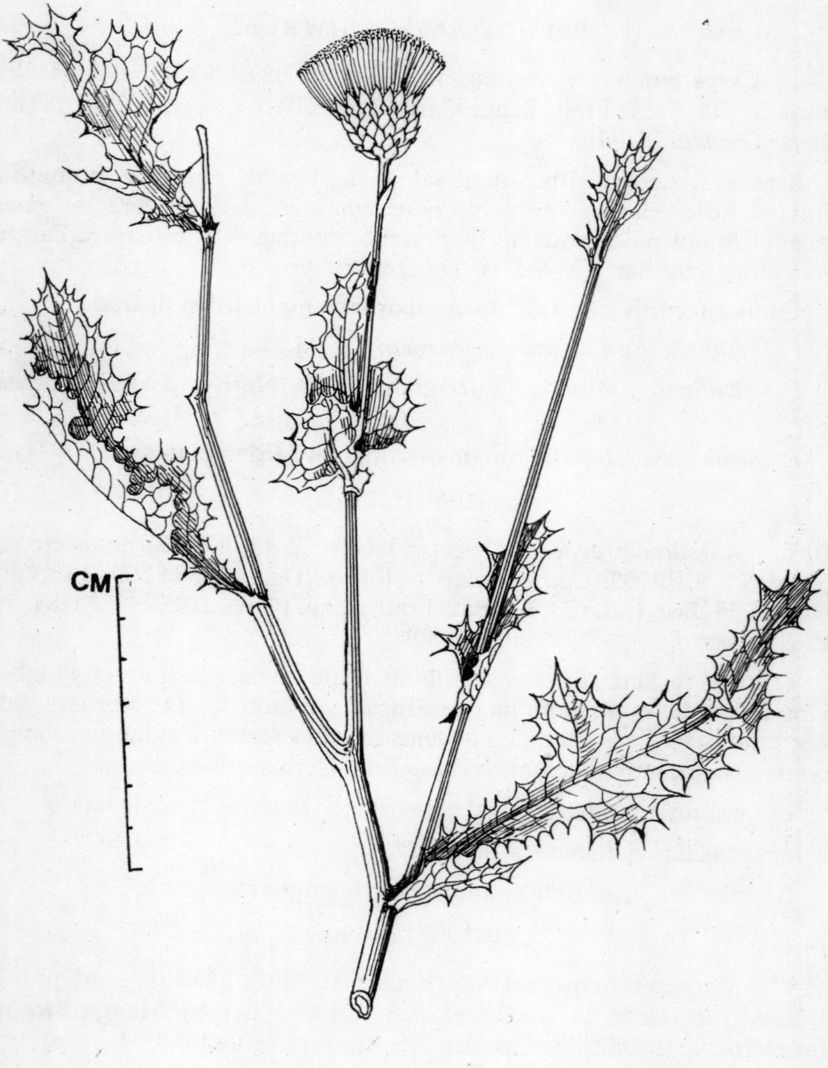

Fig.30. CIRSIUM ARVENSE (L.) Scop.
Flowering twig

12. *COSMOS* Cav.

12.1. ***Cosmos bipinnatus*** Cav. Icon. 1: 10.t. 14. Woodr., Journ. Bomb. Nat. 11: 65.1898.*(Vern. Name: Cosmos).*

Exotic, horticultural, erect herb. Leaves opposite 2-3 pinnetisect. Flower-heads large, heterogamous, long-stalked; bracts invlucre uniseriate, ray florets sterile. Achenes depressed from the top. Pappus retrosely barbellate awans. Fls. & Frts.: December -April.

Commonly planted as an ornamental garden plant in the flower garden.

Field Collection: Specimen not preserved.

Distributions: Common throughout India. Native of Mexico.

Economic Importances: Ornamental garden plants. Flowers yeild an orange-red dye.

13. *COTULA* L.

13.1. ***Cotula anthemodies*** L. Sp. Pl. 891.1753; Hook.f. Fl. Brit. Ind. 3. 316.1881; Prain Beng Plants 1: 456.1903 (Rep. ed. 1963).

Small, prostrate, annual herb, usually glabrous. Leaves upto 2 cm long, 1-2 pinnatisect or 1-2 pinnatifid. Heads solitary on the naked filiform peduncles. Achenes of dic-florets with thick narrow wing. *Fls. & Frts.:* January-April.

Frequently found in waste places.

Field Collection: Santoshpur/near Budge, Budge, *K.N. - 1964.*

Distributions: Almost throughout India. In the Tropics.

Economic Importances: Flowers after boiling with oil are used as an external use for rheumatism.

14. *CYATHOCLINE* Cass.

14.1. ***Cyathocline purpurea*** (D.Don) Kuntze, Rev. Gen. Pl. 333. 1891; Naskar Aq.& Semi-Aq. Plants 252.1990.

Erect annual herb; Naskar (l.c.) described this taxon after collecting from Baruipur, the lower Ganga Delta.

15. *ECHINOPS* L.

15.1. ***Echinops echinatus*** Roxb . Fl. Ind. 3: 447.1832; Hook f. Fl. Brit. India 3: 358.1881; Prain Beng. Plants 1: 457.1903 (Rep. ed. 1963).

Erect, much branched white, tomentose herb. Leaves alternate pinnatifid. Flower-head globose white, stipitate, bracts of the hermaphrodite flowers stiff spinescent. Cypsela elongate with connate bristles. *Fls. & Frts.:* January-April.

Commonly found in waste places; usually in the sandy areas.

Field Collection: Falta *K.N. - 1992.*

Distributions: Throughout India and Afghanistan, Pakistan.

Economic Importances: Entire plant has the diuretic and nerve tonic properties. Largely used in cough, indigestion and opthalmia. Root powder is used to the wounds of cattles to destroy maggits.

16. *ECLIPTA* L. *nom. cons.*

16.1. ***Eclipta alba*** (L.) Hassk. Pl. Jav. Rar. 528.1848; Naskar Aq. & Semi Aq. Plants 152.1990.*(Vern. Name: Kesut).*

Naskar (l.c.) described this taxon in respect of its taxonomy, ecology and economic importances. It has many medicinal values.

17. ELEPHANTOPUS L.

17.1. ***Elephantopus scaber*** L. Sp. Pl. 814.1753; *emend* Philipson, J. bot., 76. 302.1928; Watt Dict. Econ. Prod. 3: 208.1890; Hook.f. Fl. Brit. India 3: 242.1881; Prain Beng. Plants 1: 433.1903 (Rep. ed. 1963). *E. scabber* L. var. *typicus* Koster, Blumea 1: 458.1935. *(Vern. Name: Samdulun).*

Rigid herb. Leaves alternate. Heads heterogamous of 2-5 flowers, collected into a head like clusters. Involucre compressed; bracts about 8, dry; stiff, alternately flat and conduplicate. Florets purple, all equally 4-lobed and cleft on one side and with the lobes spreading somewhat palmately. Achenes truncate,10-ribbed, pappus bristly, rigid shining, dilated below. *Fls. & Frts.:* August-March.

Commonly grows in the orchards.

Field Collection: Basanti, *K.N. - 1283.*

Distributions: Almost throughout India. Pantropic or in the Tropics. An American weed; also in Australia and Tropical Asia.

Economic Importances: Used for curing urinary discharges.

18. *EMILIA* Cass.

18. 1. *Emilia sonchifolia* (L.) DC.,Wt. Contr. bot. 24.1834; Naskar Aq.& Semi-Aq. Plants 153.1990. *(Vern. Name: Sadi-modi).*

Naskar (l.c.) described this taxon from the lower Ganga delta.

19. *ENHYDRA* Lour.

19.1. *Enhydra fluctuans* Lour. Fl. Cochinch. 511.1790; Naskar, Aq. Semi-Aq. Plants 154.1990.*(Vern. Name: Hingcha).*

Profusely branched repent aquatic herb. Leaves are laxative and having several medicinal values. Naskar (l.c.) described this taxon after collecting from the lower Ganga delta.

20. *ERIGERON* L.

20.1. ***Erigeron sublyratus*** DC., Wight, Contrib. 9: 1834. *E. asteroides* Roxb. Fl. Ind. 3: 432.1832; Naskar Aq. & Semi-Aq. Plants 155. 1990. Naskar (l.c.) described this taxon after collecting from the lower Ganga delta.

21. *EUPATORIUM* L.

21.1. ***Eupatorium odoratum*** L. Syst. (ed.10) 1205.1763; Hook.f. Fl. Brit. India 3: 244.1881; Prain Beng. Plants 1: 434.1903 (Rep. ed. 1963).

Exotic coarse herb. Leaves opposite, densely gland-dotted, ovate, acuminate serrate. Heads in terminal corymbose, pale-blue, pale-lavender or white. Fruits 5 angled. Pappus rigid hairy, dull white, minutely barbellate. *Fls. & Frts.:* March-June.

Common on the road-sides, scrub-jungles and waste places.

Field Collection: Subuddhipur/Baruipur, *K.N.* - 2.

Distributions: Almost throughout India. Pantropic, but the Tropical America.

22. *GLOSSOGYNE* Cass.

22.1. ***Glossogyne pinnatifida*** DC., Wt. Contr. 19.1834; Hook.f. Fl. Brit. India: 310.1881; Prain Beng. Plants 1: 453.1903 (Rep. ed. 1963). *Zinnia bidens* Roxb. Fl. Ind. 3: 435.1832. (*Vern. Name: Barangom* Santal).

Perennial glabrous herb, with almost naked stems and branches. Stems numerous arising from a woody root-stock. Leaves mostly radical, 2-5 cm long, pinnatifid, triangular with linear or linear cuneate, acute, segments. Flowering branches mostly naked 10-30 cm long, stiff, grooved, with a small pinnatisect leaf at each fork. Heads few, long-peduncled, yellow, ray-florets female, fetile, ligules spreading; disc-florets 2-sexual, fertile. Achenes deeply groomed, black. *Fls. & Frts.:* September-October.

Frequently found in sandy grounds.

Field Collection: Tapsia, *K.N. - 1757.*

Distributions: In the Gangetic Plain; Sri Lanka.

Economic Importance: A preparation from the root is said to be remedy for snake bite and scorpion stings.

23. *GNAPHALIUM* L.

Key to the Species

1a	Heads leafy	(2)
1b	Heads in leafless, corymbose clusters, golden yellow	2. *G.luteo-album* L.
2a (1)	Heads arranged in simple or branched leafy spikes	1. *G. indicum* L.
2b (1)	Heads in rounded axillary or terminal leafy clusters	3. *G. pulvinatum* Delile

23.1. ***Gnaphalium indicum*** L. Sp. Pl. 852.1753; Naskar Aq. Semi-Aq. Plants 154.1990; Naskar & Guha Bakshi Mangr. Swamps of Sundarbans 154.1987.

Erect annual herb. Naskar (l.c.) and Naskar & Guha Bakshi (l.c.) described this taxon after collecting from the lower Ganga delta.

23.2. ***Gnaphalium luteo-album*** L. Sp. Pl. 861.1753; Hook.f. Fl. Brit. India 3: 283.1881; Watt Dict. Econ. Prod. 3: 517.1890; Prain Beng. Plants 1: 442.1903 (Rep. ed. 1963).

Erect, annual herb. Leaves oblanceolate. Flower heads in leafless, corymbose clusters, yellow; anther base sagittate. Fruits obovoid achene. Pappus hairy. *Fls. & Frts.:* December-February.

Common on the dry waste land, in the road-side thickets.

Field Collections: Lakshmikantapur, *K.N. - 1774.*

Distributions: Throughout India. Cosmopolitan, most hot and warm temperate countries.

23. 3. ***Gnaphalium pulvinatum*** Delile, Descript. de l 'Egypte 266. t. 44.1812; Naskar Aq. Semi-Aq. Plants 156.1990.

Naskar (l.c.) described this taxon after collecting from the lower Ganga delta.

24. *GRANGEA* Adans.

24.1. ***Grangea maderaspatana*** (L.) Poir. Lamk. Encycl. Method. Suppl. 2: 825.1812; Naskar Aq. Semi-Aq. Plants 156.1990.*(Vern. Name: Namuti).*

Naskar (l.c.) described this taxon from the lower Ganga delta. Leaves is used for stomach troubles, in irregular menses, infusion, hysteria and for relieving pain. Juice of the leaves is utilized in earache.

25. *GUIZOTICA* Cass. *nom. cons.*

25.1. ***Guizotica abyssinica*** (L.f.) Cass., Dict. Sc. Nat. 59: 248.1829; Hook. f. Fl. Brit. India 3: 308.1881; Watt Dict. Econ. Prod. 4: 186. 1891; Prain Beng. Plants 1: 451.1903 (Rep. ed. 1963). *Polymnia abyssinica* Linn. f. Suppl. 383.1781. *Guizotica oleifera* DC. Sept. Not. Pl. Rar. Jard. Genev. 5. *(Vern. Name: Ram-Til* or *Surguja).*

Exotic and cultivated, stout, erect annual crop. Leaves opposite at base, alternate at upper portion. Flower-heads heterogamous, yellow; cypsela glabrous, dorsally compressed, apex round. Pappus absent. *Fls. & Frts.:* February-April.

Cultivated winter annuals.

Field Collection: Basanti, *K.N. - 2278.*

Distributions: Almost throughout India. Native of Tropical America.

Economic Importances: Seeds produces an essential oil.

26. *HELIANTHUS* L.

Key to the Species

1a	Annual herb; roots not tuberous	1. *H. annuus* L.
1b	Perennial herbs; roots tuberous	2. *H. tuberosus* L.

26.1. ***Helianthus annuus*** L. Sp. Pl. 904.1753; Watt Dict. Econ. Prod. 4: 209.1891; Prain Beng. Plants 1: 450.1903. *(Vern. Name: Surjyamukhi).*

Horticultural annual herb. Leaves opposite, large, entire. Flower-heads heterogamous, ray florets uniseriate,yellow; disc black, cypsela large, oblong, bristles caducous. *Fls. & Frts.:* October-February.

Cultivated in the ornamental gardens. public gardens; it is also cultivated for oil seeds in the agricultural fields.

Field Collection: Block Farm/Minakhan, *K.N. - 1126 .*

Distributions: Throughout India. Native of North America.

Economic Importances: Seeds produce edible oil and oilcake.

26.2. ***Helianthus tuberosus*** L. Sp. Pl. 905.1753; Watt Dict. Econ. Prod. 4: 211.1891; Prain Beng. Plants 1: 451.1903 (Rep. ed. 1963). *(Vern. Name. Brahmokhra).*

Perennial garden herb with tuberous roots. Flower heads heterogamous, anther base entire. Fruit obovate compressed achenes. *Fls. & Frts.:* November-January.

Cultivated in the gardens.

Field Collection: Bamanghata/Salt Lake Area, *K.N. - 2306.*

Distributions: Native of North America.

Economic Importance: Much appreciated culinary vegetable.

27. *LACTUCA* L.

27.1. ***Lactuca sativa*** L. Sp. Pl. 795.1753; Hook. f. Fl. Brit. India 3: 404.1781; Watt Dict. Econ. Prod. 4: 578.1890; Prain Beng. Plants 1: 462.1903 (Rep. ed. 1963). *(Vern. Name: Salad).*

Stout succulent annual herb. Stem leafy, radical leaves compact, many. Flower heads yellow, in long irregular panicles. Fruit black achene, broadest towards apex. *Fls. & Frts.:* January-April.

Grown in the gardens; but not very common.

Field Collection: Titagarh/Rahara, *K.N. - 2218.*

Distributions: Cultivated throughout India; cosmopolitan.

Economic Importances: Green leaves are eaten raw in the form of 'salad'.

28. *LAGASCEA* Cav. *nom. cons.*

28.1. ***Lagascea mollis*** Cav. Anal. Cienc. Nat. 6: 331. t. 44. 1803; Hook. f. Fl. Brit. India 3: 302.1881; Prain Beng. Plants 1: 446.1903.

Tall slender herb. Stem pale, striate, terete, pubescent. Leaves ovate, acute or. acuminate, crenate, densely pubescent. Heads in clusters, silky-villous. Flowers white, peduncles long, slender, pubescent. Involucre bracts lanceolate, very acute, ciliate. Corolla tube short, limb 5-fid, the segments linear, acute. Pappus a short fimbriate cup. Achenes cuneate. *Fls. & Frts.:* July-October.

Frequent on the roadside thickets.

Field Collection: Satzelia, *K.N. - 2157.*

Distributions: An introduced Mexican weed in India. Native of Central America.

29. *LAGGERA* Schultz ex Oliver

Key to the Species

1a	Stems rather robust, more or less winged; leaves decurrent on the stem at their base; florets purplish or pinkish	1. *L. aurita* Sch.-Bip.
1b	Stems very slender, not winged; leaves half-embracing the stem at their base; florets yellow	2. *L. flava* (DC.) Benth. ex Clarke

29.1. ***Laggera aurita*** (Wild.) Sch.-Bip., Schweinf-Beitr. Flor. Aethiop.; Hook.f. Fl. Brit. India 3: 271.1881; Watt Dict. Econ. Prod. 4: 584. 1891; Prain Beng. Plants 1: 439.1903 (Rep. ed. 1963). *Blumea aurita* DC. Prodr. 5: 449. 1836.

Slender annual herb. Stem winged, not entire or continuous but shortly decurrent from the leaves. Leaves alternate, decurrent, auricled at base. Flower-heads heterogamous pinkish; cypsela small. Pappus caducous. *Fls.& Frts.:* January to April.

Occasionally on the village thickets.

Field Collection: Titagarh/ Khardah, K.N. - 2225.

Distributions: Throughout India. Bangladesh, Burma and Tropical Africa.

Economic Importances: Strongly scented plant and possesses the odour of turpentive; leaves are reported to be used by Mundas to stop bleeding from cuts.

29.2 ***Laggera flava*** (DC.) Bentham ex C.B. Clarke, Comp. Ind. 90.1876; Hook. f. Brit.India 3: 270.1881. *Blumea flava* DC. Prodr. 5.439.1836. *Blemeopsis flava* (DC.) Gagnep., Bull. Mus. Hist. Nat. Paris. 26: 76.1920.

Herbs, upto 40 cm tall, stem terete, grooved, glabrous. Leaves 8 X 4 cm, obovate-oblong, cordate-sagittate and amplexicual base, serrate-dentate. Heads in panicles; flowers yellow. Achenes ribbed. *Fls. & Frts.:* December-May.

Occasionally on the dry waste lands.

Field Collection: Budge Budge, *K.N. - 744.*

Distributions: Almost throughout India and Sikkim, Bhutan.

30. *LAUNAEA* Cass.

Key to the Species

1a	Heads terminal or racemose on the particularly branched flowering stem	1. *L. asplenifolia* DC.
1b	Heads solitary or fascicled at the nodes of the flagelliform, rooting, flowring stem	2. *L. sarmentosa* Alston

30.1. ***Launaea asplenifolia*** DC. Prodr. 7: 181.1838; Naskar & Guha Bakshi Mangr. Swamps Sundarbans 92.1987. *(Vern. Name: Tik-Chana).*

Naskar & Guha Bakshi (l.c.) described this taxon from the lower Ganga delta region. Root is useful lactagogue stimulating, the milk secretion of the milking mother.

30.2. ***Launaea sarmentosa*** (Willd.) Alston.; Naskar & Guha Bakshi, Mangr. Swamps Sundarbans 92.1987.*(Vern. Name: Tik-Chana).*

Naskar Guha Bakshi (l.c.) described this taxon from the lower Ganga delta. This Plant is said to possesses tonic, soporific, diuretic and asperient properties; and used as a substitute for Taraxacum; leaves are used or edible as famine food.

31. *MIKANIA* Willd. *nom. cons.*

31.1. ***Mikania cordata*** (Burm.f.) Robinson, Contrib. Grey Herb. 104: 35.1934; Naskar Aq. & Semi-Aq. Plants 157.1990. *(Vern. Name: Halkalmi).*

Naskar (l.c.) described this taxon after collecting from Baruipur, *K.N.- 7.*

This plant possesses many medicinal properties.

32. *PARTHENIUM* L.

32.1. ***Parthenium hysterophorus*** L. Sp. Pl. 988.1753; Adams Fl. Pl. Jamaica 751.1972; Wlh. India 7: 268.1966. (Fig.31).

Herb, 1.0-1.5 m height. Stem whitish green, hairy, angular, longitudinally grooved, profusely branched. Leaves alternate, simple but pinnately and irregularly much dissected, tips acute, margin entire, pubescent on both sides; length varies from 2-8 cm, breadth upto 5 cm. Inflorescence white in colour, terminal or axillary, peduncle upto 1 cm, slightly hairy, heads with an involucre of usually 5, outer bracts each subtending a female flower with two male flowers, one on either side; head upto 4 mm long, 4-5 mm in diam. Fruits ellipsoid, obovate, dark, bristle with remnants of cup shaped corolla, appendages and style

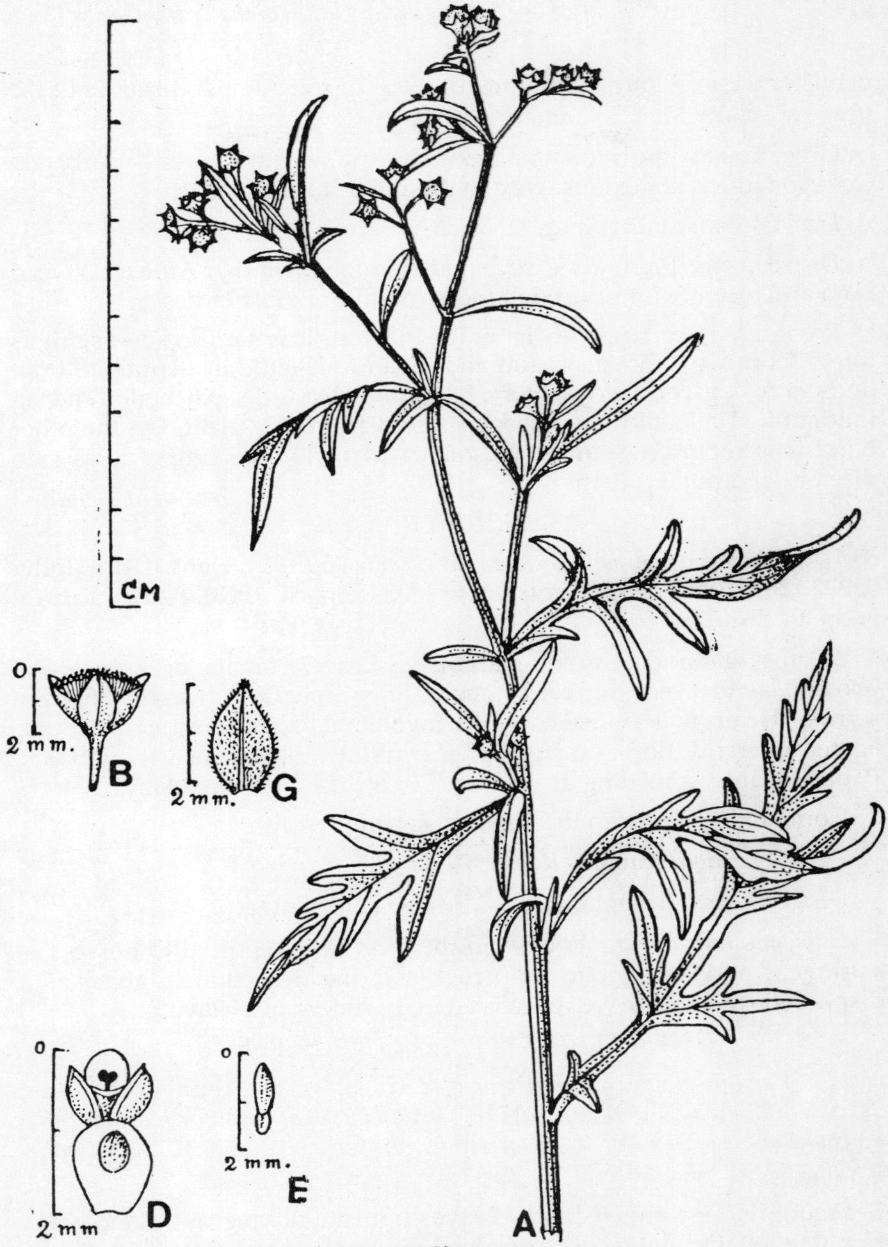

Fig.31. PARTHENIUM HYSTEROPHORUS Linn.

A. Flowering twig, B. Capitulum, C. Bracts, D. Flowers

persistent, 2 mm long, 1 mm broad. *Fls. & Frts.:* Almost throughout the year, especially March-August.

Gregariously grow on the roadsides, railway tracks and gurbages. Occasionally as obnoxious weed in the agricultural fields.

Field Collections: Baruipur, *K.N. - 331.*

Distributions: Presently distributed throughout India. An exotic weed flora and recently introduced in India. Native of West Indies.

Economic Importances: In its native habitat, it is said to be useful as tonic, febrifuse, emmenagogue and as an analgenic in neuralgia. Root decoction is given in dysentery. A bitter glycoside, parthenin and an unidentified alkaloid have been reported from this plant. On the other hand it is reported as most poisonous plant for the cattle; cause skin diseases and pollen allergy.

33. *PLUCHERA* Cass.

33.1. ***Pluchera indica*** (L.) Less., Linneea 6: 150.1831; Hook. f. Brit. India 3: 272.1881; Prain Beng. Plants 1: 440 1903 (Rep. ed. 1963). *Baccharis indica* L. Sp. Pl. 861.1753.

Shrubs, glabrous or sub-glabrous. Leaves sessile or sub-sessile, obovate, with a narrow base, acute, often apiculate at apex, coarsely serrate. Heads in terminal corymbs, involucral bracts many seriate, outer acute, inner ray florets many; achens minute, ribbed, nearly glaborus. Pappus scanty, spreading dull white. *Fls. & Frts.:* February-May.

Commonly planted in the villages, as hedge plant.

Field Collection: Amtala, *K.N. - 379.*

Distributions: Throughout India. Malaya and China.

Economic Importances: Roots and the leaves are reported to possess as astringent and antipyretic properties and the decoction is given as a disphoretic in fever; leaves juice used in dysentery in Malaya.

34. *PSEUDO-ELEPHANTOPUS* Rohr.

34.1. ***Pseudo-elephantopus spicatus*** (B.) Juss. ex Abul. Vahl, Rohr., Skriv. Nat. Selak. Kioeb 2: 214.1792; Dutta & Mitra, Ind. For. 87: 308.1961. *Elephantopus spicatus* B. Juss. ex Abul., Fl. Guian. 2: 808.1775.*Distreptus spicatus* Cass. Dict. Sc. Nat. 13: 367.1819.

Erect, rigid perennial herb. Leaves obovate-oblong or oblanceolate, crenulate or undulate, base narrowed into a short petiole; upper leaves much reduced, linear or linear lanceolate. Glomerules 1-4 headed, sessile in the upper axils. Heads 4-flowered, involucre cylindrical. Achenes puberulous; pappus bristles. *Fls. & Frts.:* January-June.

Very rare on the river banks.

Field Collection: Falta, *K.N. - 1994.*

Distributions: Throughout India. Afghanistan, Persia to North Africa.

Economic Importances: The dried plant is reported to be applied in the treatment of bruises.

35. *PULICARIA* Gaern.

35. ***Pulicaria crispa*** Sch. Bip. Webb. & Benth, Phyt. Canar 2: 223; Naskar & Guha Bakshi, Mang. Swamps Sundarbans 93.1987.

Stout, shrubby perennials, very leafy. Naskar & Guha Bakshi (l.c.) described this texon after collecting from Diamond Harbour.

36. *SENECIO* L.

36.1. ***Senecio linifolius*** (DC.) Clerke, Comp. Ind. 232.1876; Hook.f. Fl. Brit. India 3: 343.1881.*Doronicus linifolium* DC. 6: 322.1837.

Erect or ascending glabrous herb. Leaves sessile, 4-15 cm long, linear or linear-lanceolate, distantly serrate, acute or acuminate. Heads few, bracteolate, in lax corymbs; achenes slender; glabrous, equalling the red pappus. *Fls. & Frts.:* January-July.

Frequently meet within the garden beds and in waste lands.

Field Collection: Titagarh, *K.N. - 2219.*

Distributions: Less common in India. Tropical weeds.

37. *SONCHUS* L.

Key to the Species

1a	Achenes not transversely rugose	1. *S. asper* (L.) Garsault
1b	Achenes transversely rugose	(2)
2a (1)	Involucre with glandular bristles	2. *S. brachyotes* DC.
2b (1)	Involucre without glandular-bristles	3. *S. oleraceus* L.

37.1. ***Sonchus asper*** (L.) Garsault, Fig. Pl. Anim. Med. 4: 332. t. 565. 1767; Hook.f. Fl. Brit. India 3: 414. 1881; Prain Beng. Plants 1: 463. 1903 (Rep. ed. 1963). *S. oleraceus* var. *asper* L. Sp. Pl. 794.1753.

Annual herb with milky sap. Leaves with rounded auricles, sessile, semi-amplexicaul, ovate, ovate-lanceolate, elliptic oblong. Flower heads crowded, achenes distinctly compressed. *Fls. & Frts.:* December-April.

Frequently found in the waste places and also as weed of the crop fields.

Field Collection: Suryapur, *K.N. - 294.*

Distributions: Almost throughout India. Cosmopolitan and all warmer countries of the world.

Economic Importances: Young roots are edible as salad; seeds are also edible; used as emollient; also applied to wounds and boils.

37.2. ***Sonchus brachyotes*** DC. Prodr. 7: 186.1838. *S. arvensis* Hook.f. Fl. Brit. India 3: 414.1881; Watt Dict. Econ. Prod. 6 (3): 275.1893; Prain Beng. Plants 1: 463.1903 (Rep. ed. 1963). *(Vern. Name: Ban-Palang).*

Perennial herb with milky sap. Leaves toothed, auricles rounded, shining above and glaucous beneath, glabrous. Flower head lax; achenes hardly compressed, pale brown, ribbed, rugulose. Pappus white, silky. *Fls. & Frts.:* December-February.

Common on the waste and low-lands.

Field Collection: Garfa, *B. Safui-26* (CAL.).

Distributions: Almost throughout India. Europe and West Asia.

Economic Importances: Plant contains soporific properties; roots employed in the treatment of cough, bronchitis, asthma and pertussis; leaves used as application on swellings; latex used in eye troubles.

Remarks: S. brachyotes has been very much confusing with S. *arvensis;* so many authors considered it as a synonym of the latter species.

37.3. ***Sonchus oleraceus*** L. Sp. Pl. 794.1753; Hook.f. Fl. Brit. India 3: 414.1881; Watt Dict. Econ. Prod. 6 (3): 275.1893; Prain Beng. Plants 1: 463.1903 (Rep. ed. 1963). *S. ciliatus* Lamk. Fl. France 289.1805.

Annual herb with milky sap. Leaves with acute or acuminate auricles. Flower heads crowded. Achenes distinctly compressed, 3-ribbed in each side muricate. *Fls. & Frts.:* December-May.

Commonly found on waste fields and also on cultivated lands as weed.

Field Collection: Sodepur, *K.N. - 2061.*

Distributions: Throughout India; introduced in India. All temperate and tropical countries of the world.

Economic Importances: Leaves and roots used in indigetion and as a febrifuge; stems are sedative and tonic; roots are vermifuge.

38. *SPHAERANTHUS* L.

38.1. ***Sphaeranthus indicus*** L. Sp. Pl. 927.1753; Hook.f. Fl. Brit. India 3:275.1881; Naskar Aq. & Semi-Aq. Plants 158.1990. *(Vern. Name: Murmuria).*

Naskar (1.c.) described this taxon in details. Decoction of the plant are used as diuretic in urethral discharges. The watery juice of the plant is used as laxative in the stomach troubles.

39. *SPILANTHES* L.

39.1. ***Spilanthes acmella*** (L.) Murr. Syst. Veg. (ed.13) 610. 1774; Hook.f. Fl. Brit. India 3:307.1881; Watt Dict. Econ. Prod. 6(3):329.1893; Prain Beng. Plants 1:451.1903 (Rep. ed. 1963). *Spilanthes calya* DC. Prodr. 5:625. 1834. *(Vern. Name: Marhatatiga).*

Annual, erect or ascending herb. Both stem and branches hairy. Leaves opposite, ovate, acute or sub-obtuse, irregularly crenate-serrate, glabrous. Heads ovoid, solitary or sub-panicled. Involucre bracts oblong-lanceolate, sub-acute, pubescent. Pappus absent. Achenes oblong, much compressed, nearly-glabrous. *Fls. & Frts.:* November-February.

Common on the road-side waste lands, garden edges and railway tracks.

Field Collection: Piyali Nagar/Fultala, *K.N. - 103.*

Distributions: Throughout India. Sri Lanka, all warmer countries of the world.

Economic Importances: Flower-heads when chewed have a hot burning taste and cause profuse salivation.

Remarks: Prain (1903) stated about its distribution in Chhota Nagpur, N. Bengal and Chittagong–but now this species is common in this lower West Bengal.

40. *SYNEDRELLA* Gaertn. *nom. cons.*

40.1. ***Synedrella nodiflora*** (L.) Gaertn., Fruct. 2:456, t. 171. f. 7.1791; Hook.f. Fl. Brit. India 3:308.1881; Prain Beng. Plants 1:452.1903 (Rep. ed. 1963); Wlh. Ind. 10:92.1976. *Verbesina nodiflora* L.; Juslenius, Cent. I.Pl. 28.1755.

Annual pubescent or villous herb. Stem branched, glabrous-terete. Leaves opposite. Flower heads heterogamous, in forks of branches; anther sub-entire bases. Fruit dorsally compressed achene, black or dark brown. *Fls. & Frts.:* November-February.

Common in the waste lands.

Field Collection: Falta, *K.N. - 2003.*

Distributions: A ubiquitous pantropic weed of American origin. Indo-Malayasia; native of Tropical America.

Economic Importances: Leaves are used as poultice for sore legs and for the preparation of embrocation for rheumatism; leaves juice also used for earache; tender leaves are laxative.

41. *TAGETES* L.

41.1. ***Tagetes patula*** L. Sp. Pl. 887.1753; Watt Dict. Econ. Prod. 6(3): 402.1893; Prain Beng. Plants 1:445. 1903 (Rep. ed. 1963). *(Vern. Name: Chhota-genda).*

Annual garden herb. Leaves pinnatisect. Flower heads-homogamous or heterogamous; anther base entire, obtuse. Fruit linear cypsela, faintly ribbed. Pappus few. *Fls. & Frts.:*November-March.

Commonly planted in the boarder line of the ornamental gardens.

Field Collection: South Garia, *K.N. - 821.*

Distributions: Cultivated almost throughout India. Native of Mexico.

Economic Importance: Common ornamental garden plants.

42. *TRIDAX* L.

42.1. ***Tridax procumbens*** L. Sp. Pl. 900. 1753; Hook.f. Fl. Brit. India 3:311.1881; Prain Beng. Plants 1:455.1903 (Rep. ed. 1963).

Week perennial herb. Leaves pinnatisect. Flower heads heterogamous, long peduncled; anthers with basal auricles. Fruit turbinate or oblong achene. Achene brown of numerous unequal short and long feathery bristles. *Fls. & Frts.:*All round the year.

Abundantly found in the waste lands; on the road-sides and in almost all situations.

Field Collection: Subuddhipur/Baruipur, *K.N.* - 22.

Distributions: Introduced into India about the year 1830 as an ornamental plant. It is now wide-spread over tropical India. Very common on old walls, in grasslands and road-side and waste lands. Native of Tropical America, now Pantropic.

43. *VERNONIA* Schrab. *nom. cons.*

Key to the Species

1a	Robust, upto 1 m tall herb; achenes ribbed, pubescent	1. *V. anthelmintica* Willd.
1b	Small herb, upto 75 cm high; achenes not ribbed, appressedly pubescent	2. *V. cinerea* (L.) Less.

43.1. ***Vernonia anthelmintica*** Willd. Sp. Pl. 3: 1634.1890; Naskar Aq. & Semi-Aq. Plants 159.1990. *(Vern. Name: Somraj).* (Fig.32).

Naskar (1.c.) described this taxon after collecting from the lower Ganga delta (Bhangankhali, *K.N. - 149).*

43.2. ***Vernonia cinerea***(L.) Less, Linnaea 4: 291.1829; Naskar Aq. Semi-Aq. Plants 159.1990. *(Vern. Name: Kuk-shim).*

Naskar (1.c.) described this taxon from the lower Ganga delta, Subuddhipur, *K.N. - 1.*

Plant decoction help prompt perspiration in fever; seed anthelmintic.

44. *VICOA* Cass.

44.1. ***Vicoa vestita*** (Wall.ex DC.) Benth. ex Hook.f. Fl. Brit. India 3: 297.1881; Naskar Aq. Semi-Aq. Plants 160.1990; Naskar & Guha Bakshi Mangr. Swamps Sundarbans 93.1987.

Naskar and Guha Bakshi (1.c.) and Naskar (1.c.) described this taxon from the lower Ganga delta after collecting from Kakdwip, *K.N. - 1926.*

45. *WEDELIA* Jacq. nom cons.

Key to the Species

1a	Climbing shrub; leaves ovate or elliptic-ovate; heads 3 together at the ends of the branches	2. *W. biflora* (L.) DC.
1b	Herbs; leaves obovate or oblanceolate; heads always solitary, axillary	1. *W. chinesis* (Osbeck.) Merr.

45.1. ***Wedelia chinensis*** (Osbeck.) Merr., Philipp. Journ. Sci. (Bot.) 12:111. 1917; Naskar Aq. Semi-Aq. Plants 160.1990. *(Vern. Name: Kesaraj).*

Naskar (1.c.) described this taxon after collecting from Magrahat, K.N.-250. This plant is considered as tonic and alternative and useful in cough, skin-diseases and alopecin.

Fig.32. VERNONIA ANTHELMINTICA Willd.
A flowering twig.

45.2. ***Wedelia biflora***(L.) DC. Wight Contrib. 18.1834; Naskar Aq. & Semi-Plants 160.1990; Naskar & Guha Bakshi Mangr. Swamps Sundarbans 93. 1987. *(Vern. Name: Bhimraj).*

Naskar & Guha Bakshi (1.c.) described this taxon from the lower Ganga delta (Bireshwarpur, *K.N. - 1762).*

46. *XANTHIUM* L.

46.1. ***Xanthium indicum*** Koen. Roxb., Hort. Beng. 67.1814 & Fl. Ind. 3:601. 1832. *X. strumarium* L. Sp. Pl. 987. 1753; Hook.f. Fl. Brit. India 3:303. 1881; Prain Beng. Plants 1: 446.1903 (Rep. ed. 1963). *(Vern. Name: Ban-Okra).*

Rigid unarmed annual. Leaves alternate palmately lobed. Flower-heads monoecious; anther base obtuse. Fruit ovoid, covered with bristles. Pappus absent. *Fls. & Frts.:*January-August.

Abundantly found in waste places along the railway tracks.

Field Collection: Madarhat, *K.N. - 78.*

Distributions: Throughout India. Sri Lanka, native of South America.

Economic Importances: Used as diaphoretic, sedative, an agent inducing swelling and an agent increasing the flow of saliva. Useful in long-standing cases of malaria. Roots is bitter and tonic, employed in the glandular tumor and cancer. Fruits are useful as cooling agent in allaying irritation, given in small pox.

47. *ZINNIA* L.

Key to the Species

1a	Horitcultural herb, commonly planted	1. Z. *elegans* Jacq.
1b	Wild herb; occasionally planted	2 Z. *pauciflora* L.

47.1. ***Zinnia elegans*** Jacq. Coll. Suppl. 152.1796; Wlh. India 11: 107.1976. *(Vern. Name: Zinnia).*

Erect, annual herb, 1m tall; stem pilose-strigose, greenish, striate, leaves clasping, cordate-ovate or elliptic. Heads white, yellow, orange, pink; both single or double; flat globular, florets ribbon-like; ray florets reflexed, disc florets yellow or orange. *Fls. & Frts.:* February-May.

Common horticultural herbs; planted in the gardens everywhere.

Field Collection: Bamanghata, *K.N. - 2305.*

Distributions: Planted almost throughout India; introduced. Native of Mexico, at present pantropical in distribution.

Economic Importances: Contain saponin, nicotine, nornicotine etc.

47.2. ***Zinnia pauciflora*** L. Sp. Pl. (ed.2) 1269.1762; Prain Beng. Plants 1:447. 1903 (Rep. ed. 1963). *(Vern. Name: Zinnia).*

Annual garden herb. Leaves opposite. Flower heads heterogamous; anther-base entire. Achene striate, narrow. *Fls. & Frts.:* January-April.

Occasionally on the village shrubberies; rarely planted.

Field Collection: Durganagar/near Dum Dum Cant., *K.N. - 1714.*

Distributions: Wild almost throughout India. Native of Tropical America.

Economic Importance: Ornamental garden plants.

LXXV. CAMPANULACEAE

1. *WAHLENBERGIA* Schrad. *nom.cons.*

1.1. ***Wahlenbergia marginata*** (Thunb.) DC. Mon. Camp. 143. 1830; Naskar Aq. Semi-Aq. Plants 161.1990.

Naskar (1.c.) described this taxon from the lower Ganga delta, after collecting from Diamond Harbour, *K.N. - 1949.*

Beside this, *Lobelia alsinoides* Lamk. and *L. terminalis* Clarke have been reported by the earlier workers; but the present author could not collect these flora from the lower Ganga delta.

LXXVI. SPHENOCLEACEAE

(Splitted out from Campanulaceae)

1. *SPHENOCLEA* Gaertn. *nom. cons.*

1.1. ***Sphenoclea zeylanica*** Gaertn., Fruct. 1:113, t. 24, f. 5.1788; Naskar Aq. & Semi-Aq. Plants 162.1990. *(Vern. Name: Jhil-marich).*

Naskar (1.c.) described this taxon after collecting from lower Ganga delta (Canning, *K.N. - 193).*

LXXVII. PLUMBAGINACEAE

Key to the Genera

1a Styles 5, distinct, stigmas sub-capitate 1. *AEGIALITIS* R. Brown

1b Style 1, filiform divided into 5. stigmatose branches 2. *PLUMBAGO* L.

1. *AEGIALITIS* R. Brown.

1.1. ***Aegialitis rotundifolia*** Roxb. Fl. Ind. 2:111.1824; Hook.f. Fl. Brit. India 3:479.1882; Prain Beng. Plants 1:470. 1903 (Rep. ed. 1963); Naskar & Guha Bakshi Mangr. Swamps Sundarbans 142. 1987. *(Vern. Name: Tora).*

Shrub upto 3 m tall or low tree; without any aerial root system, but the base of the trunk is enlarged and fluted with spongy tissue. Leaves alternate, shining above, broad 7-9 cm long and 6-8 cm broad, broadly ovate, coriaceous, petioles dilated at the base and stem clasping, tubular leaf sheath. Flowers pentamerous, many flowered in terminal branches; white, calyx 1.3 -1.5 cm long, tubular, 5-lobed apically. Petals 5, white, upto 0.8 cm long, imbricate. Fruit an elongated, bluntly pointed, longitudinally dehiscent capsule, 4-5 cm long with persistent calyx. Seeds germination epigeal, crypto-viviparous. *Fls. & Frts.:*January-May. (Fig.33).

Grow on the river flat and inter-tidal ridge forest.

Field Collection: Sajnakhali, *K.N. - 1506.*

Distributions: Distributed in other coastal mangrove forests in India and also in Andaman & Nicobar Islands. It is the Old-World mangrove species and very common in the S.E. Asian Tropical Mangrove forests.

Economic Importances: Produces the fire woods and the tannin content in the stem-bark varies between 12.2-15%. It is one of the pioneer species of the mangrove forests, of the newly silted up soils.

2. *PLUMBAGO* L.

2.1. ***Plumbago zeylanica*** L. Sp. Pl. 151.1753; Hook.f. Fl. Brit. India 3: 480.1882; Prain Beng. Plants 1:471.1903 (Rep. ed. 1963). *(Vern. Name: Chita).*

Sub-scandent, perennial herb or spreading shrub. Stems rather woody, striate. Leaves 4-10 cm long, ovate, entire, thin, glabrous. Flowers white, in long, spiciform racemes; bracteoles much shorter than the calyx; calyx sticky, densely glandular persistent. Capsules oblong, enclosed within the persistent calyx; seeds oblong. *Fls. & Frts.:* September-March.

Frequently found amongst bushes and other waste places.

Field Collections: Bahuru, *K.N. - 1626.*

Distributions: Almost throughout India. In the tropics.

Economic Importances: The alkaloid, plumbagin has the stimulant properties and stimulate secretion of sweet gland, urine and bile etc. Roots

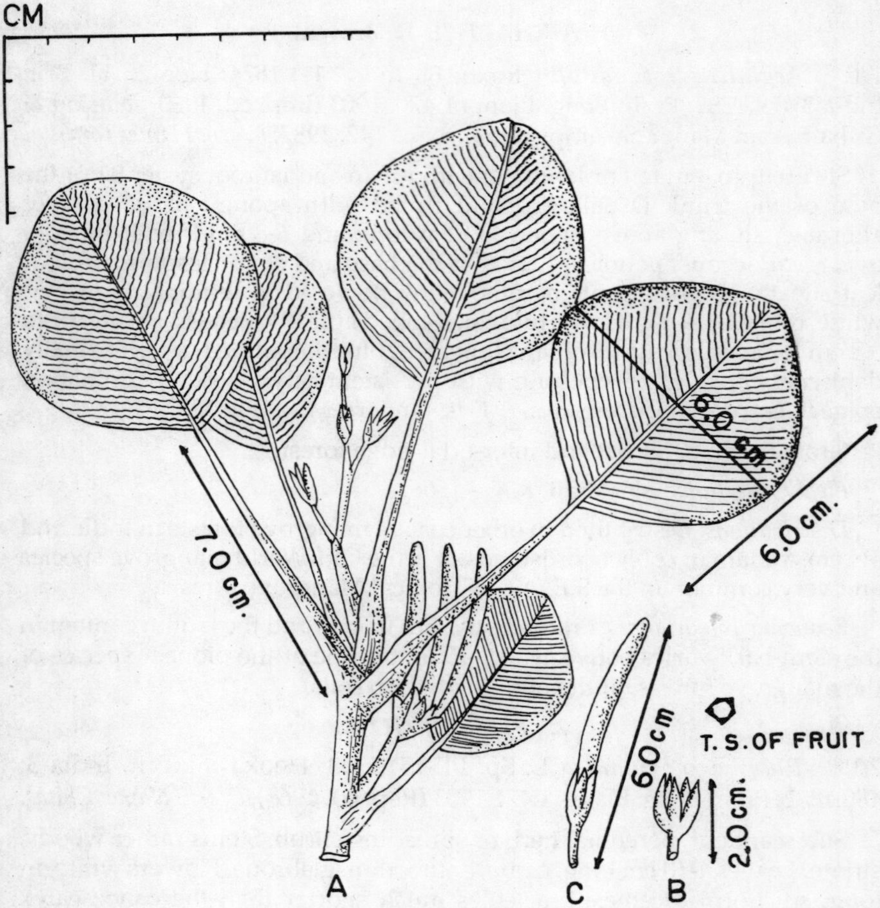

Fig.33. AEGIALITIS ROTUNDIFOLIA Roxb.

A. Flowering & Fruiting twig, B. Single flower, C. Single fruit.

act as appetizer and used also in the skin diseases, diarrhoea and bleeding piles. Milky latex is useful in scabies and unhealthy ulcers.

LXXVIII. PRIMULACEAE

Key to the Genera

1a	Corolla lobes imbricate, leaves radical	2. *PRIMULA* L.
1b	Corolla lobes twisted, leaves cauline	1. *ANAGALLIS* L.

1. *ANAGALLIS* L.

Key to the Species

1a	Leaves sessile, cordate, ovate or lanceolate, sometime gland-dotted; flowers blue	1.*A. arvensis L.*
1b	Leaves sub-sessile, broadly ovate or obovate; flowers white	2. *A. pumila* Sw.

1.1 ***Anagallis arvensis*** L. Sp. Pl. 148.1753; Naskar Aq. Semi-Aq. Plants 163.1990. *(Vern. Name: Jon Kamari).*

Erect or procumbent annual herb. Naskar (1.c.) described this taxon after collecting from the lower Ganga delta (*Dhakuria, K.N. - 1968).* Medicinally this plant is said to be used for cerebral affections, hydrophobia, leprosy, dropsy, epilepsy and in mania.

1.2. ***Anagallis pumila*** Sw., Prod. Veg. Ind. Occid. 1:40. 1788; Naskar Aq. & Semi-Aq. Plants 164.1990.

Erect branched glabrous annual herb. Naskar (1.c.) described this taxon after collecting from the lower Ganga delta (Baguihati, K.N.-1886).

2. *PRIMULA* L.

2.1. ***Primula umbrellata*** (Lour.) Bentv. Fl. Males 6(2): 191.1962; Naskar Aq. Semi-Aq. Plants 164.1990.

Naskar (1.c.) described this delicate, glandular pubescent annual herb after collecting from Nilgung, *K.N. - 2196.*

LXXIX. MYRSINACEAE

Key to the Genera

1a	Fruit globose; seeds albuminous	2. *ARDISIA* Swartz.
1b	Fruit cylindric; seeds exalbuminous	1. *AEGICERAS* Gaertn.

1. *AEGICERAS* Gaertner

1.1. ***Aegiceras corniculatum*** (L.) Blanco, Fl. Filip. 79.1837; Naskar & Guha Bakshi 138.1987. *Rhizophora corniculatum* L. Sp. Pl. 444.1753. *(Vern. Name: Khalsi).* (Fig.34).

Small tree or shrub, 2-6 m high, branches horizontally spreading, dense canopy; bark smooth dark grey. No aerial roots. Leaves alternate or spirally arranged, stipule absent, coriaceous, glabrous on both sides, pale green above and glabrous below, 5-7 cm long, 2.5-3.5 cm broad, elliptic, obovate, apex rounded or slightly emerginate. Flowers fragrant, perfect, white 1-2 cm diam., pentamerous, pointed in bud condition, axillary umbell. Sepals 5, free imbricate; petals 5, white pointed end; corolla twisted. Fruit 5-8 cm long, adult fruit curved with persistent calyx; 1-seeded, capsule pointed apically. Germination epigeal, crypto-viviparous. *Fls. & Frts.:* December-May.

Common on the mid-estuary, in the river flats inter-tidal ridge forests.

Field Collection: Rangabelia, *K.N. - 1526.*

Distributions: All the coastal and tidal forests of India and Andaman & Nicobar Islands. Sri Lanka, Burma and other S.E. Asian countries.

Economic Importances: Flower nector is very important for the honey bee, *Apis dorsata;* produces fire-woods for the rural people. The tannin percentage of the stem bark is estimated-7.0% only.

2. *ARDISIA* Swartz. *nom. cons.*

2.1. ***Ardisia solanacea*** (Poir) Roxb. Cor. Pl. 1:27.1795; Benthall Trees Calc. 284. 1946. *Anguillaria solanacea* Poir, Encycl. 7:688. 1806. *Ardisia humilis* Vahl Symb. 3: 40; Clarke, Hook.f. Fl. Brit. India 3:529.1882; Watt Dict. Econ. Prod. 1:290.1889; Prain Beng. Plants 1:475.1903 (Rep. ed. 1963). *(Vern. Name: Ban Jam).*

Large, much-branched shrub or small tree. Leaves 10-25 cm long, obovate, oblanceolate or oblong-elliptic, tapering to an acute base, acuminate or sub-obtuse at the base, entire, fleshy, sub-cariaceous, lateral nerves many; petiole about 1.25 cm long winged. Flowers pinkish, in axillary, corymbose racemes. Fruits sub-globose, with a depressed apex *Fls. & Frts.:* October-January.

Frequently found in the dampy shady places.

Field Collection: Santoshpur/near Budge Budge, *K.N. - 1952.*

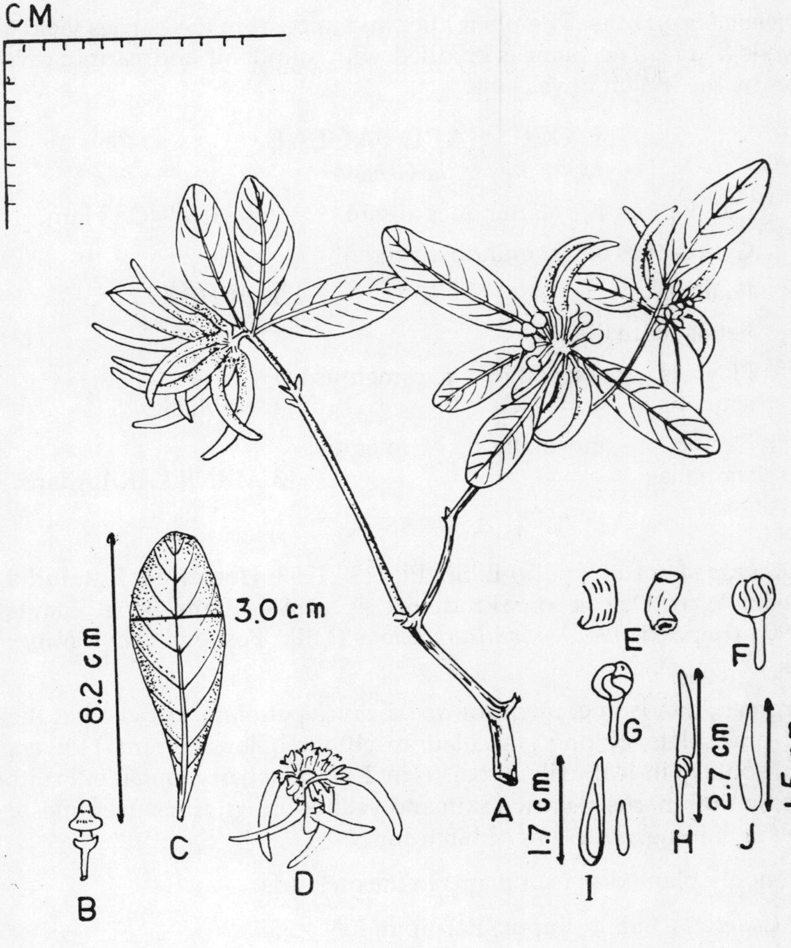

Fig.34. AEGICERAS CORNICULATUM (L.) Blanco

A. Flowering twig, B. Single enlarged flower, C. Single leaf, D. Inflorescence branch with fruits, E. Petals, F. Single flower bud (Side view), G. Single flower bud (Top view), H. Single fruit with peduncle, I. T.S. of fruit, J. Fruit with peduncle.

Distributions: Throughout India, Sri Lanka, Malaya and China.

Economic Importances: The plant juice and juice from the berries yield a good yellow dye. The plant is credited with stimulant and carminative qualities by the Indian Physicians.

LXXX. SAPOTACEAE

Key to the Genera

1a	Calyx-lobes 4-5; stamindoes absent	2. *MADHUCA* Ham.
1b	Calyx-lobes 6-8; stamindoes present	(2)
2a (1)	Petals-6	1. *ACHRAS* L.
2b (1)	Petals more than 6	(3)
3a (2)	Flowers normally tetramerous, stamens-8	4. *MIMUSOPS* L.
3b (2)	Flowers normally trimerous; stamens-6	3. *MANILKARA* Adans.

1. *ACHRAS* L.

1.1. ***Achras sapota (zapota)*** L. Sp. Pl. 1190.1753; Hook.f. Fl. Brit. India 3:534.1882; Watt Dict. Econ. Prod. 6(1):80. 1889; Prain Beng. Plants 1:477.1903 (Rep. ed. 1963). *Manilkara achras* (Mill.) Fosberg. *(Vern. Name: Sabeda).*

Evergreen, handsome, medium tree. Leaves petiolate, crowded at the ends of branchlets, oblong lanceolate or elliptic oblong, entire. Flowers whitish, solitary in leaf axils, sepals 6 in 2 whorls, hairy, outside. Fruits globose, with thin, rusty-brown skin and yellow-brown translucent flesh; seeds black shining. *Fls. & Frts.:* March-June.

Commonly planted as fruit plants in the orchards.

Field Collection: Subuddhipur/Baruipur, *K.N. - 355.*

Distributions: Planted almost throughout India; it was introduced into India by the Portuguese in the 16th Century. A native of Tropical America.

Economic Importances: Ripe fruits are delicious and edible. The most important product of the tree is 'chicle' gum, which is used in the manufacturing of chewing gum. Medicinally, the stem bark is said to be useful as an astringent, a tonic and a febrifuse, while the seeds are aperient.

2. *MADHUCA* Ham.

2.1. ***Madhuca longifolia*** (Koen.) Macbride Contr. Gray Herb. Univ. n.s. 53:17.1911. *Brassia longifolia* Koen., L. Mant. Act. App. 563.1771. *M. indica* Gmel. Syst. Nat. 2:299. 1791; Wlh. Ind. 6:207. *Bassia latifolia* Roxb. Pl. Cor. 1. 20, t. 19, 1795; Hook.f. Fl. Brit. India 3:544.1882; Watt Dict. Econ. Prod. 1:406. 1893; Prain Beng. Plants 1:478.1903 (Rep. ed. 1963). *(Vern. Name: Mahua).*

Large deciduous tree. Leaves clustered at the ends of the branches, 10-25 cm long, elliptic or oblong elliptic, obovate or broadly lanceolate, rigid. Flowers cream coloured, on long rusty-tomentose pedicels clustered at the ends of the usually leaflets branches. Berry globose or ovoid, fleshy. *Fls. & Frts.:* March-July.

Occasionally planted and frequently found wild in the villages.

Field Collection: Bakkhali, *K.N. - 1906.*

Distributions: Mainly in the central parts of India. Mostly in the Tropics.

Economic Importances: The sweet scented fleshy corolla are commonly edible by the Santals. The wood is good and is largely used for oil mills; wine is also extracted from the flowers. Seed oil is useful and oil cake is a good poison as well as good fertilizer to fish pond; now it is an essential article to the pisciculturists. Bark heals wounds and is a remedy for rheumatism; fruits cure blood diseases.

Van Royen (Blumea 10(1):53.1960) revised the taxonomy and nomenclature of the genus-Madhuca of the Malayasian area; he merged *M. indica* Gmel. and *M. longifolia* (Koening) Macb. under the latter name and distinguishes two varieties *viz.*, var. *longifoila* and var. *latifolia* (Roxb.) Chevl.

3. *MANILKARA* Adans.

3.1. ***Manilkara hexandra*** (Roxb.) Dub., Ann. Mus. Col. Marseille (Ser.3) 3:9.1915. *Mimusops hexandra* Roxb. Pl. Cor.1:16, t.15,1795 & Fl. Ind. 1:238.1820; Hook.f. Fl. Brit. India 3: 549.1882; Watt Dict. Econ. Prod. 5:251.1891. *M. indica* DC. Prodr. 8:205. 1844. *(Vern. Name: Bilati Bakul).*

Large evergreen tree, with a dense leafy crown. Leaves 5-10 cm long, oblanceolate-oblong, elliptic or elliptic-obovate. Flowers dull white, 1-6 axillary, sepals 6. Fruit ellipsoid about 1.5 cm long, reddish-yellow when ripe, tasty; 1-seeded. *Fls. & Frts.:* Novemebr-May.

Rarely planted in the gardens. Few plants are noticed in the core forest area of the tidal mangrove forest of the Sundarbans, at Netidhopani. The origin of these plants is uncertain and the presence of this plants in such tidal mangrove forests is also interesting.

Field Collection: Narendrapur, *K.N. - 1981*.

Distributions: Mostly in the Western Peninsula. Sri Lanka and Burma.

Economic Importances: The wood is tough and used for various purposes; bark is used as medicine and the fruits are eaten during scarcity; used as root-stock for apota.

4. *MIMUSOPS* L.

4.1. ***Mimusops elengi*** L. Sp. Pl. 349.1753; Hook.f. Fl. Brit. India 3:548.1882; Watt Dict. Econ. Prod. 5:249.1891; Prain Beng. Plants 1:479.1903 (Rep. ed. 1963). *(Vern. Name: Bakul).*

Medium tree, evergreen. Leaves 5-15 cm long, elliptic, usually acuminate, glabrous, shining, margin entire or undulated. Flowers white, star like on pubescent pedicels. Fruits variable, usually narrowly ovoid, orange-coloured, 1-seeded. *Fls. & Frts.:* March-August.

Frequently planted on the road-side as avenue tree and in the parks and gardens for its fragrant flowers and shades.

Field Collection: Gosaba, *K.N. - 646*.

Distributions: Native of Western Peninsula of India and now common throughout India. Tropics of both hemispheres.

Economic Importances: Ripe fruits are occasionally edible by the children. Flowers are fragrant and used for religious purposes. Nector collected by the honey bee. Timber of this plant is durable.

LXXXI. EBENACEAE

1. *DIOSPYROS* L.

Key to the Species

1a	Medium tree; leaves elliptic, seeds 2-3	1. *D. chloroxylon* Roxb.
1b	Small to moderate tree; leaves obovate or oblong; seeds 4-8	(2)
2a (1)	Flowers white, male in small cymes and female solitary	(3)

2b (1)	Flowers whitish, males in axillary cymes and females solitary or few together	4. *D. malabarica* Kostel
3a (2)	Fruits globose, glabrous	2. *D. cordifolia* Roxb.
3b (2)	Fruits ellipsoid, pubescent	3. *D. discolor* Willd.

1.1. ***Diospyros chloroxylon*** Roxb. Cor. Pl. 1:38.1795; Clarke, Hook.f. Fl. Brit. India 3:560.1882; Watt Dict. Econ. Prod. 3:137.1890; Prain Beng. Plants 1:481.1903 (Rep. ed. 1963). *D. capitulata* Wight, Icon. tt. 1224. 1588 (bis). *(Vern. Name: Anduli).*

Medium tree, sometime spinescent; bark deeply cracked, branchlets tomentose. Leaves alternate, thinly coriaceous, elliptic ovate or elliptic-oblong, obtuse, more or less pubescent above, rusty tomentose; base acute or obtuse. Inflorescence densely fulvous-pubescent; flowers white, 4-merous. Fruit globose 1 cm in diam. Seeds 2-3; albumen not ruminate. *Fls. & Frts.:*June- October.

Sometimes cultivated as ornamental tree.

Field Collection: Bahuru, *K.N. - 1627.*

Distributions: Native of parts of Orissa as well as West and South India. In the S.E. Asian countries, occasionally.

Economic Importances: Produce fuel wood and medium timber. Stem barks and fruits contain high percentage of tannin.

1.2. ***Diospyros cordifolia*** Roxb. Pl. Cor. 1:38, t.50.1795; Clarke, Hook.f. Fl. Brit. India 2:538. 1882; Benthall Trees Calc. 293.1946. *D. montana* Roxb. var. *cordifolia* (Roxb.) Hiern. Monog. Ebenaceae Trans. Canbr. Phil. Soc. 12(1): 222.1873; Prain Beng. Plants 1:481.1903 (Rep. ed. 1963). *D. montana* Clarke, Hook.f. Fl. Brit. India 3:555.1882.*(Vern. Name: Ban-Gab).*

Small tree often with scattered thorns; young shoots pubescent. Leaves ovate-oblong or elliptic-oblong, obtuse to acuminate at apex, rounded to cordate base. Male flowers usually in 3-flowered, axillary cymes; female flowers solitary, bracteate. Berry globose, yellow when ripe. *Fls. & Frts.:* March-October.

Commonly on the village thickets as wild plant.

Field Collection: Duttapukur, *K.N. - 1712.*

Distributions: Most parts of India. Very common in Burma, Malaya and Australia.

Economic Importance: Wood is reddish or yellow-white and fairly hard; fruits are poisonous, but said to be useful in boils.

1.3. ***Diospyros discolor*** Willd. Sp. Pl. 4:1108.1806; Clarke, Hook.f. Fl. Brit. India 3:654.1882; Prain Beng. Plants 1:482.1903 (Rep. ed. 1963); Li et al., Fl. Taiwan 4: 100.t.930.1978.*(Vern. Name: Bilati-Gab).*

Moderate evergreen tree. Leaves alternate oblong, acute. Flowers whitish, male in cymes and female solitary. Fruit ellipsoid, pubescent, reddish velvetry-tomentose. *Fls. & Frts.:* March-June.

Occasionally planted for its ornamental foliages.

Field Collection: Sodepur, *K.N. - 2059.*

Distributions: Introduced into India in the early 18th Century. A native of Philippines and Malaya.

Economic Importances: Fruits are edible; stem bark used in tanning fish nets.

1.4. ***Diospyros malabarica*** (Desr.) Kostel., Allg.Med. Pharmac. Floras 3:1099.1834. *Garcinia malabarica* Desr., Lam. Enc, Meth. Bot. 3:701.1789. *Diospyris embryopteris* Pers., Syn. Pl. 2:624.1807; Clarke, Hook.f. Fl. Brit. India 3: 556.1882; Prain Beng. Plants 1:481.1903 (Rep. ed. 1963). *D. peregrina* (non Gaertn.) Guerke, Engl. & Prantl., Pfam, 4(1): 164.1897. *Embryopteris peregrina* Gaertn. Fruct. 1: 145.t.29. *(Vern. Name: Gab).*

Medium to large, spreading evergreen tree. Leaves 10-25 cm long, oblong or ovate-oblong, obtuse or bluntly acuminate, rounded at the base. Coriaceous, petiole wrinkled or twisted. Flowers white, fragrant. Fruits gobose, covered with deciduous red flakes. *Fls. & Frts.:* April-November.

Village shrubberies in general.

Field Collection: Falta, *K.N. - 1294.*

Distributions: Indigenous; throuhgout India.Malaya and Australia.

Economic Importances: Sometimes the fruits are edibe. Both rind and bark are used medicinally for the cure of diarrhoea. The juice of the bark is used in tanning the fishing nets.

Ramarks: The other species of the genus-Diospyros L. in India are as follows: *D.chenum Koen., D. exsculpta* Buch - Ham., *D. ferrea* (Willd.) Bakh., *D. kaki* Linn.f., *D. lotus* L., *D. marmorata* Parker, *D. montana* Roxb., etc. Most of these species are spred over in the drier parts of India.

LXXXII. OLEACEAE

1. *JASMINUM* L.

Key to the Species

1a	Climbing shrubs	(2)
1b	Sub-erect shrub; leaves simple	3. *J. sambac* (L.) Ait.
2a (1)	Leaves compound, trifoliate with small lateral leaflets	1. *J. auriculatum* Vahl
2b (1)	Leaves simple; velvet-pubescent; bracts green or absent	2. *J. multiflorum* (Burm.f.) Andr

1.1. ***Jasminum auriculatum*** Vahl, Symb. Bot. 3:1.1794; Clarke, Hook.f. Fl. Brit. India 3: 600.1882; Prain Beng. Plants 1: 486.1903 (Rep. ed. 1963) *(Vern. Name: Jui* or *Yothika).*

Scandent shrub. Leaves trifoliate with very small lateral leaflets, which are sometime absent. Flowers white, star-shaped, in terminal cymes; calyx lobes 5, short; corolla 1.5 cm long, lobes about 8, oblong. *Fls. & Frts.:* April-November.

Planted in the gardens or lawns.

Field Collection: Narendrapur, *K.N. - 1980.*

Distributions: South India, cultivated throughout India. A Pan-Tropic species.

Economic Importances: Produces perfumes for hair oils. The flowers are said to be useful in convultion.

1.2. ***Jasminium multiflorum*** (Bum.f.) Andr., Bot. Rep. 8,t. 496.1807; Watt. Dict. Econ. Prod 4:544.1890; Wlh. India 5: 284.1959. *Nycthanthes multiflora* Burm.f., Fl. Ind. 5, t. 3. f. 1,1768. *Jasminum pubescens Willd.* Sp. P. 1:37.1797; Clarke, Hook.f. Fl. Brit. India 3:592. 1882;Prain Beng. Plants 1:486.1903 (Rep. ed. 1963). *(Vern. Name: Kunda).*

Scandent, densely pubescent shrub. Leaves simple, ovate, thick, cordate, acute or acuminate, base rounded. Flowers white, slightly scented, in terminal, many-flowered umbels. Ripe carpels globose. *Fls. & Frts.:* March-December.

Planted in gardens for its beautiful blossoms.

Field Collection: Bamanghata/Salt Lake, *K.N. - 1734.*

Distributions: Throughout India and common in the Sub-Himalayan tracts. Native of Indo-Malayasia.

Economic Importances: Exceedingly ornamental. Dried leaves of the plant are reported to be employed in poultice for indolent ulcers.

1.3. ***Jasminum sambac*** (L.) Ait. Hort. Kew.1:8.1789; Clarke, Hook.f. Fl. Brit. India 3:591.1882; Watt Dict. Econ. Prod 4: 544.1890; Prain Beng. Plants 1: 486.1903 (Rep. ed. 1963) . *Nyctanthes sambac* L. Sp. Pl. 6.1753. *(Vern. Name: Ban Mallika* or *Bel Phul).*

Erect or scandent shrub. Leaves elliptic or broadly ovate, glabrous, shining. Flowers white, pinkish-violet in bud. Usually double, fragrant, in 3-many flowered clusters. *Fls. & Frts.:* April-September.

It is cultivated commercially in the local gardens and also for ornamental gardens.

Field Collection: Chaital, *K.N. - 725.*

Distributions: Cultivated or as wild almost throughout India. Native of Indo-Malaysia.

Economic Importances: Liked by the gardens for its sweet scented flowers.

Remarks: Other ornamental and purely cultivated species of the Genus–*Jasminum* L. are as follow: *Jasminum angastifolium* Vahl, *J. arborescens* Roxb, *J. flexile* Vahl, *J. grandiflorum* L., *J. humile* L. J. *malabarica* Wight and *J. officinale* L.

LXXXIII. NYCTANTHACEAE

1. *NYCTANTHES* L.

1.1. ***Nyctanthes arbor-tristis*** L. Sp. Pl. 6.1753; Clarke, Hook.f. Fl. Brit. India 3:603.1882 Watt Dict. Econ. Prod. 5: 434.1892; Prain Beng. Plants 1:486.1903 (Rep.ed. 1963). *(Vern. Name: Sephalika/Seoli).*

Shrub or small deciduous tree with 4-angular branches. Leaves 12-20cm long ovate, acuminate or acute, very scabrous, margin entire or with distant teeth. Flowers white with orange coloured centre, very fragrant. Capsules 1.5 -2.0 cm long, flat, narrowed at the base, broadly at the apex; 1-seeded. *Fls. & Frts.:* September-December.

Commonly planted in the gardens, bunglows and temple sides.

Field Collection: Surjyapur, *K.N. - 280.*

Distributions: Throughout India. A native of Indo-Malaysia.

Economic Importances: Ornamental flower trees. Used in remittent fever; billious fever due to irregularities of the liver function and also in malarial fever.

LXXXIV. SALVADORACEAE

1. *AZIMA* Lamk.

1.1. ***Azima tetracantha*** Lamk. Encycl. 1:343.1783; Hook.f. Fl. Brit. India 3: 620.1882; Watt Dict. Econ. Prod. 1:361.1883; Prain Beng. Plants 1:489.1903 (Rep. ed. 1963). *Monetia tetracantha* Salisb. Prodr. 65.1796.*(Vern. Name: Trikanta-Gati).*

Armed shrubs. Branches obtusely 4-angular, glabrous expubescent young parts. Leaves opposite, upto 5-4 cm with a solitary or paired 1-2 cm long spines, ovate or elliptic-ovate or obovate, acute or round, tipped with a spinous point, cuneate to rounded at base. Petioles upto 3 mm long. Flowers greenish white or greenish yellow, sessile. Berry 5 mm diam., white. Seeds black. *Fls. & Frts.:* May-September.

Rarely on the hedges and thickets of the Sundarbans areas.

Field Collection: Piyalinagar/Baruipur, *K.N. - 107.*

Distributions: Western Peninsula and Sri Lanka, Tropical and South Africa; Madagascar.

Economic Importance: Medicinally this plant has emmence value.

LXXXV APOCYNACEAE

Key to the Genera

1a	Plants tree in nature		(2)
1b	Plants herbs or shrubs in nature		(7)
2a (1)	Leaves alternate		(3)
2b (1)	Leaves opposite or whorled		(5)
3a (2)	Corolla funnel shaped		(4)
3b (2)	Corolla salver shaped; flowers yellowish white or yellowish red or pink	11. *PLUMERIA* L.	
4a (3)	Flowers yellow	14. *THEVETIA* L. *nom. cons.*	
4b (3)	Flowers white	6. *CERBERA* L.	
5a (2)	Leaves whorled; flowers white in sub-terminal corymbose cymes; grow as an ornamental; fusiform follicles stalked and beaked	3. *ALSTONIA* R. Brown *nom.cons.*	

5b (2)	Leaves purely opposite	(6)
6a (5)	Corolla tube long; fruit of 2-terete, elongated, spreading and in curved follicles	7. *HOLARRHENA* R. Brown
6b (5)	Corolla tube short; fruit of 2 linear, erect connate or distinct follicles	16. *WRIGHTIA* R. Brown
7a (1)	Plant erect shrubs	(8)
7b (1)	Plants climbing or twining shrubs or herbs	(12)
8a (7)	Leaves opposite	(9)
8b (7)	Leaves whorled	(11)
9a (8)	Anthers free from the stigmas, always included within the corolla, the anther cells with round base	(10)
9b (8)	Anthers adhering to the stigma by a point on the connective; anther cells produced downwards into a subulate empty spur; corolla salver-shaped, overlapping in the right	1. *AGANOSMA D. Don*
10a (9)	Flowers white in terminal or sub-terminal cymes; salver-shaped; corolla-lobes overlapping to the left	13. *TABERNAEMONTANA* L.
10b (9)	Flowers in terminal and axillary peduncle, 3-chotomous cymes, corolla cylindric tube, lobes overlapping to the right, ellipsoid or globose	4. *CARISSA* L. nom. cons.
11a (8)	Corolla salver-shaped, corolla lobes overlapping to the right	12. *RAUVOLFIA* L.
11b (8)	Corolla funnel-shaped, corolla lobes overlapping to the right	9. *NERIUM* L.
12a (7)	Plants only shrubby in nature	(13)
12b(7)	Plants erect herb, leaf opposite; flowers variously rosy or white	5. *CATHARANTHUS* G. Don
13a (12)	Flowers small or minute	(14)

13b (12)	Flowers medium or large	(15)
14a (13)	Corolla salver-shaped; anthers more or less exserted; connective not thickened	10. *PARSONSIA* R. Brown
14b (13)	Corolla salver-shaped; anthers included	8. *ICHNOCARPUS* R. Brown *nom. cons.*
15a (13)	Flowers yellow or violet, fruit ovate compressed, spinescent capsule	2. *ALLAMANDA* L.
15b (13)	Flowers white, fruits an oblong capsule	15. *VALLARIS Burm.*

1. AGANOSMA D. Don

1.1 ***Aganosma caryophyllata*** (Roxb. ex Sims.) G. Don. Gen. Syst. 5(4): 77.1837; Hook.f. Fl. Brit. India 3: 664.1882; Watt Dict. Econ. Prod 1: 129.1889; Prain Beng. Plants 2: 503.1903 (Rep.ed.1963). *(Vern. Name: Malati* or *Gandha Malati).*

Twining shrub. Leaves opposite with reddish veins. Flowers white, large, corolla salver-shaped, irregularly twisted, fragrant. Follicles densely yellow, tomentose when young. *Fls. & Frts.:* August-April.

Sometimes planted as ornamental twinners; but most often grow widely as climbing/twining shrub on the bushes.

Field Collection: Satzelia, *K.N. - 2158.*

Distributions: A native of Lower Bengal; and planted throughout India. In the tropical countries.

Economic Importances: Useful in bronchitis; leprosy, skin diseases, ulcers, inflammations, purulent discharges from the ear and the diseases of mouth; leaves cure biliousness.

Notes: D.Prain (1903) has reported this taxon from Bihar only; though Cooke, Fl. Bombay 2: 203.1904 mentioned this taxon as a native of Lower Bengal.

2. *ALLAMANDA* L.

2.1. ***Allamanda cathartica*** L. Mant. 2: 214.1771; Watt Dict. Econ. Prod. 1: 168.1889; Prain Beng. Plants 2: 493.1903 (Rep. ed. 1963). *A. subletii* Pohl, Ph. Bras. 1: 75.1827.*(Vern. Name: Har-kakra).*

Small climbing shrub. Leaves in whorls of 3 or 4. Flowers yellow, funnel shaped in terminal panicles. Sepals 6, ovate-lanceolate, petals campanulate, lobes orbicular spreading from a wide throat, 3-6 cm across. *Fls. & Frts.:* October-February.

Frequently planted in the ornamental gardens, but sometimes selfgrown on the bushes.

Field Collection: Diamond Harbour, *K.N. - 1951.*

Distributions: Introduced into India from Guiana, now planted as ornamental plants almost throughout India. A native of Tropical South America.

Economic Importances: Stem bark used as hydrogogue in ascities. Leaves are valuable cathartic in moderate doses; sometimes used in snake-bite.

Notes: D. Prain (1903a) has reported this taxon from the Central and Eastern parts of Bengal, as naturalized flora.

3. *ALSTONIA* R. Brown *nom. cons.*

3.1. ***Alstonia scholaris*** (L.) R. Br., Mem. Wern. Soc. 1: 75.1810; Hook.f. Fl. Brit. India 3: 642.1882; Watt Dict. Econ. Prod. 1: 97. 1889; Prain Beng. Plants 2: 499.1903 (Rep. ed. 1963). *Echites scholaris* L. Mant. 55.1767.*(Vern. Name: Chhatim).*

Tall evergreen tree, up to 30 m high. Leaves 4-7 in a whorl, ovate or elliptic-oblong, whitish beneath, base acute, tip- obtusely rounded or acuminate. Cymes peduncled or sessile; flowers capitate, greenish white, pubescent, strong odour. Follicles very long and slender. *Fls. & Frts.:* December-June.

Commonly planted in the ornamental gardens, and as avenue tree. Sometimes self-grown in the village thickets; flowers highly scented.

Field Collection: Baruipur, *K.N. - 117.*

Distributions: Widely planted throughout India; mostly in the deciduous forests. Sri Lanka, Burma, and throughout East Indies.

Economic Importances: Stem bark is an astringent tonic in anthelmintic; serves as a valuable medicine in chronic diarrhoea and the advanced stages of dysentery; used in catarrhal fever. The milky juice is applied to ulcer after mixing with oil; also used in restoring the tons of the stomach. Dried bark is powerful tonic for fevers, especially for malaria fever associated with diarrhoea, dysentery and inspired digestion.

4. *CARISSA* L. nom. cons.

Key to the Species

1a	Leaves always glabrous on both surfaces, rounded or obtuse	1. *C. carandas* L.
1b	Leaves glabrous or finely puberulous beneath; acute	2. *C. spinarum* L.

4.1. ***Carissa carandas*** L. Mant. 1: 52.1767; Hook.f. Fl. Brit. India 3: 630.1882; Watt Dict. Econ. Prod. 2: 165.1889; Prain Beng. Plants 2: 495.1903 (Rep.ed.1963). *Carissa congesta* Wight.(*Vern. Name: Karamcha).*

Large shrub, dichotomously branched with strong, forked or simple thorns. Leaves opposite, obtuse. Flowers whitish with pink tinge. Fruits small ellipsoid, sour drupe. Eight seeded. *Fls. & Frts.:* February-July.

Commonly planted on the garden edges or as hedge plants.

Field Collection: Sagar Islands/Mansadeep, *K.N. - 454.*

Distributions: Throughout India, mostly on the sandy and rocky soil in a wild stages; a native of India. Sri Lanka, Java and Timor.

Economic Importances: The fruit is anti-pyretic and useful in the diseases of brain; causes indigestion and make one lethargic; it also diminishes sexual power. Unripe fruit is useful in bilious complaints. It is also edible either raw or cooked.

4.2. ***Carissa spinarum*** L. Mant. 2: 559.1771; Hook.f. Fl. Brit. India 3: 631.1882; Watt Dict. Econ. Prod. 2: 166.1889; Prain Beng. Plants 2: 495.1903 (Rep.ed.1903). *C.diffusa* Roxb. Fl. Ind.1: 639.1820.*C.hirsuta* Roth Nov. Pl. Sp. 128.1821. *C. villosa* Roxb. Fl. Ind. (ed. Carey & Wall.) 2: 525.1824.

Undershrubs, with diffused 2-3 chotomous branches, thorns divaricate, straight. Leaves ovate to orbicular-ovate, shining, glabrous, acute apiculate at apex rounded or sub-cordate at base, lateral nerves conspicous combined just below the margin from an intramarginal nerves; cymes sessile, pedicels very short. Berry ellipsoid, usually 4-seeded. *Fls. & Frts.:* July-December.

Common in the village thickets.

Field Collection: Sagar Islands, *A.K. Mukherjee-5913* (CAL).

Distributions: Major parts of India. Common in Sri Lanka and Burma.

Economic Importance: Fruits edible and sweeter than *C. carandas.*

5. *CATHARANTHUS* G. Don

Key to the Species

1a	Leaves lanceolate; flowers 0.7 cm diam.	1. *C. pusillus* G. Don
1b	Leaves ovate to obovate; flowers 2.5-5.0 cm diam.	2. *C. roseus* G. Don

5.1. ***Catharanthus pusillus*** G. Don. Gen. Syst. 4:95. 1837. *Lochnera pusilla* K. Schum. Engl. & Prantl, Pflanzenf. 4 (2): 145.1895; Wlh. Ind. 6: 162.1962. *Vinca pusilla* Murr. Nov. Comm. Soc. Reg. Sc. Gott. 3: 66. 1773; Hook.f. Fl. Brit. India 3: 640.1882; Watt Dict. Econ. Prod. 6(4): 243.1893; Prain Beng. Plants 2: 966.1903 (Rep.ed.1963).

Erect annual herb, upto 30 cm tall; branches numerous spreading from the base, quadrangular. Leaves lanceolate, acute, glabrous, with rough margins, base tapering; axillary glands long, subulate. Flowers white, small, solitary, axillary ; pedicels short; corolla tube slightly enlarged near the top, mouth narrow, hairy. Follicles slender straight pointed, membranous, glabrous, striate, diverging. Seeds cylindric, black when ripe. *Fls. & Frts.:* April-October.

Often meet in the waste lands and village thickets.

Field Collection: Kazipara/Barasat, *K.N. - 930.*

Distributions: Western Himalaya, upper Gangetic Plain and the Western Penninsula. Sri Lanka.

Economic Importances: Toxic to the cattle, causing temporary blindness with urtiarial rashes all over the body; a decoction of the dried plant and after boiling with oil is rubbed in lambago.

5.2. ***Catharanthus roseus*** (L.) G. Don. Syst. 4: 96.1837. *Vinca rosea* L. Syst. Nat. (ed.10) 944.1759; Hook.f. Fl. Brit. India 3: 640.1882; Watt Dict. Econ. Prod. 6(4): 244.1893; Prain Beng. Plants 2: 497.1903 (Rep.ed. 1963). *Lochnera rosea* (L.) Reichb., Cons. Reg. Veg. 134.1828.*(Vern. Name: Nayantara).*

Cultivated erect annual or perennial herb. Leaves opposite, entire elliptic. Flowers reddish, pink or white. Follicles 2-3 cm long, narrowly cylindric. *Fls. & Frts.:* Almost throughout the year.

Commonly planted in the flower gardens, often met with as an escape in the waste places.

Field Collection: Kholapota/Basirhat, *K.N. - 1136.*

Distributions: Almost throughout India; probably first introduced as an ornamental plant. Might be the native to America, especially West Indies; but others believd to be originated in Madagascar.

Economic Importances: The plant is used in the remedy for diabetes. Infusion of the leaves administered in menorrhagia. Alkaloid, vinblastrine acts as a heart poison; this is the source of important medicine for the treatment of cancer/therapy.

6. *CERBERA* L.

6.1. ***Cerbera odollam*** Gaertn. Fruct. 2: 193.1791; Naskar & Guha Bakshi. Mangrove Swamps Sundarbans 84.1987. *(Vern. Name: Dabur).*

Small tree in the coastal beach/swamp forests. Naskar & Guha Bakshi (l.c.) described this taxon after collecting from the lower Ganga delta (Gocharan/Sarberia, *K.N.* - 237). Bark of this plant and the nuts are narcotic , purgative and poisonous; all these are used in hydrophobia.

7. *HOLARRHENA* R. Brown

7.1. ***Holarrhena pubescens*** (Buch. Ham.) Wall, Cat No. 1673.1829. *Echites pubescens* Buch. Ham Trans. Linn. Soc. 13: 424.1822. *Holarrhena antidysenterica* (Heyne ex Roth) DC. Prodr. 8: 413.1844; Naskar & Guha Bakshi Mangrove Swamps Sundarbans 84.1987. *(Vern. Name: Kurchi).*

Small deciduous tree in the coastal mangrove forest areas. Naskar & Guha Bakshi (l.c.) described this taxon from the lower Ganga Delta (Kakdwip, K.N.-1926). Medicinally this plant is very important; useful in malarious fever; bark used in dysentery and in dropsy; seeds used in dysentery, diarrhoea and intestinal worms. Kurchi bismuth iodide is a specific pharmacopocial drug in the treatment of amoebic dysentery.

8. *ICHNOCARPUS R.* Brown *nom. cons.*

8.1. ***Ichnocarpus frutescens*** (L.) R. Brown, Mem,. Wern. Soc. 1: 62.1899; Hook.f. Fl. Brit. India 3: 669. 1882; Watt Dict. Econ. Prod. 4: 326.1890; Prain Beng. Plants 2: 504.1903 (Rep.ed. 1963). *Apocynum frutescens* L. Sp. Pl. 213.1753. *(Vern. Name: Shyamalata* or *Dudhilata).*

Large evergreen climber with rusty tomentose branches; leaves opposite, elliptic-oblong; glabrous above, slightly pubescent beneath. Flowers white in terminal pedunculate cymes. *Fls. & Frts.:* September-April.

Abundantly found on the hedges, thickets; on the small tree, prefer moist situations.

Field Collection: Beliachandi, *K.N. - 266.*

Distributions: Upper Gangetic Plains from Delhi to Bengal. Sri Lanka, China, Java, Australia and Iran.

Economic Importances: Root cures thirst, vomitting , fever, biliousness and the diseases of blood. The stalk and leaves are used in the form of decoction in fever.

9. *NERIUM* L.

9.1. ***Nerium indicum*** Mill., Gard. Dict. (ed.8) No.2.1768; Naskar & Guha Bakshi Mangr. Swamps Sundarbans 85.1987.*(Vern. Name: Karabi).*

Large evergreen shrub or small tree. Naskar & Guha Bakshi (l.c.) described this taxon from the lower Ganga Delta after collection from Sagar/Mansadwip (*K.N. - 426*).

10. *PARSONSIA* R. Brown

10. 1. ***Parsonsia spiralis*** Wall. Cat. 1631.1828; Hook. f. Fl. Brit. India 3: 650.1882; Prain Beng. Plants 2: 501.1903 (Rep. ed.1963). *Heligme rheedei* Wight, Icon. t. 1303. *Aganosma concanensis* Hook. Icon pl. t. 841.

Perennial, evergreen twining shrub. Leaves 7.8-15.0 cm X 4.0 -8.0 cm, elliptic-lanceolate, acuminate, glabrous, base rounded; main veins 6-8 pairs. Flowers greenish, in paniculately corymbose axillary cymes. Fruits coriaceous, cylindric, sharply pointed, straight, glabrous. *Fls. & Frts.:* April-October.

Occasionally in the village thickets.

Field Collection: Sundarbans, D. Prain *s.n.* (CAL).

Distributions: Assam, Lower Bengal and W. Peninsula. Lower Burma, Sri Lanka, Malaya and China.

Economic Importances: Juice of the plant used internally in insanity.

11. *PLUMERIA* L.

Key to the Species

1a	Leaves obtuse with narrowly revolute margin; flowers white	1. *P. alba* L.
1b	Leaves obovate, oblong, but more or less tapering both the ends, entire flowers red	(2)

2a (1)	Flowers red with golden yellow inside, fragrant, in terminal corymbose cymes	2. *P.rubra* L.
2b (1)	Flowers white or cream coloured with yellow spot in the centre	3. *P.rubra* L. forma *acuminata* Woodson

11. 1. ***Plumeria alba*** L. Sp. Pl. 210. 1753. *(Vern. Name: Sada Gour Champa).*

Small tree. Leaves darkgreen, obtuse, with narrowly revolute margin. Flowers subspicate on the branches of corymbose; panicle pure white. *Fls. & Frts.:* March-August.

Planted as ornamental garden plants or as avenue tree; for its beautiful foliages and flowers.

Field Collection: Baguihati,*K.N. - 1887.*

Distributions: Almost throughout India. A native of America.

Economic Importances: Applied to ulcers, herps and scabies; purgative, alterative, detergent and syphilitic ulcers.

11.2. ***Plumeria rubra*** L. Sp. Pl. 209.1753; Benthall Trees Calc. 304.1946.*(Vern. Name: Gour Champa).*

Small tree. Leaves obovate, oblong, but more or less tapering both the ends, 20-30 cm long, alternate; entire. Flowers red, golden yellow inside, fragrant, in terminal corymbose cymes. *Fls. & Frts.:* June-September.

Planted in the ornamental gardens and roadside lawns.

Field Collection: Ichhapur, *K.N. - 1800.*

Distributions: Planted almost throughout India. Introduced into India in the year 1841. A native of America, distributed from Mexico to Venezuela.

Economic Importance: The latex is applied in ulcers.

11.3. ***Plumeria rubra*** L. forma ***acuminata*** (Ait.) Woodson, Ann. Missouri Bot.Gard. 25: 211.1938; Benthall Trees Calc. 304.1946. *P. acuminata* Ait., Hort. Kew (ed. 2) 2: 70. 1811. *P. acutifolia* Poir, Lamk. Encycl. Suppl. 2: 667.1912; Hook.f. Fl. Brit. India 3: 741.1882;Prain Beng. Plants 2: 496.1903 (Rep. ed. 1963). *(Vern. Name: Gour Champa).*

Small tree. Leaves obovate, alternate, entire. Flowers white or cream coloured, with yellow spot in the centre. *Fls. & Frts.:* June-September.

Planted in the ornamental gardens.

Field Collection: Thakurpukur, *K.N. - 1677.*

Distributions: Planted almost throughout India. Indigenous to tropical America. Introduced into India in the year 1841.

Economic Importances: A stick well coated with the fresh latex is applied to the mouth of the uterus for bringing abortion.

12. *RAUVOLFIA* L.

Key to the Species

1a	Leaves glabrous	1. *R. serpentina* (L.) Benth. ex Kurz
1b	Leaves pubescent	2. *R. tetraphylla* L.

12.1. ***Rauvolfia serpentina*** (L) Benth. ex Kurz, For. Fl. Brit. Burm. 2: 171.1877; Hook.f. Fl. Brit. India 3: 632.1882; Watt Dict. Econ. Prod. 6(1): 398.1892; Prain Beng. Plants 2: 497.1903 (Rep. ed. 1963). *Ophioxylon serpentinum* L. Sp. Pl. 1043.1753. *(Vern. Name: Sarpagandha).*

Erect small herb. Leaves elliptic-lanceolate or obovate to oblanceolate. Flowers pink in terminal or lateral, long peduncled cymes. Drupes polished green becoming red and finally turning black. *Fls. & Frts.:* Almost throughout the year; specially April-June.

Almost common on the hedges, roadsides, jungles and on the railway tracks.

Field Collection: Madarhat/Baruipur, *K.N. - 71.*

Distributions: Throughout Bengal and Bihar. Burma, Sri Lanka and Java.

Economic Importances: The reserpine, a new alkaloid isolated from the resinous extracts is supposed to be therapeutically more effective. It is an effective sedative and reduce blood pressure. It possesses wonderful hypnotic effect. Reserpine is a strong and long duration sedative. It is very useful against insomnia and epilepsy. The dried root is an antidote to insect bites.

12.2. ***Rauvolfia tetraphylla*** L. Sp. Pl. 208.1753. *R. canescens* L. Sp. Pl. (ed.2) 303.1763; Prain Beng. Plants 2: 497.1903 (Rep. ed. 1963).

Small shrub. Leaves whorled, pubescent. Flowers in umbellate cymes. Fruits drupaceous. *Fls. & Frts.:* Throughout the year.

On dampy waste places, railway tracks and roadside thickets.

Field Collections: Bahuru, *K.N. - 1628.*

Distributions: Almost throughout India. In the Tropics; a native of West Indies.

Economic Importances: Plant extract and castor oil is useful for chronic skin diseases. Fruit juice produce black dye.

13. *TABERNAEMONTANA* L.

13.1. ***Tabernaemontana divaricata*** (L.) R. Brown Roem. & Schult. Syst. 4: 427.1819. *Nerium divaricatum* L. Sp. Pl. 209.1753. *N. coronarium* Jacq. 1: 138.1786. *Tabernaemontana coronaria* Willd. Enum. Hort. Berol 275. 1909; Hook.f. Fl. Brit. India 3: 646.1882; Watt Dict. Econ. Prod. 6(3): 401.1892; Prain Beng. Plants 2: 498.1903 (Rep. ed. 1963). *(Vern. Name: Tagar).*

Cultivated, much branched, glabrous shrub. Leaves eiliptic lanceolate or oblong. Flowers white in terminal or sub-terminal cymes. Fruits 3 ribbed follicles. *Fls. & Frts.:* May- October.

Commonly planted in the gardens.

Field Collection: Gocharan/Sarberia, *K.N. - 236.*

Distributions: Naturalised in the upper Gangetic plain; Garhwal hills, Khasia Hills and Assam; indigenous in the foot-hills of the Himalaya.

Economic Importances: Roots are useful in the diseases of blood; after chewed the root relieves toothache. Milky juice is useful in the eye diseases. The wood is refrigerent.

14. *THEVITIA* L. *nom cons.*

14.1. ***Thevetia peruviana*** (Pers.) Schum. Pfam. 4(2): 159.1895; Naskar & Guha Bakshi Mangr. Swamps Sundarbans 86.1987. *(Vern. Name: Kalkephul).*

Naskar & Guha Bakshi (1.c.) described this taxon after collecting from Garia *(K.N. - 311).* Medicinally it is very important in the Ayurvedic tincture.

15. *VALLARIS* Burm.

15.1. ***Vallaris solanacea*** (Roth) O. Kuntze, Rev. Gen. 2: 417.1891. *Peltanthera solanacea* Roth, Nov. Pl. Sp. 132.1821. *Vallaris heynei* Spreng., Syst 1: 635.1824; Hook.f. Fl. Brit. India 3: 650.1882; Watt Dict. Econ. Prod. 6(4): 219.1892; Prain Beng. Plants 2: 500. 1903. *(Vern. Name: Haper Mali).*

Large twining shrub. Leaves opposite, pellucid, punctate, oblong lanceolate, elliptic acuminate. Flowers white, fragrant, in lax pubescent cymes; calyx ciliate; stigma pointed. Fruits 10-15 cm long. Seeds ovoid, beaked. *Fls. & Frts.:* April-January.

Rarely found in the village thickets and hedges.

Field Collection: Magrahat, *K.N. - 271.*

Distributions: Himalaya, more or less throughout India. Sri Lanka and Burma.

Economic Importances: Said to be useful in the old sores. Bark bitter in taste and astringent and useful for fixing loose teeth. An ingredient for Vishagarbha oil. Flowers and fruits are occasionally edible.

16. *WRIGHTIA* R.Brown

Key to the Species

1a	Fruits without white tubercles, not grooved at the junction of the carpels	1. *W. tinctoria* R. Brown
1b	Fruits rough with white tubercles, grooved at the junction of the carpels	2. *W. tomentosa* Roem. & Schult.

16.1. ***Wrightia tinctoria*** R. Brown, Men. Wesn. Soc. 1: 47.1811; Hook.f. Fl. Brit. India 3: 653.1882; Watt Dict. Econ. Prod. 6(4): 316.1892; Benthall Trees Calc. 313.1946. *(Vern. Name: Indra Jau).*

Small tree with milky latex. Bark smooth, light grey, branches terete, glabrous, whorled. Leaves 4-8 in a whorl, oblong-lanceolate, coriaceous, entire, obtuse, deep green above, whitish beneath. Flowers greenish white small, fragrant, many in umbellate cymes. Follicles terete, pendulous. Seeds linear oblong, papillose. *Fls. & Frts.:* November-August.

Rarely cultivated, sometime wild.

Field Collection: Mollakhali, *K.N. - 1856.*

Distributions: Native of Central and Western India. Tropical.

Economic Importances: Bark is used as a healing agent. It is also useful in malaria fever and chronic diarrhoea. Juice of fresh fruit is used for coagulating milk.

16.2. ***Wrightia tomentosa*** Roem. & Schult. Syst. 4: 414.1819; Hook.f. Fl. Brit. India 3: 653. 1882; Watt Dict. Econ. Prod. 6(4): 317.1893; Prain Beng. Plants 2: 499.1946. *(Vern. Name: Dudh Karabi).*

Small tree with milky juice; branches divaricate; bark smooth, yellowish-grey; branchlets pubescent. Leaves elliptic-oblong; acuminate, tomentose on both sides. Flowers in short dense erect terminal corymbose, tomentose cymes, white when on the trees. Fruits cylindric; seeds slender. *Fls. & Frts.:* March- August.

Very occasionally planted in the public gardens; according to Benthall (1.c.) this taxon seems to be indigenous near Calcutta.

Field Collection: Dakshin Barasat, *K.N. - 1655.*

Distributions: Throughout the Tropical India. Sri Lanka and Penag.

Economic Importances: Produce yellow-dye; barks are useful for snake-bite and scorpion sting. Leaves and fruits and edible.

Remarks: Besides these, *Alstonia macrophylla* Wall. ex Hook.f., *Wrightia coccinea* Sims., *Allamanda neriifolia* Hook., *Beaumontia grandiflora* Wall., and *Trachelospermum jasminoides* Lem. are also grown in the lower Ganga delta region. But the present author have no collections of the above mentioned species.

LXXXVI. ASCLEPIADACEAE

Key to the Genera

1a	Filament free, without interposed glands; seeds without coma	6. *FINLAYSONIA* Wall.
1b	Filaments connate into a tube; pollen waxy	(2)
2a (1)	Anthers without a membranous tip; corona single, staminal	3. *CEROPEGIA* L.
2b (1)	Anthers with a membranous tip; pollinia often pedicelled	(3)
3a (2)	Pollinia pendulous from the corpuscle	(4)
3b (2)	Pollinia erect on the corpuscle	(8)
4a (3)	Erect leafy shrub	(5)
4b (3)	Twining or sub-erect shrub	(6)
5a (4)	Corolla wide-campanulate	2. *CALOTROPIS* R. Br.
5b (4)	Corolla rotate	1. *ASCLEPIAS* L.

6a (4)	Corona double; corolla wide, rotate-campanulate	(7)
6b (4)	Corona single, or if double, the outer as well as the inner attached to the stamens	9. *PENTATROPIS* Wight. & Arn.
7a (6)	Leaves narrow, deciduous; follicles smooth	8. *OXYSTELMA* R. Brown
7b (6)	Leaves broad; follicles echinate	10. *PERGULARIA* L.
8a (3)	Corolla lobes overlapping	(9)
8b (3)	Corolla valvate	(12)
9a (8)	Corona absent; flowers small, crowded	11. SARCOLOBUS R. Brown
9b (8)	Corona staminal	(10)
10a (9)	Corolla salver-shaped; flowers medium	12. TELOSMA Coville
10b (9)	Corolla rotate	(11)
11a (10)	Flowers medium, coronal processes stellately spreading, fleshy pollinia oblong	5. *DREGEA* E. Mey. *nom. cons.*
11b (10)	Flowers very small, coronal processes rounded or depressed; pollinia globose	13. *TYLOPHORA* R. Brown
12a (8)	Corona stellately spreading thick and fleshy corolla rotate	7. *HOYA* R. Brown
12b (8)	Corolla erect membranous corolla urceolate	4. *DISCHIDIA* R. Brown

1. *ASCLEPIAS* L.

1.1. ***Asclepias curassavica*** L. Sp. Pl. 215.1753; Watt Dict. Econ. Prod. 1: 343.1889; Hook.f. Fl. Brit. India 4: 18.1883; Prain Beng. Plants 2: 511.1903 (Rep. ed. 1963).

Erect perennial herb. Leaves linear lanceolate, alternate or whorled. Flowers orange, small in umbellate cymes; sepals connate below in a 5-partite calyx; petals 5, connate in a rotate corolla, lobes often reflexed, naked, valvate. Fruits of 2- turgid, beaked, smooth follicles. *Fls. & Frts.:* March-November.

Grow in the village shade, occasionally.

Field Collection: Sonarpur, R. N. Banerjee-*s.n.* (CAL).

Distributions: Well naturalized in W. India; introduced from West Indies. A native of America.

Economic Importances: Occasionally planted as ornamental plant. Roots are used as emetic, purgative, and in treatment of piles and gonorrhoea. Leaves juice act as anthelmintic.

2. *CALOTROPIS* R. Brown

Key to the Species

1a	Corolla bluish-pink; corona lobes auricled at apex	1. *C. gigantea* (L.) R. Brown
1b	Corolla white; corona lobes not auricled at apex	2. *C. procera* (Ait.) R. Brown

2.1. ***Calotropis gigantea*** (L.) R. Brown, Ait. Hort. Kew. (ed-2) 2: 78. 1811; Hook.f. Fl. Brit. India 4: 17. 1883; Watt Dict. Econ. Prod. 2: 34. 1889; Prain Beng. Plants 2: 510.1903 (Rep. ed. 1963). *Asclepias gigantea* L. Sp. Pl. 214. 1753. *(Vern. Name: Akanda).*

Stout, large hoary-tomentose shrub or small tree; young parts and the under surface of the leaves covered with appressed white floceous tomentum, with a silky white coma. *Fls. & Frts.:* December-August.

Commonly found in the waste, open dry lands; roadsides and railway tracks. Preferably on the sandy dry soils and village thickets.

Field Collection: Baruipur, *K.N. - 50.*

Distributions: Almost throughout India; indigenous in India. S. Asia, Sri Lanka, Malay Islands and S. China.

Economic Importances: Roots are used in elephentiates, lupus, tuberculous leprosy, syphilis, syphilitic ulcerations and chronic rheumatism. It is also useful in scorpion bites. It is good medicine in sight blindness. Hindus use flowers for various ceremonies.

2.2. ***Calotropis procera*** (Ait.) R. Brown., Ait. Hor. Kew. (ed-2) 2: 78. 1811; Hook.f. Fl. Brit. India 4: 18.1883; Watt Dict. Econ. Prod. 2: 49. 1889; Prain Beng. Plants 2: 510.1903 (Rep. ed. 1963). *Asclepias procera* Ait. Hort. Kew. (ed.1) 1: 305.1786. *(Vern. Name: Sada Akanda).*

Erect or decumbent shrub, young foliage hoary. Leaves 8-17 cm long, oblong cordate, sub-amplexicaul. Flowers purplish-red, pale silvery outside, smelling, in terminal and axillary corymbose cymes, corolla lobes erect. Follicles recurved, smooth, seeds obovate with milky white corona. *Fls. & Frts.:* December- July.

Occasionally found in the waste open dry sandy areas.

Field Collection: Piyalinagar/Baruipur, *K.N. - 105.*

Distributions: Throughout India. Ava, Persia and Tropical Africa.

Economic Importances: Though the Hindu physicians consider it useful in skin diseases, enlargement of abnormal viscera, intestinal worms, cough and ascites are also treated; but the recent clinical trials do not entertain these claims.

3. *CEROPEGIA* L.

3.1. ***Ceropegia macrantha*** Wight; Hook.f. Fl. Brit. India 4: 74. 1883; Prain Beng. Plants 2: 520. 1903 (Rep. ed. 1963).

Twining herbs with tuberous roots. Leaves opposite, minute glabrous. Flowers in axillary peduncled umbellate cymes; corolla distinctly curved; tube not longer than the glabrous lobes. Fruits terete smooth, slender follicles. *Fls. & Frts.:* January-June.

Occasionally on the village thickets, climbing/twining on the bushes.

Field Collection: Suparibhanga/Sagar Islands, *K.N. - 412.*

Distributions: Northern parts of Bengal.

Notes: D. Prain (1903a) has not mentioned this taxon in Bengal.

4. *DISCHIDIA* R. Br.

4.1. ***Dischidia nummularia*** R. Brown Prodr. 461; Hook.f. Fl. Brit. India 4: 49.1883; Prain Beng. Plants 2: 518.1903 (Rep. ed. 1963). *D. orbicularis* Dene. *Collyris minor Vahl*, Act. Soc. Hafm. 6:3. *(Vern. Name: Pargachha).*

Very slender herb. Leaves whitish when dry with recurved margins. Flowers small, red in axillary fascicles, umbels. Fruits slender follicles. Seeds slender. *Fls. & Frts.:* June-November.

Epiphytic on the large tree.

Field Collection: Nilgunj, *K.N. - 2195.*

Distributions: On the major parts of India. Bangladesh, Malay Islands, Singapore and Australia.

5. *DREGEA* E. Mey. *nom.cons.*

5.1. ***Dregea volubilis*** (Linn.f.) Benth. ex Hook.f. Fl. Brit. India 4: 46. 1883; Watt Dict. Econ. Prod. 3: 193.1880; Prain Beng. Plants 2: 517. 1903 (Rep. ed. 1963). *Asclepias volubilis* Linn.f. Suppl. 170. 1781. *Wattakaka volubilis* (Linn.f.) Stapf., Curtis, Bot. Mag. t.8976.1923. *Marsdenia volubilis* T. Cooke. *(Vern. Name: Tita-Kunga).*

Brownish tomentose herb. Leaves ovate or orbicular, acuminate at apex, cordate or rounded at base, lateral nerves 4-5 pairs. Petioles 4-6 cm long. Flowers numerous in umbels, green or yellow-green, peduncles equally or slightly longer than the petiole; pedicels slender. Follicles turgid, glabrous, striate, slightly tapering towards the tip into a slunt point. Seeds flat, obovoid, pale-yellow-brown. *Fls. & Frts.:* November-August.

Common on the hedges and thickets.

Field Collection: Jugdia Benipur/near Magrahat, *K.N. - 215.*

Distributions: Major parts of India. Sri Lanka and Java.

6. *FINLAYSONIA* Wall.

6.1. ***Finlaysonia obovate*** Wall. Pl. As. Rar. 2: 48.t.162. 1831; Naskar & Guha Bakshi Mangr. Swamps Sundarbans 88.1987. *(Vern. Name: Dudhi Lata).*

Naskar & Guha Bakshi (1.c.) described this taxon from the lower Ganga delta.

7. *HOYA* R.Br.

7.1. ***Hoya parasitica*** Wall., Wight Contrib. 37; Hook.f. Fl. Brit. India 4: 57.1883; Prain Beng. Plants 2: 519.1903 (Rep. ed. 1963). *Asclepias parasitica* Roxb. (Hort. Beng. 20.1814) & Fl. Brit. India 2: 42.1824. *(Vern. Name: Pargachha).*

Plants with stout stem, glabrous; leaves 12-25 cm ovate or oblong-ovate, acute or acuminate, very thick shining 5-nerved, margins recurved, penduncles solitary or fascicled, very thick. Flowers small, corolla tube longer, red. *Fls. & Frts.:* Throughout the year.

Commonly epiphytic on the *Ficus* sp. and *Mangifera indica.*

Field Collection: Gobardanga, *G. Sen Gupta-73(CAL).*

Distributions: Assam and Sundarbans. Chittagong to Singapore, Malacca and Andaman Islands.

Economic Importance: Cause damage to the host plant.

8. *OXYSTELMA* R. Br.

8.1. ***Oxystelma secamone*** (L.) K. Schum, Engler. & Prantl. Nat. Pfam. 4(2): 229.1895; Naskar Aq. & Semi-Aq. Plants 165.1990. *Periploca secamone* L. Mant. 216.1771. *O. esculentum* R. Br. Mem. Wern. Soc. 1: 40.1810. *(Vern. Name: Kirmi).*

Naskar (1.c.) described this taxon from the lower Ganga delta.

9. *PENTATROPIS* Wight & Arnold

9.1. ***Pentatropis capensis*** (Linn.f.) Bullock, Kew Bull. 1955: 284.1955; Naskar & Guha Bakshi Mang. Swamps Sundarbans 88.1987. *Cynanchum capense* Linn. f. Suppl. 168.1781. *P. microphylla* (Roxb.) W. & A. Wight Contri. 52.1834.

Naskar & Guha Bakshi (1.c.) described this taxon from the lower Ganga delta after collecting from Canning *K.N. - 203.*

10. *PERGULARIA* L.

10. ***Pergularia daemia*** (Forssk.) Choiv. Result. Sc. Miss. Stefan. Paoli. Somal. Ital. 1: 115. 1916; Naskar Aq. & Semi-Aq. Plants 164.1990. *(Vern. Name: Chhagal Bati).*

Naskar (1.c.) described this taxon from the lower Ganga delta, after collecting from Suryapur *K.N. - 231.* This plant has emetic and expectorant properties; a decoction of the leaves is employed in some infantile complaints.

11. *SARCOLOBUS* R. Br.

Key to the Species

1a	Corolla yellow with brown dots; fruits yellow, small	1. *S. carinatus* Wall.
1b	Corolla purplish; fruits brown, large	2. *S. globosus* Wall.

11.1. ***Sarcolobus carinatus*** Wall., As. Res. 12.570., t. 4.1816; Naskar & Guha Bashi Mangr. Swamps Sundarbans 89.1987.

Naskar & Guha Bakshi (1.c.) described this taxon after collecting from Sundarbans in the lower Ganga delta (Sajnakhali, *K.N. - 609).*

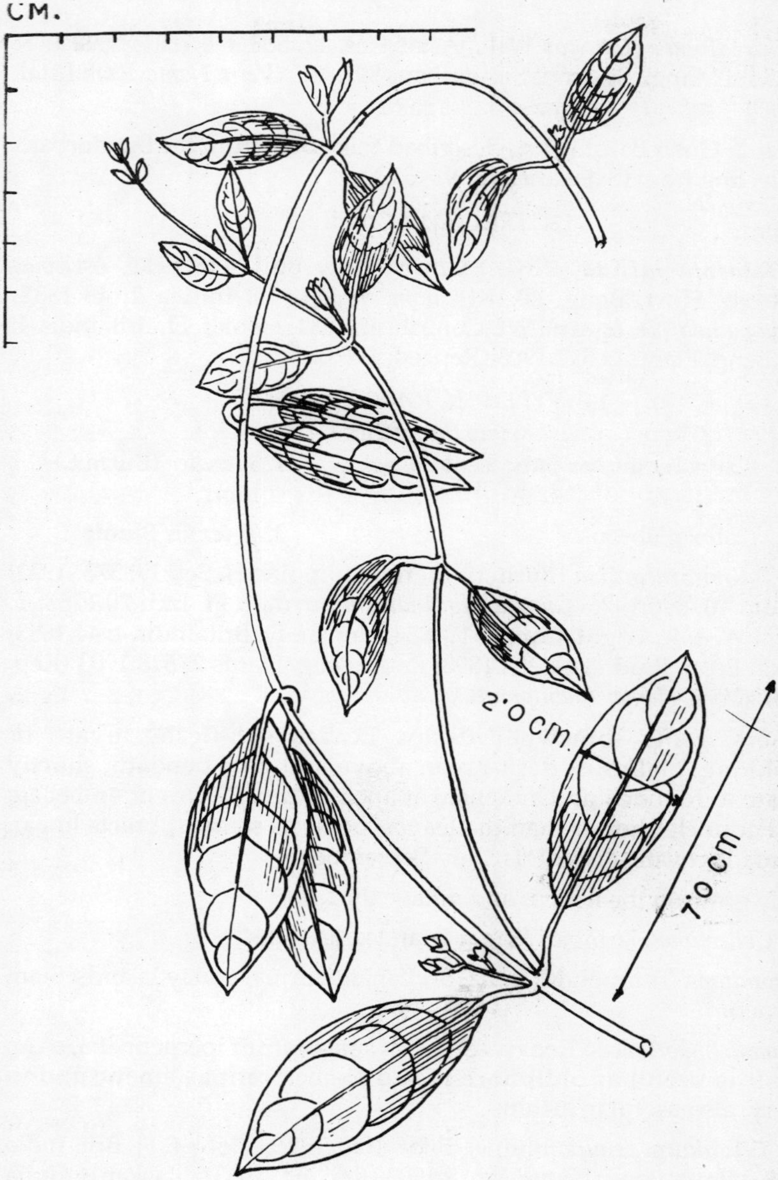

Fig.35. SARCOLOBUS CARINATUS Wall.—Flowering twig.

11.2. ***Sarcolobus globosus*** Wall., Asiat. Res. 12: 568. t. 4.1831; Naskar & Guha Bakshi Mangr. Swamps Sundarbans 89.1987. *(Vern. Name: Baoli Lata).* (Fig. 35).

Naskar & Guha Bakshi (1.c.) described this taxon from the Sundarbans after collecting from (Sajnakhali, *K.N. - 512.*)

12. *TELOSMA* Coville

12.1. ***Telosma pallida*** (Roxb.) Craib., Kew Bull. 418.1911. *Asclepias pallida* Roxb. Hort. Beng. 20.1814 *nom. nud.* & Fl. Indica 2: 48.1832. *Pergularia pallida* Wt. & Arn., Wt. Contrib. 42.1834; Hook.f. Fl. Brit. India 4: 38.1883; Beng. Plants 2: 517.1903 (Rep. ed. 1963).

13. *TYLOPHORA* R. Brown.

Key to the Species

1a	Calyx hirsute or pubescent	1. *T. indica* (Burm.f.) Merr.
1b	Calyx glabrous	2. *T. tenuis* Blume

13.1. ***Tylophora indica*** (Burm.f.) Merr., Philippines J. Sci. 19: 373. 1921; Wlh. India 10: 398.1976. *Cynanchum indicum* Burm. f. Fl. Ind. 70.1768. *T. asthmatica* W. & A. Wight, Contr. 51. 1834; Hook.f. Fl. Brit. India 4: 44.1883; Watt Dict. Econ. Prod. 6(4): 204.1893; Prain Beng. Plants 2: 518.1903 (Rep. ed. 1963). *(Vern. Name: Anantamul).*

Twiners; roots somewhat fleshy. Leaves sub-fleshy, ovate or ovate-oblong, glabrous, dark-green above and pale beneath, shortly acuminate to rounded and apiculate at apex, entire. Flowes in umbellate cymes. Peduncle shorter than the leaves; pedicels slender; bracts linear. Seeds broadly ovate. *Fls. & Frts.:* July-September.

Rarely grow on the hedges and village thickets.

Field Collection: Harinbari/Sagar Islands, *K.N. - 484.*

Distributions: Throughout India. Sri Lanka, Burma, Malay Islands, Siam and Borneo.

Economic Importances: Root is used as a substitute of ipecacunaha, as an emetic. It is useful as a diphoretic and expectorant as emetic and in dysentery; also useful in asthma.

13.2. ***Tylophora tenuis*** Blume, Bijdr. 1062.1826; Hook.f. Fl. Brit. India 4: 40.1883; Prain Beng. Plants 2: 518.1903 (Rep. ed. 1963); Naskar & Guha Bakshi Mangr. Swamps Sundarbans 90. 1987. *Asclepias tenuissima* Roxb. Fl. Ind. 2: 41. 1832.

Procumbent or twining herb. Naskar & Guha Bakshi (1.c.) described this taxon after collecting from the Sagar Islands.

Economically this taxon is useful in urticaria, excessive perspiration, bilious swelling and small-pox. Considered as valuable alexipharmac.

LXXXVII. PERIPLOCACEAE

(Splitted out from Asclepiadaceae)

Key to the Genera

1a	Corolla small, rotate	(2)
1b	Corolla large, funnel-shaped, the lobes overlapping	2. *CRYPTOSTEGIA* R. Brown
2a (1)	Corolla lobes overlapping	1. *CRYPTOLEPIS* R. Brown
2b (1)	Corolla lobe valvate	3. *HEMIDESMUS* R. Brown

1. *CRYPTOLEPIS* R. Br.

1.1. ***Cryptolepis buchanani*** Roem. & Schult., Syst 4: 409.1819; Hook.f. Fl. Brit. India 4: 5.1883; Watt Dict. Econ. Prod. 2: 642.1880; Prain Beng. Plants 2: 507.1903 (Rep. ed. 1963). *Nerium reticulatum* Roxb. Fl. Ind. 2: 8. 1824. *(Vern. Name: Karanata).*

Shrub, twining in nature with milky juice. Leaves sub-coriaceous, shining, dark-green above and pale beneath, elliptic-oblong or oblong-lanceolate. Flowers in axillary cymes; bracts ovate lanceolate; calyx glabrous; corolla pale-yellow or greenish-yellow. *Fls. & Frts.:* June-December.

Hedges and village shrubberies in general.

Field Collection: Dhosha/Chandaneswar, *K.N. - 613.*

Distributions: Throughout India. Sri Lanka.

Economic Importance: Produce fibre.

2. *CRYPTOSTEGIA* R. Brown

2.1. ***Cryptostegia grandiflora*** R. Brown, Bot. Reg. 435.1819; Hook.f. Fl. Brit. India 4: 6.1883; Watt Dict. Econ. Prod. 2: 625.1889; Prain Beng. Plants 2: 507.1903 (Rep. ed. 1963).

Climbing shrub. Leaves glossy elliptic 7.5-10.0 cm long. Flowers large, rose coloured, showy. Follicles 3-winged. *Fls. & Frts.:* June-November.

Planted in the gardens.

Field Collection: Dhakuria, *K.N. - 1969.*

Distributions: Usually cultivated throughout India. A native of Madagascar.

Economic Importances: Milky juice supply a fairly good caoutcho, which is an article of commerce. Used for tissue paper and very thin paper of commercial importance.

3. *HEMIDESMUS* R. Br.

3.1. ***Hemidesmus indicus*** (L.) R. Brown Aiton, Hort. Kew (ed-2) 2: 75.1811; Hook.f. Fl. Brit. India 4: 5.1883; Watt Dict. Econ. Prod. 4: 219.1890; Prain Beng. Plants 2: 508.1903 (Rep. ed. 1963). *Periploca indica* L. Sp. Pl. 211. 1753. *(Vern. Name: Ananta Mul).*

Twining or prostrate shrub; root stock aromatic. Leaves 5-10 cm long, elliptic oblong to linear lanceolate, obtuse or acute, apiculate, most variable in shape. Flowers greenish outside but purplish within, clustered in the leaf-axils; corolla rotate, lobes thick, ovate, acute. Follicle 10-12 cm long. Seeds flattened. *Fls. & Frts.:* August-December.

Commonly found under shades and mostly in moist localities.

Field Collection: Baruipur, *K.N. - 11.*

Distributions: North Western part of India. Sri Lanka.

Economic Importances: The decoction of root is given for cure of chronic cough and diarrhoea. In the form of paste, the root is effectively used for the treatment of swelling, rheumatic pains and boils.

LXXXVIII. LOGANIACEAE

Key to the Genera

1a	Corolla imbricate	1. *BUDDLEJA* L.
1b	Corolla valvate; capsule dehiscent	2. *CYNOCTONUM* Gmel.

1. *BUDDLEJA* L.

1.1. ***Buddleja asiatica*** Lour., Fl. Cochinch. 72.1790; Clarke, Hook.f. Fl. Brit. India 4: 82.1883; Watt Dict. Econ. Prod. 1: 546.1889; Prain Beng. Plants 2: 522.1903 (Rep. ed. 1963). *B. discolor* Roth, Nov. Pl. Sp. 83.1821. *(Vern. Name: Nimda/Bad-bhota).*

Large evergreen shrub. Leaves 5-15 cm long, lanceolate, acute; acuminate crenate or serrulate, sometimes entire, glabrous above. Flowers small, white, fragrant, in terminal or axillary. Capsules long, ovoid or ellipsoid, glabrous. *Fls. & Frts.:* March-June.

Grown as an hedge plants.

Field Collection: Bamanghata, *K.N. - 2307.*

Distributions: Throughout India. Malaya, Cochin and China.

Economic Importance: Ornamental white fragrant blooms.

Notes: Prain (1903a) has not mentioned this taxon from Lower Bengal.

2. *CYNOCTONUM* Gmel.

2.1. ***Cynoctonum mitreola*** (L.) Britton, Mem. Torr. Bot. Clus 5: 258.1894; Naskar Aq. & Semi-Aq. Plants 166.1990. *Ophiorrhiza mitreola* L. Sp. Pl. 150.1753. *Mitreola oldenlandioides* Wall. ex DC. Prods. 9: 9. 1846; Clarke, Hook.f. Fl. Brit. India 4: 79.1883. *Mitreola paniculata* Wall. Cat. 4349.

Naskar (1.c.) described this taxon from the lower Ganga delta. The collection was made from Ichhapur, *K.N. - 1818.*

Remarks.: The other taxon *Strychnos nux-vomica* L. Sp. Pl. 189.1753 is not wild in this lower Ganga delta, but occasionally planted in the ornamental gardens.

LXXXIX. MENYANTHACEAE

(*nom. alt.* Gentianaceae)

Key to the Genera

1a	Aquatic floating herbs with orbicular cordate leaves	4. *NYMPHOIDES* Hill.
1b	Semi-aquatic or terrestrial herbs	(2)
2a (1)	Flowers regular, rosy; stamens 5	2. *CENTAURIUM* Hill.
2b (1)	Flowers irregular	(3)
3a (2)	Ovary two celled; flowers blue, stamens 4-5	3. *EXACUM* L.
3b (2)	Ovary one celled; flowers white or rosy, stamens 4	1. *CANSCORA* Lamk.

1. *CANSCORA* Lamk.

Key to the Species

1a	Calyx winged; flowers white	1. *C. decussata* (Roxb.) R. & S.
1b	Calyx not winged; flowers rose	2. *C. diffusa* (Vahl) R. Br.

1.1. ***Canscora decussata*** (Roxb.) Roem. & Schult. Syst. Veg. Mant. 1: 229.1822; Clarke, Hook.f. Fl. Brit. India 4: 104.1883; Prain Beng. Plants 2: 536.1903 (Rep. ed. 1963). *Pladera decussata* Roxb. Fl. Ind. (ed. Carey) 1: 418.1820. *(Vern. Name: Dankuni).*

Erect branching, annual terrestrial herb. Stems 4-winged, dichotomously branched. Leaves 1.76-4.0 cm long, oblong-lanceolate, sessile, 3-nerved. Flowers white, irregular, calyx 4- winged. Capsule oblong, shorter than the calyx. *Fls. & Frts.:* April-December.

Grassy places, paths and waste grounds, everywhere.

Field Collection: Basiarhat/Kholapota, *K.N. - 576.*

Distributions: Throughout India; Sri Lanka, Burma and Tropical America.

Economic Importances: This plant is used as tonic, laxative and in fever; also used in nervous complaints.

1.2. ***Canscora diffusa*** (Vahl) R. Brown Prodr. Fl. Nov. 451.1810; Clarke, Hook.f. Fl. Brit. India 4: 103.1883; Prain Beng. Plants 2: 525.1903 (Rep. ed. 1963). *Pladera diffusa;* Roxb. Fl. Indica 1: 418.1820.

Erect filiform dichotomous herbs with pale rose flowers. Under moist conditions; the basal leaves become larger than normal, but those larger ones are early caducous. *Fls. & Frts.:* November-March.

Grassy places and side of the paths, everywhere, in fallow field.

Field Collection: Behala, *Dutta-64(C.C.H.)*

Distributions: Throughout India. Sri Lanka, Malaya, Australia and Tropical Africa.

Economic Importance: Useful in epilepsy and nervous debility.

2. *CENTAURIUM* Hill.

2.1. ***Centaurium centaurioides*** (Roxb.) Rao & Kammathy, J. Bombay nat. Hist Soc. 67: 357.1970; Naskar Aq. & Semi-Aq. Plants 167.1990. *(Vern. Name: Gima).*

Naskar (1.c.) described this taxon from the Lower Ganga delta (Mollakhali, *K.N. - 1855.*) This plant is bitter in taste and much useful in stomach trouble, as a substitute for chiretta.

3. *EXACUM* L.

Key to the Species

1a	Plants 20 cm or less, usually 3 nerved leaves	1. *E. pedunculatum* L.
1b	Plants above 20 cm, leaves 5 nerved	2. *E. tetragonum* Roxb.

3.1. ***Exacum pedunculatum*** L. Sp. Pl. 112. 1753; Clarke, Hook.f. Fl. Brit. India 4: 97.1883; Watt Dict. Econ. Prod. 3: 306.1890; Prain Beng. Plants 1: 524.1903 (Rep. ed. 1963); Naskar Aq. & Semi-Aq. Plants 168.1990.

Naskar (1.c.) described this taxon after collecting from Sarisha/near Diamond Harbour, *(K.N. - 1942.)*

3.2 ***Exacum tetragonum*** Roxb. Fl. Indica 1: 413.1820; Naskar Aq. & Semi-Aq. Plants 169.1990. (*Vern. Name: Kuchari*).

Naskar (l.c.) described this texon after collecting from Ichhapur (K.N.-1801).

4. *NYMPHOIDES* Seguier.

Key to the Species

1a	Corolla lobes entire; seeds scabrous	1. *N. hydrophylla* Kuntze
1b	Corolla lobes fimbriate; seeds smooth	2. *N. indicum* (L.) Kuntze

4.1. ***Nymphoides hydrophylla*** (Lour.) Kuntze, Rev.Gen. Pl. 2: 429.1891; *Menyanthes hydrophylla* Lour., Fl. Cochinch. 129.1790. *N. cristatum* Kuntze, Rev. Gen. Pl. 2: 429. 1891. *M. cristata Roxb. Pl. Cor. 2: 3, t.106.1798. Limnanthemum cristatum* Griseb., Gen. & Sp. Gent. 342.1839. (*Pan-chuli*).

Naskar (Aquatic & Semi-Aquatic Plants of Lower Ganga delta 170. 1990) described this taxon after collecting from Mallickpur *(K.N. - 90)*.

Economically the stalks and leaves are pounded with oil and applied to ulcers and bites; a decoction is used as a wash for parasitic skin affections; seeds anthelmintic.

4.2. ***Nymphoides indicum*** (L.) Kuntze, Rev. Gen. Pl. 429.1891; Naskar Aq. & Semi-Aq. Plants 171.1990.

Naskar (1.c.) described this taxon after collecting from Sashan *(K.N. - 1179)*. Plants is used as substitute for Chiratha.

XC. POLEMONIACEAE

1. *PHLOX* L.

1.1. ***Phlox drummondi*** Hook.; Prain Beng. Plants 2: 528.1903 (Rep. ed. 1963).

Annual garden herb. Leaves opposite. Flowers red or white, solitary or in terminal cymes. Fruit ovoid, 3-nerved capsule. Seeds ovoid. *Fls. & Frts.:* January-April.

Commonly planted as ornamental plants in the public gardens.

Field Collection: Bamanghata, *K.N. - 2308*.

Distributions: Cultivated throughout India. Native of North America.

Economic Importance: Ornamental garden plants.

XCI. HYDROPHYLLACEAE

1. *HYDROLEA* L. *nom. cons.*

1.1. ***Hydrolea zeylanica*** (L.) Vahl, Symb. Bot. 2: 46.1791; Naskar Aq. & Semi-Aq. Plants 171.1990. *(Vern. Name: Kasschra).*

Naskar (1.c.) described this taxon from the lower Ganga delta after collecting from Sitakundu/Baruipur *(K.N. - 55)*.

XCII. EHRETIACEAE

(Splitted out from Boraginaceae)

Key to the Genera

1a	Shrubs or small tree; styles terminal, twice-forked	1. *CORDIA* L.
1b	Trees or shrubs; styles terminal, once forked	2. *EHRETIA* L.

1. *CORDIA* L.

Key to the Species

1a	Corolla small white; tube hardly larger than the calyx	(2)
1b	Corolla large, orange, tube much longer than calyx	3. *C. sebestena* L.
2a (1)	Leaves beneath densely tomentose	1. *C. dichotoma* Forsst.f.

2b (1) Leaves beneath pubescent or puberulous when young; but not densely tomentose 2. *C. myxa* L.

1.1. ***Cordia dichotoma*** Forsst.f. Prodr. 18.1786 *C. obliqua* Willd., Phytogr. 4, t. 1794; Clarke, Hook.f. Fl. Brit. India 4: 139.1883; Prain Beng. Plants 2: 530.1903 (Rep. ed. 1963). *C. latifolia* Roxb. Fl. Indica 2: 330. 1824. *(Vern. Name: Chhoto Lashora).*

Small to moderate deciduous tree. Leaves coriaceous elliptic-lanceolate to broadly ovate or sub-orbicular, rounded, cordate or somewhat cuneate at base. Flowers white, fragrant, in large, lax, terminal and axillary, peduncled cymes. Drupes ovoid, apiculate. *Fls. & Frts.:* February-August.

Grow in the hedges, rarely cultivated.

Field Collection: Chemagari/Sagar Islands, *K.N. - 460.*

Distributions: Throughout India, both wild and cultivated.

Economic Importances: Fruits used as vegetable. Decoction of the leaves in combination with leaves of *Ocimum sanctum* is a very common remedy for cold, cough and influenza.

1.2. ***Cordia myxa*** L. Sp. Pl. 190.1753; Clarke, Hook.f. Fl. Brit. India 4: 136.1883; Prain Beng. Plants 2: 530.1903 (Rep. ed. 1963). *(Vern. Name: Bohuari* or *Lashora).*

Moderate, deciduous tree; branches usually glabrous. Leaves alternate, broadly obtuse, sinuate-dentate, glabrous. Flowers white, pedicels short. Drupe ovoid supported by the saucer shaped irregularly 5-10 lobed more or less longitudinally striate glabrous calyx. *Fls. & Frts.:* March-July.

Occasionally on the village thickets.

Field Collection: Kalinagar, *K.N. - 768.*

Distributions: Throughout India. Sri Lanka, Egypt, Cochin—China and Tropical Australia.

1.3. ***Cordia sebestena*** L. Sp. Pl. 190.1753; Prain Beng. Plants 2 : 531. 1903 (Rep. ed. 1963). *(Vern. Name: Bokhaf* or *Bohari).*

Small evergreen tree. Leaves alternate, ovate or elliptic. Flowers orange red in lax terminal cymes. Fruit ovoid drupe, enclosed in the persistent calyx. *Fls. & Frts.:* March-August.

Occasionally planted in the public gardens.

Field Collection: Bhakuria/ on the public garden. *K.N. - 1970.*

Distributions: Commonly cultivated throughout India. A native of West Indies and Peru; now widely planted throughout the hotter parts of the World.

Economic Importances: Ornamental garden plant.

2. *EHRETIA* L.

2.1. ***Ehretia serrata*** Roxb. Fl. Indica 1: 596.1820; Benthall Trees Calc. 326.1946. *E. acuminata* R. Br.; Clarke, Hook.f. Fl. Brit. India 4: 141. 1883; Prain Beng. Plants 2: 533.1903 (Rep. ed. 1963). *(Vern. Name: Kula-aja).*

Medium sized evergreen tree. Leaves alternate, ellipitc-oblong. Flowers white in terminal compound panicles. Fruit ellipsoid drupe. *Fls. & Frts.:* March-August.

Occasionally in the village thickets.

Field Collection: Kestour/near Baguihati, *M.K. Ghosh- 3004.*

Distributions: A native of the hills of the Northern and N.E. India. Cosmopolitan, throughout Burma to Japan and Australia.

XCIII. BORAGINACEAE

Key to the Genera

1a	Ovary deeply 4-lobed; fruits usually 4-nutlets	2. *CYNOGLOSSUM* L.
1b	Ovary entire or slightly 4-lobed; fruit drupaceous	(2)
2a (1)	Styles-2	1. *COLDENIA* L.
2b (1)	Style-1	3. *HELIOTROPIUM* L.

1. *COLDENIA* L.

1.1. ***Coldenia procumbens*** L. Sp. Pl. 125.1753; Clarke, Hook.f. Fl. Brit. India 4: 144. 1883; Prain Beng. Plants 2: 534.1903 (Rep. ed. 1963). *(Vern. Name: Triphunki).*

Prostrate, procumbent, deep-rooted herb, hairy. Leaves alternate, crisped, coarsely serrate. Flowers minute, sub-sessile, solitary, axillary white. Fruits a drupe of pyrene, 4 lobed. *Fls. & Frts.:* April-January.

Commonly found in moist places, dry paddy fields.

Field Collection: Titagarh, *K.N. - 2220.*

Distributions: Throughout the Tropical India. A Native of America, at present widely distributed in the warmer parts of the world.

Economic Importances: Fresh leaves are pested and applied to rheumatic swelling. Dried plants with equal part of fenugreek seeds is made into a fine powder and applied warm on suppuration of boils.

2. *CYNOGLOSSUM* L.

2.1. ***Cynoglossum lanceolatum*** Forssk., Fl. Aegypt. Arab 41.1775; Clarke, Hook.f. Fl. Brit. India 4:156.1883; Prain Beng. Plants 2: 535.1903 (Rep. ed. 1963). *C. lanceolatum* Desf., Tabl. Hort. Par. 220.1804. *C.racemosum* Roxb. Fl. Indica: 456.1820.

Erect hispid, annual or biennial hairy herb. Leaves alternate radical, oblong, obovate or broadly lanceolate strongly nerved beneath. Racemes elongate, ebracteate, flowers pedicelled, white; calyx deeply 5-lobed. Ovary lobes 4. Nutlets shortly ovate, all sides depressed. *Fls. & Frts.:* April-January.

Common on the waste places along the roadsides and railway tracks.

Field Collection: Hingalgunj, *K.N. - 1491.*

Distributions: Throughout India; Sikkim, Nepal, Burma, Afghanistan China and Formosa.

3. *HELIOTROPIUM* L.

Key to the Species

1a	Plants erect; comparatively larger in size; leaves more than 3 cm	2. *H. indicum* L.
1b	Plants prostrate or procumbent; comparatively smaller in size; leaves less than 2 cm long linear	(2)
2a (1)	Plants halophytic; fleshy; inflorescence downwardly curved; cymes wooly or smooth	1. *H. curassavicum* L.
2b (1)	Plants terrestrial or sub-aquatic; not fleshy; leaves very minute; inflorescence very small cymes	(3)
3a (2)	Leaves elliptic or obovate, obtuse; spikes persistently bracteate	3. *H. ovalifolium* Forssk.

3b (2) Leaves lanceolate; spikes not conspicously bracteate towards the upper part 4. *H. stigosum* Willd.

3.1. ***Heliotropium curassavicum*** L. Sp. Pl. 130.1753; Naskar & Guha Bakshi Mangr. Swamps Sundarbans 99. 1987.*(Vern. Name: Nona hatisur).*

Glaucous, prostrate villous halophytic herb. Naskar & Guha Bakshi (l.c.) described this taxon from the lower Ganga delta after collecting from Minakhan *K.N. - 1104.* Medicinally the plant is useful to sores and wounds. The root powder is applied to sores and wounds.

3.2. ***Heliotropium indiucm*** L. Sp. Pl. 130.1753; Clarker, Hook.f. Fl. Brit. India 4: 152.1753; Prain Beng. Plants 2: 532.1903 (Rep. ed. 1963).*(Vern. Name: Hatisura).*

Coarse, diffused, hairy annual herb. Leaves alternate or sub-opposite, petioled, ovate or ovate-oblong, obtuse or sub-acute. Flowers violet, 2 ranked, in elongated, bristly, ebracteate spikes. Fruits deeply 2-lobed, each lobe in bluntly 4-ribbed. *Fls. & Frts.:* All round the year.

Common on the waste places, rubbish heaps, everywhere.

Field Collection: Khodarbazar/Baruipur, *K.N. - 116.*

Distributions: Common throughout India. America, Sri Lanka, Malaya and Tropical Africa.

Economic Importances: Plants diuretic, leaves applied to boils, sores and the stings of insects and reptiles.

3.3. ***Heliotropium ovalifolium*** Forssk., F. Aegypt. -Arab. 38.1775; Clarke, Hook.f. Fl. Brit. India 4: 150.1883; Prain Beng. Plants 2: 532.1903 (Rep. ed. 1963). *H.coromandelianum* Retz. Obs. 2: 9.1781.*(Vern. Name: Nag-phuli).*

Prostrate, villous sub-aquatic herb, usually with an erect central stem and spreading lateral ones or decumbent. Leaves elliptic or obovate, constracted at the base into a slender petiole. Flowers white, in compact, hairy, usually forked spikes. Calyx deeply 5-partite. Nutlets-4. *Fls. & Frts.:* February-August.

Commonly found on roadsides and railway tracks.

Field Collection: Amtala, *K.N. - 364.*

Distributions: Bengal and Western Peninsula. Tropics of the Old World; a native of West Indies.

Notes: Prain (1903a) did not mention this taxon from the lower Bengal.

3.4. ***Heliotropium strigosum*** Willd. Sp. Pl. 1: 743.1798; Clarke, Hook.f. Fl. Brit. India 4: 151.1890; Prain Beng. Plants 2: 532.1903 (Rep. ed. 1963). *(Vern. Name: Chiti-phul).*

Small, much-branched, usually procumbent, strigose, perennial herb. Branches slender, several from the woody base. Leaves lanceolate, or linear-lanceolate, acute at the apex, entire. Flowers small, white, sessile or lower ones pedicelled. 2- rancked,in simple spikes; bracts small, linear, the lower bracts larger, leafy. Fruit subglobose, depressed, nutlets 4, hairy minute, more or less united, slightly hairy, nutlets. *Fls. & Frts.:* March-August.

Waste places and railway tracks.

Field Collection: Ghoshpur/Kulti Lock Gate, *K.N. - 1423.*

Distributions: Throughout India. Bangladesh, Burma, Malaya, China, Australia and North Africa.

Economic Importances: Raw juice of the plant is used in diuretic cases. Plant is laxative and the plant juice is applied to sore eyes, gum boils and sores.

Notes: Prain (1903a) did not report this taxon from the lower Bengal.

XCIV. CONVOLVULACEAE

Key to the Genera

1a	Plants not so much woody or stout		(2)
1b	Plants mostly stout, woody, scandent perennial twinners		(8)
2a (1)	Styles -2 or deeply 2-partites		(3)
2b (1)	Styles-1, undivided upto the stigma		(4)
3a (2)	Corolla sub-entire, infundibuliform; capsules 4-valved, 4- grained	3. *EVOLVULUS* L.	
3b (2)	Corolla lobed, campanulate; capsules 2-valved, 1 or 2, sometimes with 3 grains	11. *VOLVULOPSIS* Roberty	
4a (2)	Fruits indehiscent; leaves densely white tomentose beneath (except *A. hookeri*)	1. *ARGYREIA* L.	

4b (2)	Flowers usually dehiscent by valves; leaves never densely white tomentose beneath	(5)
5a (4)	Flowers yellow	7. *MERREMIA* Demsh. Emend. Hallie
5b (4)	Flowers white, pink or violet, never yellow	(6)
6a (5)	Flowers numerous, in large panicles	9. *PORANA* Burm.
6b (5)	Flowers solitary or in cymes	(7)
7a (6)	Stigma with rounded lobes; ovary 2-4 celled	5. *IPOMOEA* L.
7b (6)	Stigmas with ovate or elongate lobes; ovary 1-2 celled	2. *CONVOLVULUS* L.
8a (1)	Fruits surrounded by enlarged calyx; leaves variable in shape	4. *HEWITTIA* Wt. & Arn.
8b (1)	Fruits not surrounded by enlarged calyx	(9)
9a (8)	Woody twiners, seeds pubescent	10. *STICTOCARDIA* Halier .f.
9b (8)	Stout, scandent climbing villous herbs; seed glabrous	(10)
10a (9)	Plants without milky juice	6. *LEPISTEMON* Bl.
10b (9)	Plants with milky juice	8. *OPERCULINA* Manso

1. *ARGYREIA* L.

Key to the Species

1a	Leaves glabrous above, white tomentose beneath broadly ovate-cordate	2. *A. nervosa* *(Burm.f.)* Boj.
1b	Leaves glabrous both on the above and beneath, not white tomentose	1. *A. hookeri* Clarke

1.1. ***Argyreia hookeri*** Clarke, Hook.f. Fl. Brit. India 4: 186.1883. *A. malaberica* Woodr., J. Bombay nat Hist Soc. 12: 170.1898.

Huge climbing shrub; stem grooved, glabrous or slightly pubescent. Leaves broadly ovate, acute or acuminate, glabrous, base cordate with rounded lobes; main nerves 10-12 pairs. Flowers in long pedunculate, few-flowered cymes; peduncles 15-20 cm long, grooved more or less hairy rose coloured. Berry 2 cm diam, globose. *Fls. & Frts.:* August-November.

Very rare, but occasionally planted in the hedges as ornamental plant.

Field Collection: Baruipur, *K.N. - 14.*

Distributions: Assam, W. Peninsula and Andamans & Nicobar Islands. Nepal, Sikkim Himalaya and Bhutan.

Notes: Prain (1903a) did not report this taxon from Bengal.

1.2. ***Argyreia nervosa*** (Burm.f.) Boj, Hort. Maurit. 224.1837. *Convolvulus nervosus* Burm. f. Fl. Ind. 48, t. 20, f. 1.1768. *Argyreia speciosa* Sweet, Hort. Brit. (ed.1) 289.1827; Clarke, Hook.f. Fl. Brit. India 4: 185.1883; Watt Dict. Econ. Prod. 1: 310.1889; Prain Beng. Plants 2: 551.1903 (Rep. ed. 1963). *Ipomoea speciosa* Pers. Syn. Pl. 1: 183.1805. *(Vern. Name: Guguli* or *Bich-tarak).*

Large woody, handsome climber. Leaves, ovate, cordate, acute. Flowers 5-10 cm in diam., white woolly outside and bright purple within, in very long peduncled, many flowered cymes. *Fls. & Frts.:* August-December.

Commonly planted in gardens, climbing on the walls.

Field Collection: Magrahat, *K.N. - 1202.*

Distributions: Assam, Bengal, Mysore; cultivated elsewhere. Java and other Asian countries.

Economic Importances: Ornamental garden plant. Roots are used as tonic, alterative and also beneficial in rheumatism and nervous diseases.

2. *CONVOLVULUS* L.

2.1. ***Convolvulus arvensis*** L. Sp. Pl. 153.1753; Clarke, Hook.f. Fl. Brit. India 4: 219.1883; Watt Dict. Econ. Prod. 2: 518.1889.

Trailing or twining herb; more or less angular glabrous or pubescent stem. Leaves 2.5-8.0 cm long, breadth variable, ovate or oblong, obtuse and apiculate at the apex, hastate at the base; lower leaves often lobed. Peduncles axillary, solitary, slender. Corolla broadly infundibuliform, pink or nearly white, glabrous. Stamens unequal. Ovary glabrous, stigmas filiform. Capsules globose, glabrous. *Fls. & Frts.:* November-January.

Rare in distribution, occasionally on the village thickets.

Field Collection: Garia Rly. Station, *M.K.Ghosh- 2765*(CAL).

Distributions: Throughout India. Cosmopolitan.

Notes: Prain (1903a) did not report this taxon from Bengal.

3. *EVOLVULUS* L.

3.1. ***Evolvulus alsinoides*** (L.) L. Sp. Pl. (ed.) 392.1762; Clarke, Hook.f. Fl. Brit. India 4: 220.1883; Watt Dict. Econ. Prod. 3: 3.1890; Prain Beng. Plants 2: 539.1903 (Rep. ed. 1963). *Convolvulus alsinodeso* L. Sp. Pl. 157.1753. *Evolvulus hirsutus* Lamk. Method. 3: 538.1789.

Perennial herb. Stem branched from the base, prostrate, spreading, densely hairy. Leaves linear to broadly elliptic, obtuse apiculate at apex, acute at the base, pilose on both surface; petiole short. Flowers in axillary, 1-2 flowered cymes; capsule globose, 4-valved. Seed 3-4, glabrous. *Fls. & Frts.:* August-December.

Occasionally on the open grass-lands, waste places and road-sides.

Field Collection: Goranbose/Sundarbans, *K.N. - 1868.*

Distributions: Tropical and Sub-tropical countries of the world and throughout India.

Econorric Importances: Imporves appetite.

4. *HEWITTIA* Wt. & Arn.

4.1. ***Hewittia sublobata*** (Linn.f.) O.K. Rev., Pl. 441.1891. *Convolvulus sublobatus* Linn. f. Supl. 135.1781. *Hewittia bicolor* Wt.& Arn., Madr. J. Sc. 1 (5): 22.1827; Clarke, Hook.f. Fl. Brit. India 4: 216.1883; Prain Beng. Plants 2: 540.1903 (Rep. ed. 1963). *Shutereia bicolor* Choisy, Mem. Soc. Phys. Genev. 6 : 486.1834. *Palmia bicolor* Endl. ex Dlaz. & Gibs. Bombay Flora 163. *(Vern. Name: Jarad Kalmi).*

Perennial or annual twining herb, rarely prostrate, pubescent to nearly glabrous. Leaves shortly pubescent, or sub-glabrous, highly variable in shape, angular to shallowly lobed, ovate, acute to rounded at apex, wavy, cordate at base, firmly pubescent. Flowers axillary, solitary peduncle; peduncle pubescent, lanceolate. Capsule pilose, sub-globose, surrounded by enlarged calyx. *Fls. & Frts.:* November-February.

Commonly grow on the hedges and thickets of waste places.

Field Collection: Haroa, *K.N. - 587.*

Distributions: South East India & W. Peninsula. Tropical Africa, Polynesia, S.E.Asia and China.

5. *IPOMOEA* L.

Key to the Species

1a	Plants with tuberous roots	2. *I. batatas*(L.) Lamk.
1b	Roots not as above	(2)
2a (1)	Plants with filiform segments of leaves	11. *I. quamoclit* L.
2b (1)	Plants not as above	(3)
3a (2)	Leaves deeply bilobed	9. *I. pes-caprae* (L.) Sweet
3b (2)	Leaves not bilobed	(4)
4a (3)	Large scandent shrub	(5)
4b (3)	Herbaceous or woody twiners, trailers or climbers	(6)
5a (4)	Stem somewhat stout (root tuberous)	6. *I. mauritiana* Jacq.
5b (4)	Stem fistular	5. *I. carnea* Jacq.
6a (4)	Ornamental garden plants	10. *I. purpurea* Roth
6b (4)	Mostly wild plants	(7)
7a (6)	Aquatic herb; leaves mostly hastete	1. *L. aquatica* Forssk.
7b (6)	Terrestrial plants	(8)
8a (7)	Leaves compound	(9)
8b (7)	Leaves simple	(10)
9a (8)	Leaves usually 3-lobed	7. *I. nil* (L.) Roth
9b (8)	Leaves usually pentadigitate or palmate	3. *I. cairica* (L.) Sweet
10a (8)	Flowers pink, purple or white	(11)
10b (8)	Flowers white with yellow or green bands	(12)
11a (10)	Capsule globose, hairy	4. *I. eriocarpa* R. Br.
11b (10)	Capsule ovoid, glabrous	12. *I. sepiaria* Koening
12a (10)	Capsule ovoid, glabrous, seeds dark brown	8. *I. obscura* (L.) Ker.
12b (10)	Capsule globular, glabrous	13. *I. tuba* (Sch.) G. Don.

1.1. ***Ipomoea aquatica*** Forssk., Fl. Aegypt.-Arab. 44. 1775; Naskar, Aq. & Semi-Aq. Plants 171. 1990. *(Vern. Name: Kalmi Sak).*

Naskar (l.c.) described this taxon from the lower Ganga delta, after collecting from Khodarbazar, *K.N. - 1162.*

Economic Importances: The leaves and young stems are used as vegetable. The juice of the leaves acts as a mild purgative and is supposed to purify blood; the juice is also considered as emetic acting on the nerves of stomach, as antidote counter acting, the poisonous action of opium and arsenic. The plants is considered healthy for female, suffering from nervous and general debility.

5.2. ***Ipomoea batatas*** (L.) Lamk., Tabl. Encycl. 1: 465.1793; Clarke, Hook.f. Fl. Brit. India 4: 202.1883; Watt Dict. Econ. Prod 4: 478.1890; Prain Beng. Plants 2: 547.1903 (Rep. ed. 1963). *Convolvulus batatus* L. Sp. Pl. 154.1753. *Batatas edulis* Choisy, Convolv. Orient. Mem. Soc. Phys. Genev. 6: 435.1824. *(Vern. Name: Mitha-Alu).*

Prostrate or diffuse spreading, glabrous, tuberous rooting at the nodes. Tubers usually red or white. Leaves ovate, cordate, long petioled. Flower purple, 2 to several, on long axillary peduncles. *Fls. & Frts.:* November-March.

Normally cultivated in the sandy soils for its tuberous roots; the red variety is more frequently found in cultivation in this area.

Field Collection: Baruipur, *K.N. - 56.*

Distributions: Cultivated throughout India; introduced into India by the Portuguese in the early period of 16th Century. A native of Brazil; widely grown throughout the Tropics and some parts of the temperate regions in Africa, China, Japan, Malay Archipelago and Pacific Islands.

Economic Importances: Tubers are used as vegetable; sometimes in the preparation of sweets; leveas are green vegetable.

Notes: Cytological data show the complexity of its origin; some workers beleived that sweet potato originated from *I. tiliacea* (Wild.) Choicy–a wild species of Tropical America (Cooley, *Sc. Mon.* N. Y. 1951. 72, 325.).

5.3. ***Ipomoea cairica*** (L.) Sweet, Hort. Brit. 287.1827; Wlh. Ind. 5: 247.1959. *Convolvulus cairicus* L. Syst. (ed.10) 922.1759. *I. palmata* Forssk., Fl. Aegypt. Arab. 43.1775; Clarke, Hook.f. Fl. Brit. India 4: 214.1883 (partly). *I. pulchella* Roth, Nov. Pl. Sp. 115.1821; Prain Beng. Plants 2: 548.1903 (Rep. ed. 1963).

Very extensively twining shrub with tuberculate stems and slender long glabrous branches. Leaves usually deeply palmatey 5-7 lobed, orbicular. Flowers about 4.5 cm long, campanulate, pale, purple violet, anther purple; stigma white. Capsule sub-globose, glabrous, 2-celled. *Fls. & Frts.:* August-November.

Sometimes found climbing on bushes and even on small trees; often cultivated in gardens.

Field Collection: Bamanghata /Salt Lake, *K.N. -2310.*

Distributions: Throughout India; specially in Western India. Sri Lanka, a native of Tropical Asia and Africa.

Economic Importances: An ornamental plant for its beautiful flowers and foliages. Seeds are also used as purgative.

5.4. ***Ipomoea eriocarpa*** R. Br. Prodr. 424.1810; Clarke Hook.f. Fl. Brit. India 4: 204.1883; Watt Dict. Econ. Prod. 4: 485.1890. *Convolvulus hispidus* Vahl, Symb. Bot. 3: 29.1794. *I. hispida* Roem. & Schult. Syst. 4: 238.1819 (non Zucc. 1809); Prain Beng. Plants 2: 545.1903 (Rep. ed. 1963). *I. sessiliflora* Roth, Nov. Pl. Sp. 116.1821.

Slender twining villous annual shrub. Leaves ovate or oblong, lanceolate to deltoid, base cordate with rounded lobes. Flowers pink or purplish, in short stalked, dense, axillary heads, corolla campanulate. Capsules globose, hairy, 4-seeded. *Fls. & Frts.:* August-November.

Frequently found twining on shrubs.

Field Collection: Ichhpur, *K.N. - 1803.*

Distributions: Throughout India. Sri Lanka, Afghanisthan, Tropics of the Old World.

Economic Importances: Leaves and tubers are cooked as vegetable.

5.5. ***Ipomoea carnea*** Jacq. subsp. *fistulosa* (Mart. ex Choisy) Austin, Taxon 26: 237.1977. *I. fistulosa* Mart. ex Choisy, DC. Prodr. 9: 249.1845; Naskar Aq. & Semi-Aq. Plants 74.1990. *Batatas crassicaulis* Benth.Voy. Sulph.134. 1844. *I. crassicaulis* (Benth.) Robins, Proc. Amer. Acad. 51: 539.1916. *(Vern. Name: Dal Kalmi).*

Large straggling shrub with milky juice. Naskar (l.c.) described this taxon after collecting from Sonarpur, *K.N. - 333.*

A native of America and probably introduced in India very recently. Used as hedge plants in the village gardens and often very obnoxious weed in the agriculture fields and ponds.

5.6. ***Ipomoea mauritiana*** Jacq., Collect. 4: 216.1791. *I. digitata* L. Syst, Nat. 924.1759; Clarke, Hook.f. Fl. Brit. India 4: 202.1883; Watt Dict. Econ. Prod. 4: 484.1890; Wlh. Ind. 5: 248.1959. *I. paniculata* R. Brown, Prodr. 486.1819 (nom. Burm. f. 1768); Prain Beng. Plants 2: 547.1903 (Rep. ed. 1963). *(Vern. Name: Bhui Kumra).*

Perennial large glabrous twiner or scandent shrub; roots tuberous; stems somewhat stout. Leaves broad than long, usually 3-9 palmate lobed, lobes ovate, lanceolate or elliptic, shortly acuminate or acute at apex. Flowers many in corymbosely paniculate cymes; peduncles; bracts minute, deciduous. Capsule ovoid, 4-valved, surrounded by slightly cottony detachable hairs. *Fls. & Frts.:* July-January.

Commonly grow on the grassy gounds and scrub-jungles.

Field Collection: Kakdwip, *K.N. - 1927.*

Distributions: Throughout India. A pantropic species.

Economic Importances: Tubers contain a resin, similar to jalap resin and is considered tonic alterative, approdisiac, demulent,lactagogue and cholagogue. It is recommended for children emaciation.

5.7. ***Ipomoea nil*** (L.) Roth, Cat. Bot. 1: 36.1797; Clarke, Hook.f. Fl. Brit. India 4: 199.1883; Watt Dict. Econ. Prod. 4: 485.1890; Prain Beng. Plants 2: 546.1903 (Rep. ed. 1963). *Convolvulus nil* L. Sp. Pl.(ed.2) 219.1762. *(Vern. Name: Nil Kalmi).*

Hairy, twining annual herb. Leaves ovate, cordate, 3-lobed; lobes ovate, acuminate. Flowers deep blue tinged with pink; capsules 3-celled, sub-globose or ovoid, glabrous; seeds 6 or 4 glabrous, puberulent. *Fls. & Frts.:* August-November.

In the gardens, sometimes in the waste lands as wild.

Field Collection: Duttapukur, *K.N. - 1713.*

Distributions: Throughout India. Probably a native of America; naturalized in the Old World.

Economic Importances: Dried seeds are purgative; used as substitute for jolap.

5.8. ***Ipomoea obscura*** (L.) Ker. Gawl., Bot. 3: 239.1817; Clarke, Hook.f. Fl. Brit. India 4: 207.1883; Watt Dict. Econ. Prod. 4: 488.1890; Prain Beng. Plants 2: 547.1903(Rep. ed. 1963). Wlh.India 5: 251. 1959. *Convolvulus obscurus* L. Sp. Pl. (ed.2) 220.1762.

Slender, twining, annual herb. Stem hairy or glabrous, often tinged with purple. Leaves broad ovate, cordate, acute or acuminate, entire. Flowers white with yellow bands, usually one or 2-3 flowered peduncles. Capsule 2-valved, ovoid, glabrous. Seeds dark brown. *Fls. & Frts.:* September-February.

Twining on hedges or thickets.

Field Collection: Dhakuria, *K.N. - 1971.*

Distributions: Throughout India. Tropical Asia, Africa and Malaya.

5.9. ***Ipomoea pes-caprae*** (L.) Sweet, Hort. Suburb. 35.1818; Prain Beng. Plants 2: 547.1903 (Rep. ed. 1963). *Convolvulus pes-caprae* L. Sp. P. 159.1753. *I.biloba* Forssk. Fl. Aegypt.-Arab. 44.1776; Clarke, Hook.f. Fl. Brit. India 4: 212.1883. *(Vern. Name: Chagar Kuri).*

Prostrate week stemmed, much branched herb on the sea side sandy areas. Act as soil binder. Medicinally leaves used in rheumatism; astringent, stomachic alteration, tonic, diuretic and laxative. Naskar & Guha Bakshi (1987) described this taxon after collecting from Sagar Island, *(K.N. - 435).*

5.10. ***Ipomoea purpurea*** Roth, Bot. Abhand. 27.1787; Clarke, Hook.f. Fl. Brit. India 4: 200. 1883; Prain Beng. Plants 2: 446.1903 (Rep. ed. 1963); Wlh. India 5: 252.1959.

Stout hirsute twiner. Leaves ovate-cordate entire. Flowers 3-5 cm long; bracts minute; sepals without ligulate tips; peduncles almost always exceedig; pedicels 2.5 cm long, deflexed in fruits; petals glabrous externally; white to dark purple. *Fls. & Frts.:* July-October.

Occasionally planted as ornamental graden plants.

Field Collection: Kalyanpur, *K.N. - 1163.*

Distributions: Planted almost throughout India; introduced from Tropical America.

Economic Importances: Used as purgative and anti- syphilitic; showy ornamental plants.

5.11. ***Ipomoea quamoclit*** L. Sp. Pl. 159.1753; Clarke, Hook.f. Fl. Brit. India 4: 199. 1883; Ooststroom, Fl. Males I, 4(4): 482.1953.*Quamoclit pinnata* (Desr.) Boj., Hort. Maurit. 224.1837; Prain Beng. Plants 2: 548.1903 (Rep. ed. 1963). *Convolvulus pinnatus* Desr. Lamk. Encycl. 3: 567.1962. *Q.vulgaris* Choisy, Mem. Soc. Phys. Geneve 6: 434.1834. *Q. quamoclit*

Britton & Brown., Fl. North. Unit. Stat. 3: 22, 1893, *nom.* illegit. *(Vern. Name: Taru-Lata).*

Glabrous, slender, twining, annual herb. Leaves pinnately, divided into filiform segments. Flowers red, on 1-few flowered peduncles; corolla narrowly funnel shaped. *Fls. & Frts.:* August- November.

Grown in the gardens for its ornamental appearances. Sometimes wild in the waste places as climbing on bushes and trees.

Field Collection: Barasat, *K.N. - 918.*

Distributions: Throughout India. In the Tropics; a native of Tropical America.

Economic Importances: Ornamental garden plants.

5.12. ***Ipomoea sepiaria*** Koening ex Roxb. (Hort. Beng. 14.1814); Roxb. Fl. Ind. 2: 90.1824; Clarke, Hook.f. Fl. Brit. India 4: 209.1883; Watt Dict. Econ. Prod. 4: 401.1890; Prain Beng. Plants 2: 547.1903 (Rep. ed. 1963).*I. maxima* auct. non. Linn. f. Don ex Sweet 1830. *(Vern. Name: Ban-Kalmi).*

Perennial slender twiner, sometimes prostrate; stem woolly with deciduous hairs. Leaves glabrous, ovate, rarely angular or hastate, acute at apex; cordate at base, 3-9 cm long and broad. Flowers many in subumballate cymes; peduncle axillary, solitary, stout, highly variable in length; bracts small, caducous. Capsule glabrous, ovoid, 4-valved, 4 or 2 seeded. Seeds grey, silky pubescent. *Fls. & Frts.:* August-November.

Common on the village thickets and waste places.

Field Collection: Ghoramara/Sagar Island, *K.N. - 403.*

Distributions: Throughout India. Sri Lanka and Malaya.

Economic Importances: Plant juice considered deobstruent and diuretic; used as an antidote to arsenic.

5.13. ***Ipomoea tuba*** (Sch.) G. Don. Gen. Syst. 4: 271.1838. *Convolvulus tuba* Sch. Linn. 6: 735. 1831. *C. grandiflorus* Jacq. Hort. Vindob. 3: 39, non Linn.f. 1781.

Glabrous, woody twining shrub, terete or angular, often longitudinally wrinkled. Leaves orbicular or ovate 5-15 X 5-10 cm, deeply cordate at the base. Inflorescence axillary, one-few flowered; peduncle terete, variable in length; white with green bands. Capsule globular, glabrous pale brown, 2-celled, 4-valved. Seeds 4, black. *Fls. & Frts.:* August-November.

Occasionally on the coastal regions of Sagar Islands.

Field Collection: Sagar Islands/Handelketki, *A.K. Mukherjee-8781.*

Distributions: Western Peninsula. Tropical America, East Tropical Africa, Mascarene Islands and Tropical continental Asia to Polynesia.

6. *LEPISTEMON* Blume

6.1. ***Lepistemon binectariferum*** (Wall.) Rev. Gen. 446. 1891; Ooststroom, Blumea 5: 341. 1943; Fl. Males. I, 4(4): 489.1953. *Lepistemon wallichii* Choisy, Mem. Soc. Phys. Geneve 6: 443. 1833; Clarke, Hook.f. Fl. Brit. India 4: 216. 1883; Prain Beng. Plants 2: 549.1903 (Rep. ed. 1963).

Stout, scandent, climbing, villous herb. Leaves cordate. Flowers in axillary cymes, dense; bracts narrow, deciduous; corolla externally hairy; carpels, connate in a 2-celled ovary; ovules 4. Fruit a 4-valved capsule; seeds glabrous. *Fls. & Frts.:* July-November.

Very rare on the village thickets, climbing on the bushes.

Field Collection: Taki Madhyampur *K.N. - 959.*

Distributions: Vere rare in the Indian territory but common in Bangladesh.

Notes: Prain (1903a) did not report this taxon from Bengal (W.B.).

7. *MERREMIA* Demsh. emend. Hallier.f.

Key to the Species

1a	Peduncle very short or nearly absent, flowers in axillary clusters	1. *M. gangetica* (L.) Cufodontis
1b	Peduncle longer	(2)
2a (1)	Sepals unequal in length; outer ones shorter, elliptic to oblong-elliptic or oblong	3. *M. hirta* (L.) Merr.
2b (1)	Sepals slightly unequal in length; outer ones broadly obovate to orbicular or broadly spathulate	2. *M. hederacea* (Burm.f.) Hallier.f.

7.1. ***Merremia gangetica*** (L.) Cufodontis, Bull. Jard. bot. Burm. Suppl. 31. 743.1961. *Convolvulus gangeticus* L., Torner. Cent. Plant 2: 9. 1756. *Evolvulus emarginatus* Burm.f. Fl. India 77, t. 30,f.1. 1768. *Ipomoea reniformis* Choisy, Convolv. Orient. Mem. Soc. Phys. Genev. 6: 446. 1834; Clarke, Hook.f. Fl. Brit. India 4: 206.1883; Watt Dict. Econ. Prod. 4: 491.1890. *(Vern. Name: Bhui Kamari).*

Prostrate, creeping herbs, rooting at each nodes. Leaves reniform or ovate-cordate, more or less 3 cm long. Flowers axillary, long peduncled, bracteate yellow. Sepals 5, slightly enlarged in fruit; petals 5, yellow with 5 vertical brands, each marked by 5 dark violet lines. *Fls. & Frts.:* August-February.

In the grassy gardens and open sandy areas.

Field Collection: Durganagar, *K.N. - 1715.*

Distributions: India, mainly in Western Peninsula. Sri Lanka, Malaya, S.E. Africa and Tropical Africa.

Economic Importances: Diuretic and is used in rheumatism.

7.2. ***Merremia hederacea*** (Burm.f.) Hallier f., Engl. Bot. Jahrb. 18: 118.1894; Prain Beng. Plants 2: 543.1903 (Rep. ed. 1963). *Evolvulus hederaceous* Burm. f. Fl. Ind. 77, t. 30, f. 2,1768. *Ipomoea chrysoides* Ker-Gawl., Bot. Reg. t. 270.1818; Clarke, Hook.f. Fl. Brit. India 4: 206.1883. *Merremia convolvulacea* Deunst. (Schluss. Hort. Malab. 12, 23,34, 1888 nomen nund.) ex Hallier f. Bot. Jahrd. 16: 552.1893. *Merremia chryseides* Hallier.f., Engl. Bot. Jahrd. 16: 552. 1893.

Twining or climbing herb. Stem angular, hairy at the nodes. Leaves ovate, acute, apiculate, often shallowly 3-lobed, deeply cordate, repand toothed, petiole upto 6 cm long. Flower yellow, in dense axillary cymes; capsules ovoid or sub-quadrangular, 2- celled; seeds trigonous. *Fls. & Frts.:* August-January.

Frequently found climbing upon small trees and bushes. It looks very beautiful when in full bloom.

Field Collection: Baruipur/Khoderbazar, *K.N. - 114.*

Distributions: Throughout India. Sri Lanka, Malaya, China, Australia and Tropical Africa.

Economic Importance: Leaves poultice is useful to chopped hands.

7.3. ***Merremia hirta*** (L.) Merr. Philip. Sc. 7. Bot. 244. 1917; Ooststroom, Fl. Males, I, 4(4): 442.1953. *Convolvulus hirta* L. Sp. Pl. 1: 159.1753. *C. reptans* L. Sp. Pl. 158. 1753. *Merremia caespitosa* Hallier.f. Bot. Jahrb. 16. 552. 1893; Prain Beng. Plants 2: 543.1903 (Rep. ed. 1963). *C. caespitosus* Roxb. Fl. Ind. 1: 483.1820. *I. linifolia* Blume, Bijdr. 721.1825; Clarke, Hook.f. Fl. Brit. India 4: 205.1883.

Very small twining herb. Stem slender 20-60 cm long, filiform, sparsely hirsute or glabrous. Leaves very variable in form, linear, linear oblong lanceolate. Flowers one to few; broadly funnel-shaped, pale yellow. Capsule broadly ovoid to globular. *Fls. & Frts.:* August-December.

Occasionally in between the stones of the railway tracks; very small and twining habitat.

Field Collection: Kalyanpur, *K.N. - 1166.*

Distributions: Throughout India. South China, Siam, Tropical Australia and Malaysia.

8. *OPERCULINA* Manso.

8.1. ***Operculina turpethum*** (L.) Manso., Enum. Subst. Brass. 16.1836; Prain Beng. Plants 2: 543. 1903 (Rep. ed. 1963). *Convolvulus turpethum* L. Sp. Pl. 155.1753. *Ipomoea turpethum* R. Br. Prodr. 485. 1810; Clarke, Hook.f. Fl. Brit. India 4: 212. 1993; Watt Dict. Econ. Prod. 4: 493. 1890. *(Vern. Name: Dudh Kolmi or Tohri).*

Perennial twiners with milky juice; stems twisted cylindric, angled, compressed narrowly 2-5 winged. Leaves very variable in shape, ovate to oblong, acute or acuminate, apiculate at apex, cordate or sub-cordate at base. Flowers 2-4 in cymes; peduncle 3-6 cm long, axillary stout. Capsule globose, 2-4 seeded. Seeds 4-6 mm, black, glabrous. *Fls. & Frts.:* November-February.

Common on the waste places and village thickets.

Field Collection: Harinbari/Sagar Islands, *K.N. - 430.*

Distributions: Throughout India. Sri Lanka, Malaya Islands, Tropical America, Mauritius, Philippines and Tropical Africa.

Economic Importance: Roots has been used as purgative.

9. *PORANA* Burm.

9.1. ***Porana paniculata*** Roxb. Pl. Cor. 3: 31, t. 235.1815; Clarke, Hook.f. Fl. Brit. India 4: 222.1883; Prain Beng. Plants 2: 538.1903 (Rep. ed. 1963).

Large, beautiful climbing, grey tomentose shrub. Leaves 4-10 cm long, petioled, ovate, cordate, acuminate entire. Flowers small, white, numerous, arranged in large axillary and terminal, drooping panicles. Capsules membranous, bristled, hairy 1-seeded. *Fls. & Frts.:* August-March.

Commonly found climbing on trees and looks very beautiful appearances when in full-blossom.

Field Collection: Bamanghata/Salt Lake, *K.N. - 2308.*

Distributions: Many parts of India. A Pantropic species.

Economic Importance: Ornamental plants.

10. *STICTOCARDIA* Hallier.f.

10.1. ***Stictocardia tilifolia*** (Desr.) Hallier.f., Bot. Jahrb. 18: 159. 1894; Prain Beng. Plants 2: 530.1903 (Rep. ed. 1963). *Evolvulus tilifolius* Desc., Lamk. Encycl. 1: 467. 1781. *Argyreia tiliaefolia* Wt. Ic. t. 1358.1850; Clarke, Hook.f. Fl. Brit. India 4: 184.1883.

Woody twiners; young parts pubescent, becoming fine; glabrescent by age. Leaves ovate-cordate, acuminate mucronate at apex. Flowers 1-3 on axillary peduncle, which is shorter than the petiole; pedicels as long as or longer than the calyx, pubescent. Capsule 2-4 cm diam., globose. Seeds upto 10 mm long black, pubescent. *Fls. & Frts.:* November-February.

Commonly climbing on bushy plants and also trailing on grounds.

Field Collection: Bakkhali, *K.N. - 1907.*

Distributions: Throughout India. Malaya to Philippines.

11. *VOLVULOPSIS* Rober.

11. ***Volvulopsis nummularia*** (L.) Roberty, Candollea 14: 28. 1852. *Convolvulus nummularius* L. Sp. Pl. 157. 1753. *Evolvulus nummularius* (L) L. Sp. Pl. (ed.2) 391.1762; Prain Beng. Plants 2: 539.1903 (Rep. ed. 1963).

Perennial herb. Stems prostrate and rooting from the nodes. Leaves entire, distichous, broadly ovate, elliptic or orbiculate. Flowers white; bracts linear or lanceolate. Capsule globular. Seeds 3 or less, black or brown. *Fls. & Frts.:* August-January.

Rapidly spreading and abundantly found in the moist lands.

Field Collection: Dehimedanmallapur/Mallickpur, *K.N. - 82.*

Distributions: Introduced into India during the last part of the 18th Century. A native of West Indies.

Economic Importances: Juice of the plant is used for cooling.

XCV. CUSCUTACEAE

(Splitted out from Convolvulaceae)

1. *CUSCUTA* L.

1.1. ***Cuscuta reflexa*** Roxb., Pl. Cor. 2:3,t.104, 1799; Clarke, Hook.f. Fl. Brit. India 4: 225.1883; Watt Dict. Econ. Prod. 2: 671.1889; Prain Beng. Plants 2: 537.1903 (Rep. ed. 1963). *(Vern. Name: Swarnalata).*

Twining, yellow, leafless, fleshy stem parasitic herb. Flowers whitish or pinkish, clustered or in racemes, corolla lobes reflexed. Capsules depressed-globose, seeds 2-4. *Fls. & Frts.:* February-April.

Common as parasitic and forming dense mass on low trees and shrubs.

Field Collection: Madarhat/Baruipur, *K.N. - 72.*

Distributions: Throughout India. Sri Lanka and Malaya.

Economic Importance: The stems and seeds have medicinal values.

XCVI. SOLANACEAE

Key to the Genera

1a	Fruits a berry or a pod-like berry, but never capsular, not dehiscent	(2)
1b	Fruits capsular	(7)
2a (1)	Anthers connivent around the style	(3)
2b (1)	Anthers not connivent around the style	(4)
3a (2)	Opening of anthers mostly by a pore or slit at or near the apex	7. *SOLANUM* L.
3b (2)	Anthers opening by a long slit	4. *LYCOPERSICON* Mill.
4a (2)	Calyx enlarging in fruit	(5)
4b (2)	Calyx not as above	(6)
5a (4)	Flowers solitary	6. *PHYSALIS* L.
5a (4)	Flowers clustered	8. *WITHANIA* Pauq. *nom. cons.*
6a (4)	Corolla rotate or bell-shaped	1. *CAPSICUM* L.
6b (4)	Corolla distinctly tubular or cylindrical	2. *CESTRUM* L.

7a (1)	Calyx nearly or quite coveing the fruit	5. *NICOTIANA* L.
7b (1)	Calyx much shorter than the fruit, mostly reflexing	3. *DATURA* L.

1. *CAPSICUM* L.

1.1. ***Capsicum frutescens*** L. Sp. Pl. 189.1753; Prain Beng. Plants 2: 557.1903 (Rep. ed. 1963). *C. minimum* Mill. Gard. Dict. (ed.6) 10. 1762; Clarke, Hook.f. Fl. Brit. India 4: 239. 1883; Watt Dict. Econ. Prod. 2: 139.1889. *(Vern. Name: Dhani Lanka).*

Perennial shrub, branches angular, upto 2 m high. Leaves broadly ovate, acuminate, usually wrinkled, more or less pubescent. Corolla white. Berries conical, acuminate, red at maturity. *Fls. & Frts.:* December-April.

Occasionally planted for its pungent fruits; very rarely as garden escape.

Field Collection: Harinbari/Sagar Islands, *K.N. - 482.*

Distributions: Cultivated throughout, the Portuguese brought this plant to India from Pernambuco. A native of Tropical America.

Economic Importances: Fruits are edible in curries. It is a powerful stimulant and carminative in flatulent dyspepsia. Remedial for painful rheumatism.

Notes: The other common species of the cultivated *Capsicum* L. are as follows:

C. annuum L. var. *annuum* L., *C. annuum* L.var. *acuminata* roxb., *C. annuum* L. var. *abbreviata* Fingerh, *C. annuum* L. var. *cerasiformis* Mill., *C. annuum* L. var. *grossa* Sendt.

2. *CESTRUM* L.

Key to the Species

1a	Corolla-lobes obtuse or rounded; reflexed	1. *C. diurnum* L.
1b	Corolla-lobes acute, not reflexed	2. *C. nocturnum* L.

2.1. ***Cestrum diurnum*** L. Sp. Pl. 191.1753; Bor & Raizada Beaut. Ind. Climb. & Shrub. 118. 1954. *(Vern. Name: Buno-Hasnu Hana).*

Medium shrub, 1.5-2m high, young shoots pubescent. Leaves whitish beneath, oblong or elliptic-oblong, both ends acute or obtuse. Flowers numerous, peduncled, axillary or terminal branches compound panicles;

corolla infundibuliform; lobes rounded or obtuse, reflexed; stamens included; filaments inserted near the base of the corolla and adnate upto the throat. Berry subglobose, dark, purple when ripe. *Fls. & Frts.:* November-May.

On the village shrubberies and railway tracks. Grow gregariously.

Field Collection: Garia, *K.N. - 336;* Joynagar, *K.N. - 864.*

Distributions: Almost throughout India. A Tropical species.

Notes: D. Prain (1903a) did not mention this taxon from Bengal.

2.2. ***Cestrum nocturnum*** L. Sp. Pl. 191. 1753; Bailey Man. Cult. Plant 874.1949. *(Vern. Name: Hasnu-Hana).*

Diffuse, glabrous shrub. Leaves thin oblong-ovate to elliptic, shining, tapering at the ends. Flowers greenish-white in axillary and terminal, umbelliform racemes; corolla tube about 1.5 cm long. *Fls. & Frts.:* July-October.

Planted in the gardens and residential areas for its sweet scented flowers, which bloom at night.

Field Collection: Barasat, *K.N. - 915.*

Distributions: Presently cultivated throughout India. A native of West Indies.

Economic Importances: It is an American contribution to oriental gardens. Widely cultivated for its fragrant flowers at night.

Notes: Prain (1903a) did not mention this taxon from Bengal.

3. *DATURA* L.

Key to the Species

1a	Leaves entire or angled with acute lobes	1. *D. fastuosa* L.
1b	Leaves sinnate, slightly pubescent	2. *D. stramonium* L.

3.1. ***Datura fastuosa*** L. Syst. Nat. (ed.10) 2: 932.1959. *D. metel* L. Sp. Pl. 179. 1753. *D. alba* Nees, Trans. L. 17: 73.1834. *D. fastuosa* L. var. *alba* Clarke, Hook.f. Fl. Brit. India 4: 243.1883; Watt Dict. 3: 36.1890; Prain Beng. Plants 2: 559. 1903 (Rep. ed. 1963). *(Vern. Name: Sada Dhutra).*

Large, perennial herb or undershrub. Leaves ovate-sinuate, unequal sided at base. Flowers large, funnel shaped, white, tinged with green below or reddish purple. Capsules globose, prickled. *Fls. & Frts.:* Almost throughout the year.

Commonly growing in waste places, near roadsides, railways tracks and sandy areas of the coastal lands.

Field Collection: Madarhat, *K.N. - 75;* Malirdhar, *K.N. - 189.*

Distributions: Throughout India. A native of Tropical America.

Economic Importances: Seeds poisonous, also useful in hydrophobia.

3.2. ***Datura stramonium*** L. Sp. Pl. 179. 1753; Gamble Fl. Madras 3: 660. 1923. *(Vern. Name: Kalo-Dhutra).*

Undershrub, petiole 1-3 cm long; leaf blade sinuate, lanceolate, acute at apex, narrow at base, 12 x 6 cm, membranous, sparsely pubescent. Calyx subglobose, spinous, subtended by calyx base, 4-valved: seeds, reniform 3 mm long. *Fls. & Frts.:* April- November.

Very rare on the village thickets but common in sandy areas.

Field Collection: Diamond Harbour. *S.K. Mukherjee- 3829*(CAL).

Distributions: Almost throughout India; introduced in India during the 18th century. A native of Asia.

Economic Importances: Used the dried plant for the treatment of asthma: used as expectorant, antispasmodic, demulcent and anodyne.

4. *LYCOPERSICON* Mill.

4.1. ***Lycopersicon lycopersicum*** (L.) Karstem, Beutschl. Pl. 966.1883. *Solanum lycopersicum* L. Sp. Pl. 185. 1753. *L. esculentum* Mill., Gard. Dict. (ed.8), n.2. 1768; Clarke, Hook.f. Fl. Brit. India 4: 237. 1883; Watt Dict. Econ. Prod. 5: 100. 1891; Prain Beng. Plants 2: 553. 1903 (Rep. ed. 1963). *(Vern. Name: Tomato* or *Gur Begun).*

Erect herb with spreading hairy branches. Leaves 15-25 cm long, odd-pinnate; leaflets ovate to oblong, acuminate, irregularly toothed. Flowers yellow in few flowered peduncled cymes. Fruits red or yellow; pulpy. *Fls. & Frts.:* October-May.

Cultivated throughout; also found as a garden escape in the moist waste lands.

Field Collection: Amtala, *K.N. - 371.*

Distributions: Extensively cultivated as winter annuals throughout India. A native of S.W. America.

Economic Importances: Green and ripe fruits are eaten as vegetable both raw and cooked; it is highly nutritious.

5. *NICOTIANA* L.

Key to the Species

1a	Plants wild, generally smaller in size; flowers pure-white; capsule very narrow	1. *N. plumbaginifolia Viu.*
1b	Plants cultivated, generally larger in size; flowers rosy-white; capsule not very narrow	2. *N. tabacum* L.

5.1. ***Nicotiana plumbaginifolia*** Viu, Planch. Pl. Hort.Dinegro 26, t.5, 1802; Naskar Aq. & Semi-Aq. Plants 176. 1990. *(Vern. Name: Ban-Tamaku).*

Common erect sticky glandular herb, founds along the canal sides or road side thickets. Naskar (1.c.) described this taxon after collecting from Dhanyakuria *(K.N. - 569),* Haroa *(K.N. - 579)* and Budge Budge *(K.N. - 736).* A native of Tropical America.

5.2. ***Nicotiana tabacum*** L. Sp. Pl. 180.1753; Clarke, Hook.f. Fl. Brit. India 4: 245. 1883; Prain Beng. Plants 2: 559. 1903 (Rep. ed. 1963). *(Vern. Name: Tamak or Tamaku).*

Erect, glandular-pubescent herb. Lower leaves large, oblong or ovate-lanceolate, upper ones smaller, amplexicaul, obovate. Flowers reddish or pinkish, in terminal panicles. Capsules ovate. *Fls. & Frts.:* November-April.

Cultivated rarely in the public gardens in the Lower Ganga delta.

Field Collection: Specimen not collected.

Distributions: Cultivated almost throughout South India. A Native of Peru.

Economic Importances: Leaves are used as tobacco.

6. *PHYSALIS* L.

Key to the Species

1a	Commonly wild; calyx in fruit, less than 2.5 cm length, ovoid or sub-globose	1. *P. minima* L.
1b	Commonly cultivated; calyx in fruit; always more than 2.5. cm long, ovoid-conical	2. *P. peruviana* L.

6.1. ***Physalis minima*** L. Sp. Pl. 183. 1753; Naskar, Aq. Semi-Aq. Plants 176. 1900. *(Vern. Name: Ban-Tipariya).*

Erect or decumbent, annual herb. Leaves ovate, acute sinuate toothed or tobulate, acuminate. Flowers yellow, solitary, on long slender deflexed pedicels. Naskar (1.c.) described this taxon after collecting from Garia *(K.N. -335),* Budge Budge *(K.N. - 732)* and Masjidbati *(K.N.-835).* Fruits occasionally edible.

6.2. ***Physalis peruviana*** L. Sp. Pl. 1670.1753; Clarke, Hook.f. Fl. Brit. India 4: 238.1883; Prain Beng. Plants 2: 558.1903 (Rep. ed. 1963). *(Vern. Name: Tipariya).*

Small shrub or herb. Leaves alternate, entire. Corolla 1.5 cm across; calyx over 2.5 cm long. Fruits berry yellow. *Fls. & Frts.:* September-February.

Frequently cultivated for its fruits.

Field Collections: Specimen not preserved.

Distributions: Cultivated almost throughout India. A native of Tropical America.

Economic Importances: Fruits are edible either cooked or raw.

7. *SOLANUM* L.

Key to the Species

1a	Plants unarmed		(2)
1b	Plants armed with prickles		(4)
2a (1)	Mostly cultivated; roots mixed with tubers; stem herbaceous; faintly angular; leaves pinnatifid; flowers large, blue	9. *S. tuberosum* L.	
2b (1)	Mostly wild; roots not mixed with tubers; stem mostly woody, round; flowers small, white		(3)
3a (2)	Leaves ovate or oblong, minutely lobed, glabrous; pedicels sub-umbellate	4. *S. nigrum* L.	
3b (2)	Leaves elliptic, acute, entire; stellately woolly, peduncles sub-terminal	10. *S. verbascifolium* L.	

4a (1)	Leaves pinnatifid or twice pinnatifid; flowers racemose, white or violet, calyx-tube inflated, deeply cup-shaped, much enlarged	5. *S. sisymbriifolium* Lamk.
4b (1)	Leaves not pinnatifid; racemes lateral; calyx-tube not inflated, shallowly cup-shaped	(5)
5a (4)	Flowers white	(6)
5b (4)	Flowers blue, berry smooth	(7)
6a (5)	Leaves very prickly on the nerves, with distinct angular lobes; corolla densely woolly	1. *S. ferox* L.
6b (5)	Leaves not prickly on the nerves, with rounded lobes; corolla sparsely pubescent	7. *S. torvum* Swarty.
7a (5)	Stem erect, stout herbs or undershrubs; leaves stellately woolly; ovate, sinuate or lobed	(8)
7b (5)	Stems not erect; leaves glabrescent	(9)
8a (7)	Racemes many flowered; calyx lobes triangular-oblong; berry small, globose	2. *S. indicum* L.
8b (7)	Racemes few-flowered; calyx lobes narrowly elliptic or linear oblong; berry large globose	3. *S. melongena* L.
9a (7)	Diffused herbs	6. *S. surattense* Burm.f.
9b (7)	Scandent or trailling shrubs	8. *S. trilobatum* L.

7.1. ***Solanum ferox*** L. Sp. Pl. (ed.2) 267. 1762; Clarke, Hook.f. Fl. Brit. India 4: 233. 1883; Watt Dict. Econ. Prod. 6(3): 257. 1893; Prain Beng. Plants 2: 555. 1903 (Rep. ed. 1963) *(Vern. Name: Ram-begun).*

Medium shrub, stout upto 1.6 m tall; densely clothed with long coarse, stellate hairs. Leaves usually 2 at a node and unequal, broadly elliptic in outline, minutely or pinnately cut into few short triangular lobes. Flowers solitary, leaf-opposed densely hairy cymes. Berry globose. *Fls. & Frts.:* June-October.

Common on the dry-waste lands.

Field Collection: Sagar Islands, *A.K.Mukherjee- 5940* (CAL).

Distributions: Bengal, Assam and W. Peninsula. A Tropical plant; Sri Lanka and China.

7.2. ***Solanum indicum*** L. Sp. Pl. 187.1753; Clarke, Hook.f. Fl. Brit. India 4: 234. 1883; Watt Dict. Econ. Prod. 6(3): 258. 1893; Prain Beng. Plants 2: 555. 1903 (Rep. ed. 1963). *(Vern. Name: Byakur.).*

Usually armed, terrestrial errect herb. Cymes extra-axillary few flowered. Flowers blue. Berry usually small, less than 8 mm wide. *Fls. & Frts.:* Throughout the year.

Common on the dry waste lands, roadsides and on the railway tracks.

Field Collection: Madarhat, *K.N. - 67;* Hasnabad, *K.N. - 1050.*

Distributions: Throughout the tropical India. Sri Lanka, Malaya, China and Philippines.

Economic Importances: Fruits are edible as vegetable. Roots are useful for cough, catarrhal affections, colic and nasal ulcers. Glyco-alkaloids– solasonine and solanine of this plant useful for the cortisone and sex-hormone.

7.3. ***Solanum melongena*** L. Sp. Pl. 186. 1753; Clarke, Hook.f. Fl. Brit. India 4: 235. 1883; Watt Dict. Econ. Prod. 6(3): 258. 1. 1893; Prain Beng. Plants 2: 255. 1903 (Rep. ed. 1963). *S. esculentum* Dunal, Hist. Solan. 208. 1813. *(Vern. Name: Begun).*

Erect or diffused undershrub, often prickly, young parts tomentose. Leaves large, elliptic, 10-18 cm long, sinuate or shallowly lobed with rounded lobes. Flowers violet or bluish. Berries globose white, yellow or dark purple. *Fls. & Frts.:* June-March.

Cultivated in the gardens; occasionally as garden escape.

Field Collection: Champadalia/Barasat, *K.N. - 918.*

Distributions: Widely cultivated throughout India. Native country is uncertain, distributed in the Old World Tropics.

Economic Importances: A common vegetable fruit.

7.4. ***Solanum nigrum*** L. Sp. Pl. 186. 1753; Clarke, Hook.f. Fl. Brit. India 4: 229. 1883; Watt Dict. Econ. Prod. 6(3): 263. 1893; Prain Beng. Plants 2: 554. 1903 (Rep. ed. 1963). *(Vern. Name: Gurki begun or Gurkamari).*

Erect, bushy, much-branched, annual herb. Leaves ovate or oblong or lanceolate-ovate, sometimes toothed or lobed. Flowers white small, in drooping sub-umbellate extra-axillary peduncles. Berries globose, black when ripe. *Fls. & Frts.:* August- July.

Frequently found as a weed of the cultivated fields and also in the moist shady waste lands.

Field Collection: Canning, *K.N. - 186.*

Distributions: Common throughout India. Sri Lanka and all the Temperate and Tropical regions of the world.

Economic Importances: The fruits are eaten by the poor villagers as vegetable. Useful for anthrax pustules, cirrhosis of liver and dropsy.

7.5. ***Solanum sisymbriifolium*** Lamk. Illust. 2: 25. 1811; Clarke, Hook.f. Fl. Brit. India 4: 236. 1883; Prain Beng. Plants 2: 555. 1903 (Rep. ed.).

Undershrub, densely prickly and glandular velutinous. Leaves alternate, deeply pinnatifid or pinnatisect. Flowers 3-6 in pedunculate cymes, white or violet; pedicels 0.5-1.0 cm long. Corolla campanulate, pubescent outside. Berry 6-12 mm diam., globose, shining red. Seeds light brown. *Fls. & Frts.:* November-May.

Common on the waste places and along the roadside hedges.

Field Collection: Dakshin Chatra, *K.N. - 522.*

Distributions: Almost throughout India. A native of America.

Notes: Prain (1903a) did not report this taxon from Bengal.

7.6. ***Solanum surattense*** Burm. f. Fl. Ind. 57. 1768 (excl. syn. Pluck. & Rai.). *S. xanthocarpum* Schrard & Wendl., Sert. Hanov. 1: 8, t.2, 1795; Clarke, Hook.f. Fl. Brit. India 4: 236.2883; Watt Dict. Econ. Prod. 6(3): 273. 1893; Prain Beng. Plants 2: 555. 1903 (Rep. ed 1963). *(Vern. Name: Kantakari).*

Diffuse or prostrate, prickly herb, sometimes woody at the base. Prickles straight, yellow, very sharp. Leaves ovate, oblong or elliptic, deeply lobed or sub-pinnatifid, dark green. Flowers bluish-purple, in extra-axillary, few-flowered cymes. Fruits globose, seeds smooth. *Fls. & Frts.:* Novemebr-July.

Commonly found in the waste lands also as weed of the crop fileds.

Field Collection: Goramara/Sagar Islands, *K.N. - 414.*

Distributions: Throughout India. Sri Lanka, S.E. Asia, Malaya, Tropical Australia and Polynesia.

Economic Importances: Roots are much used as medicines by Hindu; fruits are edible. It is used in cough, asthma and chest pain.

7.7. ***Solanum torvum*** Swarty., Prodr. Descr. Veg. 47. 1788; Clarke, Hook.f. Fl. Brit. India 4: 234. 1883; Watt Dict. Econ. Prod. 6(3): 264.1893; Prain Beng. Plants 2: 555.1903 (Rep. ed. 1963). *(Vern. Name: Gota-begun).*

Shrubby, branches sparsely prickly, covered with brown stellate tomentum. Leaves ovate, remotely stellate-tomentose. Flowers white with yellow, anthers in corymbose. Berries globose, 2 cm in diam. *Fls. & Frts.:* January-June.

On the dry waste lands, along the roadsides and thickets or hedges.

Field Collection: Baruipur, *K.N. - 24;* Canning, *K.N. - 210.*

Distributions: Indigenous in some parts of India, very common. Old World Tropics.

Economic Importances: Berries are edible occasionally. Useful in cough and act as sedative, diuretic and digestive tonic.

7.8. ***Solanum trilobatum*** L. Sp. Pl. 188.1753; Naskar & Guha Bakshi Mangr. Swamps Sundarbans 167. 1987.

Armed scandent or trailling prickles hooked herb. Naskar & Guha Bakshi (1987) described this taxon after collecting from Garia *(K.N. - 347)* and Salt Lake area of the lower Ganga delta. Its leaves are occasionally cooked as vegetable.

7.9. ***Solanum tuberosum*** L. Sp. Pl. 185. 1753; Clarke, Hook.f. Fl. Brit. India 4: 229. 1883; Watt Dict. Econ. Prod. 6(3): 265. 1893; Prain Beng. Plants 2: 555. 1903 (Rep. ed. 1963). *(Vern. Name: Alu).*

Herbaceous with underground stem, tubers. Leaves odd pinnate; leaflets entire. Flowers whitish or bluish-violet, in terminal cymes. *Fls. & Frts.: January-March.*

Cultivated during winter season.

Field Collection: Naraharipur/Sagar Islands, *K.N. - 471.*

Distributions: Cultivated almost throughout India; it was introduced into India during the latter part of the 16th century. A native of South America.

Economic Importances: Tubers are common vegetable. It is raw material for preparing of starch and ethyl alcohol.

7.10. ***Solanum verbascifolium*** L. Sp. Pl. 184. 1753; Clarke, Hook.f. Fl. Brit. India 4: 230.1883; Watt Dict. Econ. Prod. 6(3): 273.1893; Prain Beng. Plants 2: 554. 1903 (Rep. ed. 1963). *S. erianthum* D. Don Prodr. Fl. Nepal. 96. 1825. *(Vern. Name: Arasa).*

Shrub or small unarmed tree, upto 7 m high. Leaves 10-20 cm X 5- 15 cm, elliptic-lanceolate; acuminate, entire, velvety pubescent above, densely woolly beneath. Flowers numerous; corolla white, deeply divided filaments flat, glabrous. Berry globose, yellow, covered with small stellate hairs. *Fls. & Frts.:* July- December.

Very common in waste lands.

Field Collection: Budge Budge, *K.N. - 746;* Birati, *P. Basu-104*(CAL)

Distributions: Indigenous in most part of India. Malaya, Tropical Australia and Tropical America.

Economic Importances: In South India, the fruits are eaten as vegetable; dried plants with hot water applied to lessen inflammation and pain; also used in burns.

8. *WITHANIA* Pauq. nom. cons.

8.1. ***Withania somnifera*** (L.) Dunal., DC. Prodr. 19(1): 453. 1852; Naskar Aq. & Semi-Aq. Plants 177.1990. *(Vern. Name: Aswagandha).*

Erect branched undershrub, occasionally grow on the moist lands. Naskar (1.c.) described this taxon after collecting from Kestopur/near Baguihati *(K.N. - 1600).*

Economic Importances: It is considered a reputed medicine both by the Ayurvedic and Yunani practitioners in tumerous and in tubercular glands. The tuber has a bitter, sharp and acrid taste and is considered useful in inflammatory conditions, psorassis, bronchitis, ulcers and scabies, when applied locally. The powder or fresh root is very powerful nerve tonic and tones up shattered nerves, sharpens dull memory, restore debilated conditions from old age. It is very much invaluable in impotency.

Remarks: Beside all these, some other purely cultivated species of the Solanaceae are mentioned bellow; these are common in the lower parts of Bengal.

Atropa belladonna L., *Brunfelsia americana* L., *Petunia nyctaginiflora* Juss., *P. violacea* Lindl., *Solanum argenteum* Dun., *S. macranthum* Dun, *S. melongena* L. var. *esculenta* Roxb. & *S. anguivi* Lamk. and *Capsicum annum* L. var. *abbreviata* Fingerh.

XCVII. SCROPHULARIACEAE

Key to the Genera

1a	Mostly indigenous and naturalised	(2)
1b	A recently introduced, terrestrial prostrate herb; calyx 5- partite; corolla lemon-yellow with purple veins	9. *MECARDONIA* Ruiz. & Pav.
2a (1)	Corolla 5-merous, the upper tip or the 2 upper lobes outside in bud	(3)
2b (1)	Corolla 5-or 4-merous with the upper lip or the 2 upper lobes, inner most in the bud	(10)
3a (2)	Corolla saccate at the base mouth closed	1. *ANTIRRHINUM* L.
3b (2)	Corolla not saccate, mouth open	(4)
4a (3)	Corolla tube very long and narrow; flowers cymose, red	10. *RUSSELIA* Jack.
4b (3)	Corolla tube not very long; flowers axillary or racemose, not red	(5)
5a (4)	Anther cells not divaricate or separate	(6)
5b (4)	Anther-cells if perfect divaricate or separate	(8)
6a (5)	Stamens 4 perfect, included in the tube	(7)
6b (5)	Stamens only 2 perfect, staminodes-2	4. *DOPATRIUM* Buch. Ham.
7a (6)	Leaves mostly radical, flowers in subsecund racemes	8. *MAZUS* Lour.
7b (6)	Leaves mostly cauline, flowers not secund	2. *BACOPA* Aubl. *nom. cons.*
8a (5)	Stamens 4 perfect, all included in the tube	5. *LIMNOPHILA* R. Brown
8b (5)	Stamens 4, 2 perfect in the tube, 2 in the throat perfect on reduced to staminodes	(9)

9a (8)	Capsule ovoid or oblong not exceeding or linear exceeding the calyx	7. *LINDERNIA* All.
9b (8)	Capsule oblong, not exceeding the calyx	12. *TORENIA* L.
10a (2)	Corolla 4-merous; stamens 4	(11)
10b (2)	Corolla 5-merous	3. *CENTRANTHERA* R. Brown
11a (10)	Calyx more or less polysepalous	11. *SCOPARIA* L.
11b (10)	Calyx wide-campanulate, 2-fid	6. *LINDENBERGIA* Lehm.

1. *ANTIRRHINUM* L.

1.1. ***Antirrhinum majus*** L. Sp. Pl. 617. 1753; Hook.f. Fl. Brit. India 4: 253. 1884; Prain Beng. Plants 2: 564.1903 (Rep. ed. 1963).

Erect, annual or perennial herb. Leaves entire or lobed. Flowers solitary, axillary, or passing into terminal racemes, with bract-like floral leaves; bracteoles 0; sepals connate in a 5-partite calyx; lobes imbricate; petals 5, connate in a compressed, tubular corolla. Fruits an ovoid or globose, 2-celled capsule. Seeds minute. *Fls. & Frts.:* July-October.

Planted as ornamental garden plants.

Field Collection: Baruipur, *K.N. - 1239.*

Distributions: Spontaneous in some parts of India and sometimes cultivated also. A native of the Mediterranean region.

Notes: Prain (1903a) reported this taxon, but mentioned that does not grow spontaneously in Lower Bengal.

2. *BACOPA* Aubl. *nom. cons.*

2.1. ***Bacopa monnieri*** (L.) Wettstein, Pfamilien 4 (3b): 77.1891; Naskar Aq. & Semi-Aq. Plants 179. 1990. *(Vern. Name: Brahmi Sak).*

Prostrate, much branched, fleshy marshy loving herb. Naskar (1.c.) described this taxon after collecting from Canning *(K.N.-140)* & Bamanpukur *(K.N. - 1090)*. The entire plant is useful as nerve tonic; cure epilepsy and insanity.

3. *CENTRANTHERA* R. Brown

3.1. ***Centranthera tranquebarica*** (Spreng.) Merr.; Hook.f. Fl. Brit. India 4: 300. 1884; Prain Beng. Plants 2: 578.1903 (Rep. ed. 1963). *C. humifusa* Wall.; Majumder, Ind. Agr. 6: 167.1962.

Annual, rigid, more or less erect herb. Leaves usually opposite, linear, entire. Flowers axillary, peduncled, bracteoles 2, small. Sepals 5, united into a compressed tubular calyx, split on one side. Petals 5, tube is curved, limbs oblique, lobes subequal, yellow. Stamens 4, didynamous, included, one cell fertile. Carpels-2, ovary 2-celled, ovules many, style flattened at the top, passing into a flat stigma. Fruit subglobose. Capsule loculicidal, valves carrying away the placentas. Seeds many, oblong cuneate, testa lax, reticulate. *Fls. & Frts.:* June- November.

Very occasionally on the village thickets.

Field Collection: Ichhapur, *K.N. - 1804.*

Distributions: Almost throughout India. Cosmopolitan species.

4. *DOPATRIUM* Buch. Ham.

4.1. ***Dopatrium junceum*** (Roxb.) Buch.-Ham., Benth. Scrop. India 31. 1835; Naskar, Aq. & Semi-Aq. Plants 180. 1990.

Slender fleshy moist loving herb in the paddy field. Naskar (1.c.) described this taxon after collecting from Bakkhali *K.N. - 1415.*

5. *LIMNOPHILA* R. Brown

Key to the Species

1a	Leaves with pinnate nerves arching from the midrib	(2)
1b	Leaves with 3-5 nerves running from the base of the apex	(3)
2a (1)	Leaves oblong or elliptic-oblong, obtuse, crenate-serrate; flowers sessile; calyx glabrous or nearly so, calyx-lobes lanceolate-acuminate	4. *L. repens* Benth.
2b (1)	Leaves elliptic-oblong, sub-crenate, obtuse; flowers pedicelled; always axillary, solitary; calyx hirsute, lobes subulate	3. *L. pulcherrima* Hook.f.

3a (1)	Flowers sessile or very shortly pedicelled; calyx lobes ovate, shortly acuminate	1. *L. heterophylla (Roxb.) Benth.*
3b (1)	Flowers pedicelled, pedicels as long as or longer than the calyx; calyx lobes ovate acuminate	2. *L. indica* (L.) Druce

5.1. ***Limnophila heterophylla*** (Roxb.) Benth., Scrop. India 25.1836; Naskar Aq. & Semi-Aq. Plants 181. 1990.

An aquatic perennial herb of the swampy areas. Naskar (1.c.) described this taxon after collecting from Sasan/on the paddy field *(K.N. - 29).* Ghaturmore/Amtala *(K.N. - 383).*

5.2. ***Limnophila indica*** (L.) Druce, Rep. Bot. Exch. Club. Brit Is. 3: 420. 1914; Naskar Aq. & Semi-Aq. Plants 182. 1990. *(Vern. Name: Karpur).*

Partially submerged aquatic herb; common on the swampy areas. Naskar (1.c.) described this taxon after collecting from the lower Ganga delta.

The plant is considered as an antiseptic and when made into a liniment with coconut oil, is used in elephantiasis. The juice of the plant is rubbed over the body in pestilent fevers. It is also given internally with ginger, cumin and other aromatics, in dysentery.

5.3. ***Limnophila pulcherrima*** (Griff.) Hook.f. Fl. Brit. India 4: 267. 1884; Naskar Aq. & Semi-Aq. Plants 183. 1990.

Sub-aquatic annual herb; common on the moist waste lands. Naskar (1.c.) described this taxon after collecting from Mahishmari/Sundarbans *(K.N. - 1864).* A native of India.

5.4. ***Limnophila repens*** Benth., DC. Prodr. 10: 387. 1846; Naskar Aq. & Semi-Aq. Plants 184. 1990.

Annual, occasionally, moist loving herb. Naskar (1.c.) described this taxon after collecting from Durganagar/near Dum Dum Cant. *K.N. - 1717.*

6. *LINDENBERGIA* Lehm.

6.1. ***Lindenbergia indica*** (L.) Vatke, Osters, Bot. Zeitschr. 25: 10.1875; Naskar Aq. & Semi-Aq. Plants 184. 1990. *(Vern. Name: Haldi-Basanta).*

Erect, tufted annual herb; frequently found in the moist waste lands. Naskar (1.c.) described this taxon after collecting from the lower Ganga delta.

7. *LINDERNIA* All.

Key to the Species

1a	Capsule about equalling the calyx or shorter	(2)
1b	Capsule twice as long as calyx or longer	(5)
2a (1)	Leaves penni-nerved	(3)
2b (1)	Leaves parallel-nerved; perfect stamens-4	9. *L. pyxidaria* All.
3a (2)	Flowering calyx cleft to the base or nearly so	(4)
3b (2)	Flowering calyx cleft to the middle or less	4. *L. crustacea* (L.) Muell.
4a (3)	Glabrous; sepals shorter than the capsule	6. *L. multiflora* Mukerjee
4b (3)	Pubescent with spreading hairs, sepals longer than capsule	5. *L. hirsuta Wettst.*
5a (1)	Perfect stamens 4	(6)
5b (1)	Perfect stamens 2	(7)
6a (5)	Flowering calyx cleft to the middle	7. *L. numularifolia* Wettst.
6b (5)	Flowering calyx cleft almost to the base	3. *L. cordifolia* Merr.
7a (5)	Leaves oblong or oblong-lanceolate, elliptic	(8)
7b (5)	Leaves commonly lanceolate	(9)
8a (7)	Leaves sharply spinose-serrate	2. *L. ciliata* Pennell
8b (7)	Leaves distantly serrate	1. *L. anagallis* Pennell
9a (7)	Leaves multicordate	8. *L. parviflora* Haines
9b (7)	Leaves unicordate	10. *L. verbenaefolia* Pennell

7.1. ***Lindernia anagallis*** (Burm.) Pennell J. Arn. Arb. 24: 252.1943; Naskar Aq. & Semi-Aq. Plants 186. 1990.

Creeping marshy herbs; Naskar (1.c.) described this taxon after collecting from Basirhat *(K.N. - 938).*

7.2. ***Lindernia ciliata*** (Colsm.) Pennell, Brittonia 2: 182. 1936; Naskar Aq. & Semi-Aq. Plants 187.1990.

Small, erect, rigid, branched, glabrous herb on the moist areas. Naskar (1.c.) described this taxon after collecting from Chouhati/way to Minakhan, *(K.N. - 1732).*

7.3. ***Lindernia cordifolia*** (Colsm.) Merr. Enum. Philip. Pl. 3: 437.1923; Naskar Aq. & Semi-Aq. Plants 188.1990.

Small, branched, diffused, glabrous herb. Naskar (1.c.) described this taxon after collecting from the lower Ganga delta, Madhyampur *(K.N. - 963.).*

7.4. ***Lindernia crustacea*** (L.) Muell. Cens. Austral. Pl. 97.1882; Naskar Aq. & Semi-Aq. Plants 189. 1990.

Annual, glabrous, diffusely branched herb; common in the moist places. Naskar (1.c.) described this taxon after collecting from Madarhat *(K.N. - 66)* Gosaba (K.N. - 655).

7.5. ***Lindernia hirsuta*** (Benth.) Wettst., Nat. Pflauzen. f. 4: 36. 1891; Naskar Aq. & Semi-Aq. Plants 190. 1990.

Erect, succulent, herb on the waste lands. Naskar (1.c.) described this taxon from the lower Ganga delta regions.

7.6. ***Lindernia multiflora*** (Roxb.) Mukerjee, J. Ind. Bot. Soc. 24 (3): 131. 1945; Naskar Aq. & Semi-Aq. Plants 190. 1990.

Erect, branched, glabrous and small succulent herb on the moist waste lands. Naskar (1.c.) described this taxon from the lower Ganga delta after collecting this from Taki *(K.N. - 1003.).*

7.7. ***Lindernia numularifolia*** (Don) Wettst., Pfam. 4(36); 79. 1819; Naskar Aq. & Semi-Aq. Plants 190.1990.

Erect glabrous herb; common on the dampy and shady places. Naskar (1.c.) described this taxon after collecting from Bamanghata *(K.N. - 1736).*

7.8. ***Lindernia parviflora*** (Roxb.) Haines Bot. Bihar & Orissa 2: 635. 1922; Naskar Aq. & Semi-Aq. Plants 191.1990.

Erect or decumbent, small glabrous much branched herb; prefers moist places. Naskar (1.c.) described this taxon after collecting from Chandpara *(K.N. -1672).*

7.9. *Lindernia pyxidaria* All., Misc. Maxim. 3. tab. 5. 178.1766; Naskar Aq. & Semi-Aq. Plants 191.1990.

Annual, erect, glabrous branched herb; abundantly on the moist and shady places. Naskar (1.c.) described this taxon after collecting from Narendrapur *(K.N. - 1979).*

7.10. *Lindernia verbenaefolia* (Colsm.) Pennell. Acad. Nat. Sci. Phill. Monograph 5: 31. 1943; Naskar Aq. & Semi-Aq. Plants 192.1990.

Erect glabrous herb; rarely on the moist waste lands. Naskar (1.c.) described this taxon after collecting from Satzelia/Sundarbans *(K.N. - 2159).*

8. *MAZUS* Lour.

8.1. *Mazus pumilus* (Burm.f.) Van Steenis, Nova Guina n.s. 31.1958; Naskar Aq. & Semi-Aq. Plants 193.1990.

Small annual herb; moisture loving plant and on the bank of the river. Naskar (1.c.) described this taxon from the lower Ganga delta.

9. *MECARDONIA* Ruiz. & Pav.

9.1. *Mecardonia procumbens* (Miller) Small, Fl. S.E. Us. 1065. 1903; *M. dianthera* (Swattz.) Pennell, Proc. Nat. Sci. Phil. 98:87.1947.

Annual, terrestrial, prostrate, glabrous herb; a weed of the paddy fields. Naskar (1.c.) described this taxon from the lower Ganga delta and collected from Baruipur *(K.N. - 8)*; Canning *(K.N. - 185):* Maslandapur *(K.N. - 551)* and Budge Budge *(K.N. -730).*

10. *RUSSELIA* Jack.

10.1. *Russelia equisetiformis* Schlecht. & Cham. Linnaea 6: 377. 1831; Chatterjee & Bharadwaja, Bull. bot. Soc. Bengal 9(2) 132. 1955. *R. juncea* Zucc. Flora 15(2): 99.1832; Bor & Raizada 227. Pl. 82. *R. floribunda* Woodraw, J. Bombay nat. Hist. Soc. 12: 174.1898.

Shrub with rush like branches. Leaves-upper ones reduced to scales. Flowers scarlet, with tubular corolla, peduncles, 1-3 flowered on drooping branches. *Fls. & Frts.:* Almost throughout the year.

Cultivated garden plants; specimen not preserved.

Distributions: Cultivated almost throughout India. A native of Mexico.

Economic Importances: As ornamental garden plants.

11. *SCOPARIA* L.

11.1. ***Scoparia dulcis*** L. Sp. Pl. (ed.1) 116.1753; Hook.f. Fl. Brit. India 4: 280.1884; Prain Beng. Plants 2: 575.1903. (Rep. ed. 1963).

Erect, branched terrestrial herb. Stem- 4 to 6 angled, glabrous or slightly hairy. Leaves simple, often opposite, 3-nately whorled, glandular, elliptic or rhomboid, subacute or obtuse, serrate, base tapering into a short petiole. Flowers small, white, 3-6 from each whorl or leaves, often forming many flowered racemes; pedicel slender, terete. Calyx more or less polysepalous; sepals 4, oblong, acute or subacute, entire, 3-5 nerved. Corolla white, rotate, densely pilose; tube very short, lobes 4, subequal, obtuse, stamens 4, style stout, stigma exserted from the corolla. Capsule small, sub-globose; valves membranous, ultimately 2-fid. Seeds obovoid; scrobiculate. *Fls. & Frts.:* Throughout the year.

Abundantly, weed in the waste lands. Common as hedge plant.

Field Collection: Dhopagachhi, *K.N. - 40;* Dattapukur, *K.N. - 797.*

Distributions: Introduced in Bengal and India also. A native of Tropical America.

Economic Importances: Leaves infusion is used in fever, cough and bronchitis; helpful for toothache. Also used for diabetes.

12. *TORENIA* L.

12.1. ***Torenia cordifolia*** Roxb. Cor. Pl.2: 32. t.f. 161.1798; Naskar Aq. & Semi-Aq. Plants 195.1990.

Erect annual glabrous herbs. Grow on moist places. Naskar (1.c.) described this taxon from the lower Ganga delta.

XCVIII. OROBANCHACEAE

1. *OROBANCHE* L.

1.1. ***Orobanche indica*** Buch. -Ham., Roxb. Fl.Ind.3: 27.1832; Hook.f. Fl. Brit. India 4: 326.1884; Watt Dict. Econ. Prod. 5: 495. 1891; Prain Beng. Plants 2: 580.1903 (Rep. ed. 1963). *O. aegyptic*a Pers. Syn. 2: 181.1806. *Phelipaea indica* G. Don Gen. Syst. 4: 632.1837. *(Vern. Name: Bena Bau.).*

Erect, brownish, herbaceous root-parasite; scales lanceolate pubescent. Flowers bluish, violet in lax spikes; bracts ovate, acuminate pubescent; bracteoles subulate, filiform; capsule oblong. *Fls. & Frts.: . December-April.*

As a root parasite on mustard and brinjal fields.

Field Collection: Narendrapur, *K.N. - 1976.*

Distributions: Native of India. Central and Western part of India.

Notes: Other species *Aeginetia pedunculata* Wall. is also an occasional root parasite of *Vetiveria* sp.

XCIX. LENTIBULARIACEAE

1. *UTRICULARIA* L.

Key to the Species

1a	Flowers on short pedicels; peduncles with a whorl of spongy floats below the inflorescences	3. *U. inflexa* var. *stellaris* Taylor
1b	Flowers as long as pedicels; peduncles without floats	(2)
2a (1)	Large; leaves 4.0 -7.5 cm long; seeds not or scarcely winged	1. *U. aurea* Lour.
2b (1)	Small; leaves rarely exceeding 0.5 cm long; seeds with a broad corky winged	2. *U. gibbosa* spp. *exoleta* Taylor

1.1. ***Utricularia aurea*** Lour. Cochinch. 1: 26.1790; Naskar Aq. & Semi-Aq. Plants 192.1903. *(Vern. Name: Jhangi).*

Rootless, submerged but free floating aquatic herb. Naskar (l.c.) described this taxon from the lower Ganga delta after collecting from Satzelia *K.N. - 2160.*

1.2. ***Utricularia gibbosa*** L. spp. *exoleta* P. Taylor, Mitt, Bot . Staat. Munch. 4: 101.1961; Naskar Aq. & Semi-Aq. Plants 198.1990. *(Vern. Name: Chhota Jhangi).*

Small, submerged but free floating aquatic herb. Naskar (l.c.) described this taxon from the lower Ganga delta.

1.3. ***Utricularia inflexa*** Forssk. var. *stellaris* (Linn.f.) Taylor, Mitt. Bot., Staata. Munchen 4: 96.1961; Naskar Aq, & Semi-Aq. Plants 199.1990 *(Vern. Name: Jhangi).*

Rootless, submerged, free floating aquatic herb. Common in the shallow ponds, marshy places and paddy fields. Naskar (l.c.) described this taxon after collecting from Goranbose/Sundarbans *K.N. - 1877.* This weed flora largely hampered the normal growth of fish species cultured in the closed water and act as the obnoxious weed in the paddy fields.

C. BIGNONIACEAE

Key to the Genera

1a	Capsule septifragally dehiscent, the valves parallel to compressed septum	(2)
1b	Capsule loculicidally dehiscent, the valves transverse to the septum	(4)
2a (1)	Trees with 2-3 pinnate leaves	(3)
2b (1)	Climbers with di-or tri-foliate leaves and the terminal converted into a tendril; flowers showy	1. *BIGNONIA* L.
3a (2)	Flowers dark red, corolla very broad	6. *OROXYLUM* Vent.
3b (2)	Flowers white with long slender tube	5. *MILLINGTONIA* Linn.f.
4a (1)	Staminodes short or absent	(5)
4b (1)	Staminodes very long, clavate at tip; flowers blue	4. *JACARANDA* Juss.
5a (4)	Calyx not spathaceous	(6)
5b (4)	Calyx spathaceous; corolla with elongated tubular, base white	2. *DOLICHANDRONE* Seem.
6a (5)	Flowers bright yellow	7. *TECOMA* Juss. emend. H.B.&.K.
6b (5)	Flowers yellowish brown	3. *HAPLOPHRAGMA* P.Dop.

1. *BIGNONIA* L.

1.1. ***Bignonia unguiscati*** L. Sp. Pl. 623.1753. *B. gracilis* Lodd. Bot. Cab. t. 1705.1831.

Climbing shrub with small hooked tendrils. Leaflets deep green, shining, broadly oval, acuminate. Flowers bright yellow. *Fls. & Frts.:* April-July.

Sometime planted as an ornamental garden plant.

Field Collection: Garia, R.N.Banerjee-21 (CAL).

Distributions: Occasionally planted throughout India. A native of Central America to South Brazil and Argentina.

Notes: Prain (1903a) did not mention this taxon from Bengal.

2. *DOLICHANDRONE* Seem.

2.1. ***Dolichandrone spathaceae*** (Linn.f.) K. Schum. Fl. Kais- Wilbelm Land 123.1889; Naskar & Guha Bakshi Mangr. Swamps Sundarbans 99.1987. *(Vern. Name: Gorshingiah).* (Fig. 36).

Medium to large tree; occasionally grow in the mangove cleared areas of the Sundarbans. Naskar & Guha Bakshi (l.c.) described this taxon after collecting from the Sundarbans region (Rudra Nagar/Sagar Islands, *(K.N. - 444)*. A coarse fibre can be obtained from this bark.

3. *HAPLOPHRAGMA* P. Dop.

3.1. ***Haplophragma adenophyllum*** (Wall.) P. Dop., Bull. Soc. Bot. France 72: 890.1925. *Bignonia adenophylla* Wall., Cat 6502.1832, *nom. nud. Heterophragma adenophyllum* Seem. ex Benth. & Hook.f., Gen. Pl. 2: 1047. 1876 & Clarke, Hook.f. Fl. Brit. India 4: 381.1884; Prain Beng. Plants 2: 588.1903 (Rep. ed. 1963).

Medium sized to, large tree with dense crown. Leaves large, unipinnate; leaflets 5-7, broad elliptic. Flowers large, yellowish brown, rusty, woolly tomentose, in terminal panicles. Capsules cylindric, ribbed, mostly curved, seeds winged. *Fls. & Frts.:* August-February.

Sometimes planted as ornamental garden plants.

Field Collection: Specimen not preserved.

Distributions: Andaman & Nicobar Islands. Indigenous to Burma.

4. *JACARANDA* Juss.

4.1. ***Jacaranda mimosaefolia*** D. Don, Bot. Reg. t. 631.1822. *J. ovalifolia* R. Brown Bot. Mag. t. 2327.1822.

Small to moderate-sized, beautiful, dense crowned tree. Leaves bipinnate; leaflets many, acuminate. Flowers bluish in lax pyramidal panicles. Capsules orbicular, woody, compressed at right angles to the septrum. *Fls. & Frts.:* April-August.

Occasionally planted in the gardens and parks.

Field Collection: Barasat, *K.N. - 923.*

Distributions: Now planted almost throughout India; it was inroduced to India in the year 1841. A native of Brazil and North-West Argentina.

Economic Importance: Beautiful ornamental plant.

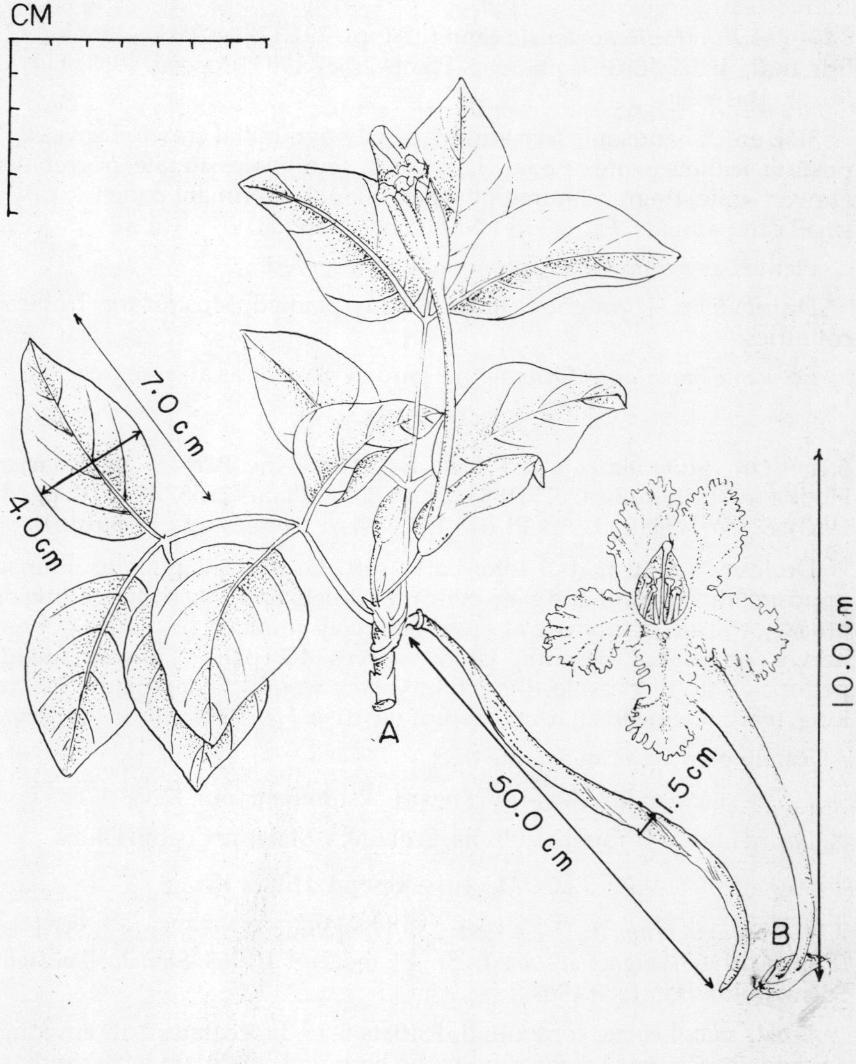

Fig36. DOLICHANDRONE SPATHACEAE (Linn. f.) Schum.
A. Twig with pod, B. Single flower.

5. *MILLINGTONIA* Linn. f.

5.1. ***Millingtonia hortensis*** Linn.f. Suppl. 291.1781; Clarke, Hook.f. Fl. Brit. India 4: 377.1883; Prain Beng. Plants 2: 587.1903 (Rep. ed. 1963). *(Vern. Name: Akas -nim).*

Tall, erect, handsome tree with narrowly pyramidal crown. Leaves 2-3 pinnate, leaflets ovate or ovate lanceolate, acuminate, sinuate, or crenate. Flower white, fragrant, numerous, pendulous, in termianl panicles; calyx small campanulate. *Fls. & Frts.:* November-March.

Planted as avenue tree. Specimen not preserved.

Distributions: Throughout India, mostly planted. Most of the Tropical countries.

Economic Importance: Ornametnal garden plant or as avenue tree.

6. *OROXYLUM* Vent.

6.1. ***Oroxylum indicum*** (L.) Vent., Dec. Gen. Nov. Paris 8.1808; Clarke, Hook.f. Fl. Brit. India 4: 377.1883; Prain Beng. Plants 2: 587.1903 (Rep. ed. 1963). *Bignonia indica* L. Sp. Pl. 625.1753. *(Vern. Name: Sona* or *Sonpatti).*

Deciduous 5-10 m tall tree; bark, soft. Leaves 2-3 pinnate. Pinnae opposite; rachis stout and pale beneath; leaflets opposite ovate-elliptic or broadly elliptic, acuminate at apex, obliquely rounded or obtuse at base nerves prominent beneath; lateral nerves 4-7 pairs. Flowers foetid; peduncles stout; capsule thick, flat; valves woody. Seeds many 5-7 cm long, margin winged all round except the base. *Fls. & Frts.:* June-January.

Rarely planted as an avenue tree.

Field Collection: Bireshwarpur/near Lakshmikantapur, *K.N. - 1763.*

Distributions: Throughout India; Sri Lanka, Malacca Cochin China.

7. *TECOMA* Juss. emend. H.B.& K.

7.1. ***Tecoma stans*** (L.) Juss. Gen. 139.1789; Prain Beng. Plants 2: 587.1903 (Rep. ed. 1963). *Bignonia stans* L. Sp. Pl. (ed.2) 871.1763. *Stonolobium stans* Seem. Journ. Bot. 1: 88.1863.

Small tree. Leaves compound, leaflets 5-15, lanceolate, 5-10 cm long, acuminate. Flowers bright yellow; the limb with undulate lobes, anthers pubescent. Capsule 12-29 cm long. *Fls. & Frts.:* Throughout the year.

Cultivated, sometime noted as garden escape.

Field Collection: Diamond Harbour, *K.N. - 1449.*

Distributions: Almost throughout India. A native of Tropical America.

Economic Importances: Sometimes planted as hedge plant; useful as garden screen.

CI. PEDALIACEAE

1. *SESAMUM* L.

1.1. ***Sesamum indicum*** L. Sp. Pl. 634.1753; Clarke, Hook.f. Fl. Brit. India 4: 387.1884; Watt Dict. Econ. Prod. 6(2) 502.1893; Prain Beng. Plants 2: 590.1903 (Rep. ed. 1963). *S. orientale* L. Sp. Pl. 634.1753. *(Vern. Name: Til).*

Erect more or less foetid and glandular, usually pubescent, annual herb. Leaves lanceolate or oblong. Flowers pinkish or whitish, solitary with a strong and bad odour. Capsules oblong, 4-gonous, erect, dehiscent from above downwards; seeds black or white, glabrous. *Fls. & Frts.:* August-November

Cultivated crop, but often as a garden escape in the waste moist lands.

Field Collection: Barasat, *K.N. - 919.*

Distributions: Cultivated almost throughout India from a very early period. Probably a native of Tropical Africa.

Economic Importances: Seeds often used in the Hidu's religious ceremonies; used in confectionary. Seed-oil is edible and used for confectionary and for making margarine, soaps, cosmetics, perfumes, insecticides, and pharmaceutical products. Oil-cake is used as cattle feed. Mucilaginous juice of the plant is used for killing hair lice.

CII. ACANTHACEAE

Key to the Genera

1a	Climbing woody shrub; calyx minute, annular or 10-15 toothed, pollen globose	19. *THUNBERGIA* Retz. *nom. cons.*	
1b	Erect or prostrate herbs or shrubs		(2)
2a (1)	Seeds not supported on hard retinacula		(3)
2b (1)	Seeds supported on hard upward curved retinacula		(5)
3a (2)	Leaves pinnately divided; corolla-lobes contorted; ovules not in 2 rows in each cell	5. *CARDANTHERA* Ham.	

3b (2)	Leaves not pinnately divided; corolla-lobes imbricate; ovules superposed in 2 rows in each cell	(4)
4a (3)	Calyx 4-partite; stamens 2	12. *NELSONIA* R. Brown
4b (3)	Calyx 5-partite; stamens 4	18. *STAUROGYNE* Wall.
5a (2)	Corolla without upper lip, lower lip large, expanded, 3-lobed; anther filaments without any excurrent process; ovules 2 in each cell	1. *ACANTHUS* L.
5b (2)	Corolla with either 2 lips or with 5 sub-equal lobes	(6)
6b (5)	Corolla -lobes twisted to the left in bud	(7)
6b (5)	Corolla lobes imbricate in buds	(11)
7a (6)	Ovules more than 2 in each cell	(8)
7b (6)	Ovules 2 - in each cell; capsules normally with 4 or fewer seeds; placentas separating elastically from the valves from the base upwards; stamens 4; bracteoles none	14. *PHAULOPSIS* Willd.
8a (7)	Corolla distinctly 2 lipped	10. *HYGROPHILA* R. Brown
8b (7)	Corolla sub-equally 5-lobed	(9)
9a (8)	Bracteoles large; capsule clavate with a solid base	(10)
9b (8)	Bracteoles small, narrow or none; anthers blunt; capsule seed bearing throughout	9. *HEMIGRAPHIS* Nees emend. Ander
10a (9)	Inflorescence few flowered, contracted	7. *DIPTERACANTHUS* Nees emend Br.
10b (9)	Inflorescence laxly cymose, many flowered	16. *RUELLIA* L. emend Bremek
11a (6)	Ovules 3-10 in each cell; capsule compressed, normally 6 or more seeded; seeds ovoid hardly compressed	3. ANDROGRAPHIS Wall. ex Ness

11b (6)	Ovules 2 or rarely 1 in each cell	(12)
12a (11)	Corolla lobes 5, sub-equal	(13)
12b (11)	Corolla distinctly 2-lipped; anther 2-celled	(14)
13a (12)	Anther 1-celled, muticous; stamens 4, didynamous	6. *CROSSANDRA* Salisb.
13b (12)	Anther 2-celled; stamens 2 or 3 perfect	4. *BARLERIA* L.
14a (12)	Placentas separating elastically from the valves from the base upwards; bracts arranged in unilateral spikes	17. *RUNGIA* Nees
14b (12)	Placentas not separating elastically from the valves	(15)
15a (14)	Anther-cells parallel, subequal, bracts	8. *ECBOLIUM* Kurz.
15b (14)	Anther-cells placed one higher up than the other	(16)
16a (15)	Anther-cells with white, basal spur-like appendage	11. *JUSTICIA* L.
16b (15)	Anther-cells not spured at the base	(17)
17a (16)	Anther cells apiculate	2. *ADHATODA* Nees
17b (16)	Anther cells muticous	(18)
18a (17)	Bracts and bracteoles small, sub-equal; corolla white	15. *RHINACANTHUS* Nees
18b (17)	Bracts larger than bracteoles; corolla rose or purple	13. *PERISTROPHE* Nees.

1. ACANTHUS L.

1.1. ***Acanthus ilicifolius*** L. Sp. Pl. 639.1753; Naskar & Guha Bakshi Mangr. Swamps Sundarbans 80. 1987. *(Vern. Name: Harkoch Kanta).*(Fig.37).

Viny shrub/herb, scarcely woody, upto 1 m high; aerial roots from the lower reclining stem. Leaves spiny to very spiny margin, pinnatifid and apex oblong or elliptic decussate, glabrous. Flowers 3.5-4.0 cm long, light blue or violet. Fruit 1 cm green and 2.5-3.0 cm long, seed 0.5-1.0 cm long, germination of seed hypogeal. *Fls. & Frts.:* November-February.

This mangrove associated flora grow both in the tidal influenced and above the tidal reaches of the mangrove forests.

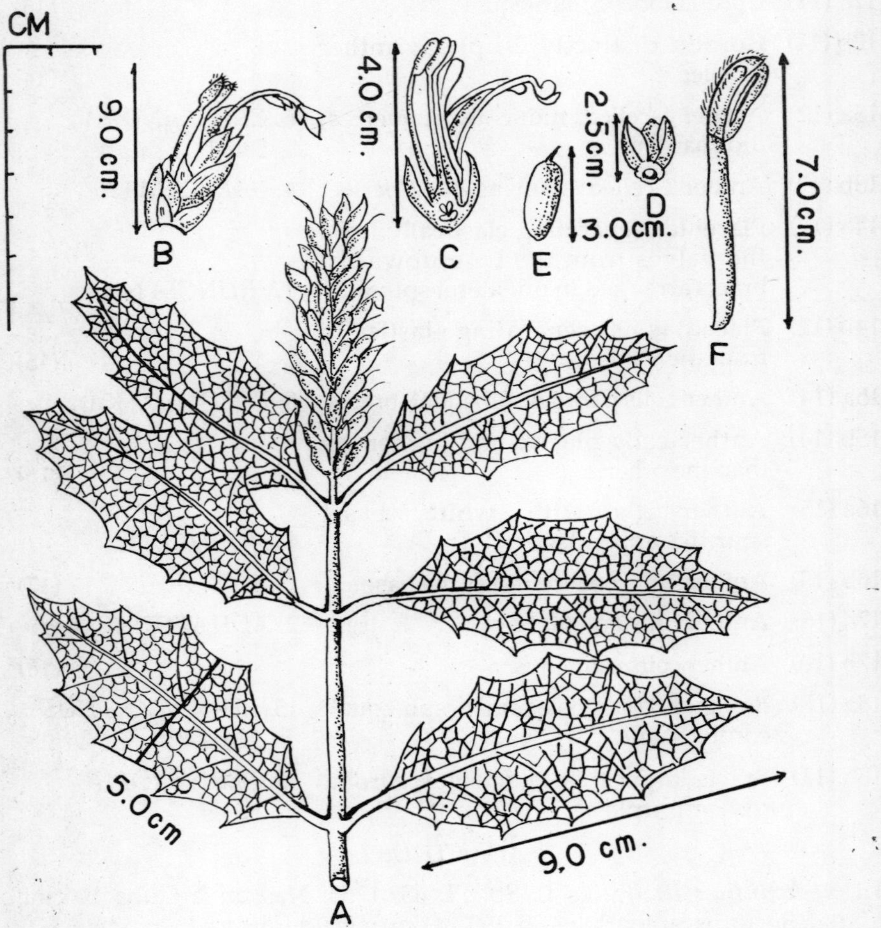

Fig.37. ACANTHUS ILICIFOLIUS L.

A. Flowering twig, B. Part of Inflorescence, C. Section of Flower, D. Sepals, E. Fruit, F. Gynoecium.

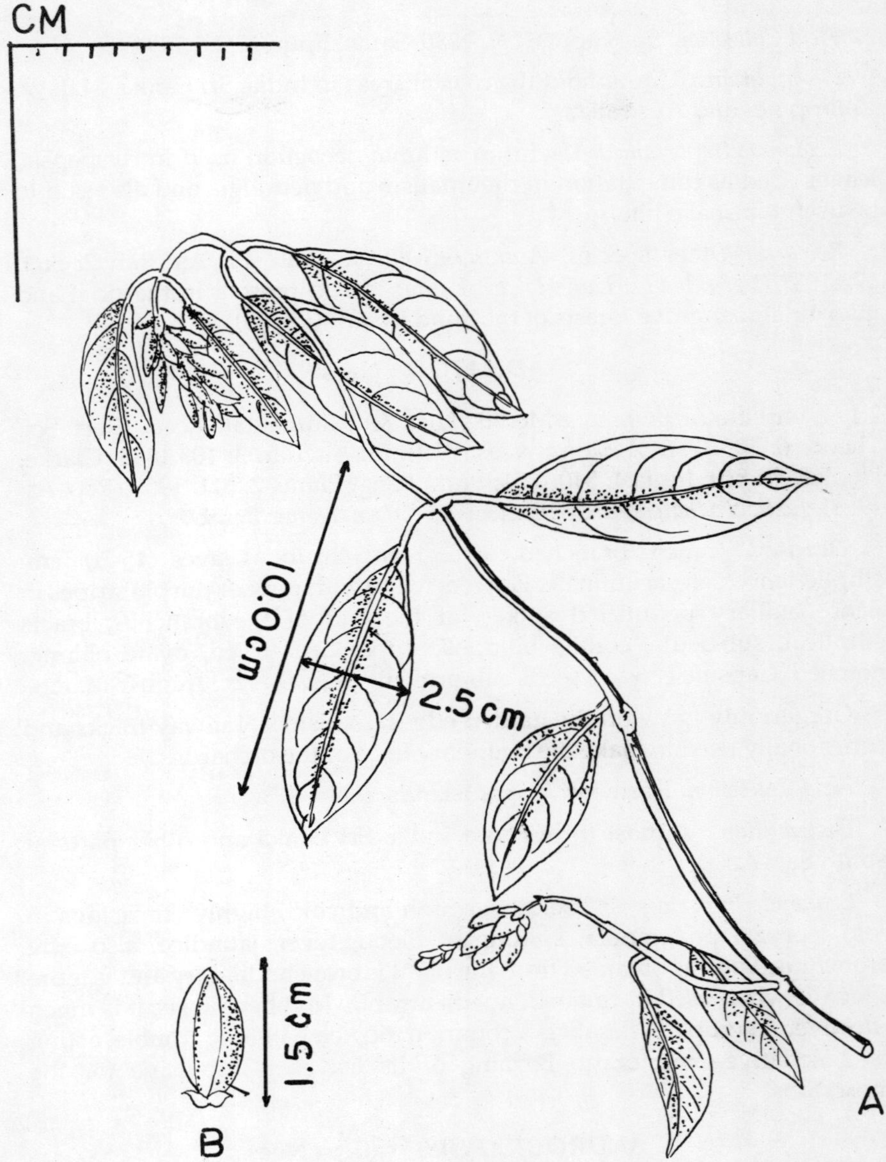

Fig.38. ACANTHUS VOLUBILIS Wall.
A. Flowering twig, B. Single fruit.

Field Collection: Surjyapur, *K.N. - 280;* Santoshpur, *K.N. - 500.*

Distributions: Throughout the coastal areas in India. Sri Lanka, Malaya, Philippines and Australia.

Economic Importances: Useful in asthma; decoction used in dyspepsia; leaves used as fomentation in rheumatism and neuralgia and also said to be useful in snake bite.

Remarks: Other species, *Acanthus volubilis* Wall. Pl. As. Rar. 2: 56.t. 172.1831 *(Vern. Name: Lata-Hokoch Kanta)* is also present in the tidal and inter-tidal mangrove forests of the Sundarbans. (Fig.38).

2. *ADHATODA* Nees

2.1. ***Adhatoda zeylanica*** Medic., Hist. & Commentat. Acad. Elect. Sci. Theod. 6: 393.1790. *A. vasica* Nees, Wall. Pl. As. Rar. 3: 103.1832; Clarke, Hook.f. Fl. Brit. India 4: 540.1885; Prain Beng. Plants 2: 611. 1903 (Rep. ed. 1963). *Justicia adhatoda* L. Sp. Pl. 15.1753. *(Vern. Name: Basak).*

Diffused, much branched, evergreen shrub. Leaves 15-20 cm, elliptic-lanceolate, acuminate. Flowers white with pink or purple stripes in dense axillary peduncled spikes, at the end of the branches; bracts elliptical, sub-acute; corolla bilipped, upper lip curved, ovate oblong, notched. Capsule clavate, seeds sub-orbicular. *Fls. & Frts.:* August-March.

Gregariouly grow on the garden edges, roadsides, railway tracks and occasionally used to make the boundary lines of the orchards.

Field Collection: Harinbari/Sagar Islands, *K.N. - 473.*

Distributions: Almost throughout India. Sri Lanka and other parts of South East Asia.

Economic Importances: Useful in cough and cold; highly efficacious in cold, coryza, pneumonia, spitting of blood, fever, jaundice, also cure bronchitis. Vaccine obtained from the plant is bronchodilatory and is more effective when used in conjuction with atropin. In tuberculosis, it is much effective. A syrup of 'Basak' is a common medicine in lung trouble, acting as a sedative-expectorant. Burning of its leaves is repulsive for the mosquitos.

3. *ANDROGRAPHIS* Wall. ex Nees

3.1. ***Andrographis paniculata*** (Burm.f.) Wall. ex Nees, Wall. Pl. As. Rar. 3: 116.1832; Clarke, Hook.f. Fl. Brit. India 4: 501.1884; Prain Beng. Plants

2: 603.1903 (Rep. ed. 1963). *Justicia paniculata* Burm.f. Fl. Ind. 9.1768. *(Vern. Name: Kalmegh).*

Erect, glabrous, annual herb. Stems 4 - angular, grooved . Leaves linear-lanceolate or lanceolate, glabrous. Flowers pink in solitary lax spreading axillary and terminal racemes; bracts opposite; calyx and corolla glandular,hairy on the outside, stamens 2. Capsule tapering at the ends, sparsely gland hairy, seeds sub-quadrate, rugose, glabrous. *Fls. & Frts.:* August-April.

Gregariously grow on the village garden, prefer shady waste lands; sometimes planted in the garden for its medicinal values.

Field Collections: Gosaba, *K.N. - 176;* Falta, *K.N. - 2018.*

Distributions: Throughout India. Sri Lanka, West Indies and some of the hot-countries.

Economic Importances: Domestically used for flatulence and diarhoea of children and it is considered anthelmintic; used for tropidity of liver, neuralgia, dyspepsia and worm. Bitter leaves juice is excellent for children sufering from irregular stools and loss of appetite.

4. *BARLERIA* L.

Key to the Species

1a	Unarmed, branches clothed with yellowish hairs, undershrub; flowers blue	1. *B. cristata* L.
1b	Spinous, branches terete, grey whitish, flowers orange yellow	2. *B. prionitis* L.

4.1. ***Barleria cristata*** L. Sp. Pl. 636.1753; Clarke, Hook.f. Fl. Brit. India 4: 488.1884; Prain Beng. Plants 2: 606.1903 (Rep. ed. 1963). *B. dichotoma* Roxb. Fl. Ind. 3: 39.1832 *B. ciliata* Roxb. Fl. Ind. 3: 38.1832. *(Vern. Name: Jati or Jhinti).*

Erect undershrub; branches clothed with yellowish hairs. Leaves elliptic oblong, acute, tapering towards the base. Flowers blue, purple both in axillary and terminal; leaf interrupted spikes; calyx hairy below, spinous; stamens 5, 2 longer with proper anthers. Capsules ellipsoid; 4-seeded. *Fls. & Frts.:* October-February.

Common on the thickets and hedges; road-side and railway tracks.

Field Collection: Madarhat/Baruipur, *K.N. - 99.*

Distributions: Almost throughout India. Distributed in the Tropics.

Economic Importances: Plants used in snake bite, infusion is given in cough; the roots and leaves are used to reduce swelling.

4.1. ***Barleria prionitis*** L.Sp. Pl. 636.1753; Clarke, Hook.f. Fl. Brit. India 4: 482.1884; Prain Beng. Plants 2: 605.1903 (Rep. ed. 1963). *(Vern. Name: Kanta-Jati).*

Bushy prickly undershrub, branching from the base. Stems terete or enlarged, grey white. Spines 3-4 or more in the leaves axil, whitish. Leaves 10 cm long, ovate to elliptic, tapering at the ends. Flowers orange yellow, in terminal spikes. Capsule black, pointed, longer than the calyx. *Fls. & Frts.:* October- June.

Commonly grow in the scrub jungles, roadside hedges or railway tracks.

Field Collection: Sitakundu/Baruipur, *K.N.* - *53.*

Distributions: Throughout India. Sri Lanka, Malacca, Tropical Asia and Africa.

Economic Importances: Root-paste is applied to cure boils and glandular swellings; chewing of leaves relieve toothache and dried bark used in cough and jaundice; leaves used in caterrhal affections of children.

Notes: Barleria lupulina Lindl. and *B. stigosa* Willd. also grow in this lower part of Bengal. But the present author could not collect these taxa and study them.

5. *CARDANTHERA* Ham.

5.1. ***Cardanthera difformis*** (L.) Druce,Rep . Bot. Each., Cl. Brit. Isles, Suppl. 2: 612.1917. *C. triflora* Buch. - Ham. ex Benth, Benth & Hook.f. Gen. Fl. 2: 1074.1874; Clarke, Hook.f. Fl. Brit. India 4: 405.1884; Prain Beng. Plants 2: 595.1903 (Rep. ed. 1963). *Ruellia triflora* Roxb. (Hort. Beng. 46.1814) & Fl. Ind. 3: 52.1832. *Adenosma triflora* Nees Wall Pl. Rar. 3: 79.1832. *(Vern. Name: Kala).*

Annual viscid herb, rooting at the base, 30-40 cm tall. Lower leaves pinnatifid, in large pectinate, membranous, cauline. Flowers 1-3 nate, in axillary whorls; sepals hairy, one longer than the others, very acute, corolla lobes twisted to left. Fruit narrow capsule, pubescent. *Fls. & Frts.:* January-May.

Common on the semi-aquatic area; on the embankment of the ponds, canals and ditches.

Field Collection: Baruipur, *K.N. - 19.*

Distributions: Throughout India. Peru and Singapore.

6. *CROSSANDRA* Salisb.

6.1. ***Crossandra infundibuliformis*** (L.) Nees, Wall. Pl. As. Rar. 3: 98. 1832. *Justicia infundibuliformis* L. Sp. Pl. 21.1753. *Ruellia infundibuliformis* Roxb. Fl. Ind. 3: 41.1832. *Crossandra undulaefolia* Salisb. Parad. Lond. t. 12. 1805; Clarke, Hook.f. Fl. Brit. India 4: 492.1884; Prain Beng. Plants 2: 605.1903 (Rep. ed. 1963).

Undershrub; leaves ovate or lanceolate, often undulate, tapering both the ends. Flowers orange or orange scarlet in linear-oblong spikes; bracts elliptic, acute, sepals 5, ovate, 2 inner smaller; corolla with narrow tube, covered at the mouth; stamens didynamous. *Fls. & Frts.:* April-December.

Palnted in the gardens as ornamental plants.

Field Collection: Santoshopur/near Budge Budge, *K.N. - 1953.*

Distributions: Western Peninsula; cultivated in the N. India and West Bengal; Sri Lanka.

Economic Importances: Ornamental and the plants are aphrodisiac.

7. *DIPTERACANTHUS* Nees emend. Bremek.

7.1. ***Dipteracanthus prostratus*** (Poir.) Nees, Wall. Pl. As. Rar. 3: 81. 1832.*Ruellia prostrata* Poir., Lamk. Encycl. 6: 349.1804; Clarke, Hook.f. Fl. Brit. India 4: 411.1884; Prain Beng. Plants 2: 598.1903 (Rep. ed. 1963). *Dipteracanthus dejectus* Nees, Wall. Pl. As. Rar. 3: 82. 1832.

Diffused, undershrub. Stem prostrate, much branched, internodes long, nodes more or less hairy and often tinged with purple, leaves ovate, elliptic, entire, deep-green above, pale beneath. Flowers pale violet, solitary or few together; corolla caducous, tube funnel shaped above. Capsules clavate, pointed, pubescent, seeds brown. *Fls. & Frts.:* June-December.

Frequently found in the bushy places, along the road-sides and railway tracks.

Field Collections: Subuddhipur/Baruipur, *K.N. - 4;* Santoshpur, *K.N. - 496.*

Distributions: Throughout India. Distributed in the Tropics.

Economic Importances: Leaves are given with liquid copal as remedy for gonorrhoea, also used as remedy for ear diseases.

8. *ECBOLIUM* Kurz.

8.1. ***Ecbolium viride*** (Forssk.) Alston. var. ***dentata*** (Klein ex Link.) Raizada, Ind. For. 84: 482.1958. *E. viride* (Forssk.) Alston., Trimen Hand-Book Fl. Ceylon 6: Suppl. 229.1931. *E. linianum* Kurz, Journ. As. Soc. Beng. 40: 275.1871; Clarke, Hook.f. Fl. Brit. India 4: 544.1885. *Justicia dentata* Klein ex Link. Jahrab 1(3): 48.1820. *Ecbolium linneanum* Kurz var. *dentata* (Klein ex Linx.) Clarke, Hook.f. Fl. Brit. India 4: 545. 1885; Prain Beng. Plants 2: 609.1903 (Rep. ed. 1963). *(Vern. Name: Udu-Jati).*

Erect herb with swollen nodes. Leaves glabrous, dark green, ovate or ovate lanceolate, acuminate at apex, acute or acuminate at base; lateral nerves 4-7 pairs. Seeds broadly ovoid, compressed, pale, base notched. *Fls. & Frts.:* November-April.

Common on the waste lands, village thickets and road-side hedges.

Field Collection: Amtala, *K.N. - 378;* Masjidbati, *K.N. - 836.*

Distributions: Major parts of India. Sri Lanka, Malacca, Arabia, Abyssinia and Madagascar.

Economic Importances: The plant is used for gout and disuria; leaves decoction is used in jaundice, menorrhagia and rheumatism.

9. *HEMIGRAPHIS* Nees emend. T. Ander.

9.1. ***Hemigraphis hirta*** (Vahl) T.Ander.,Jour. Linn. Soc. (Bot) 9: 462. 1867; Naskar Aq.& Semi-Aq. Plants 201.1990.

Small prostrate much branched herb; gregariously grow in the moist places. Naskar (l.c.) described this taxon after collecting from Budge Budge, *K.N. - 737.*

10. *HYGROPHILA* R. Brown

Key to the Species

1a	Plant armed with prominent axillary spines	1. *H. auriculata* (Sch.) Heine
1b	Plants armed	(2)

2a (1)	Flowers in terminal dense oblong spikes	3. *H. polysperma* (Roxb.) T. And.
2b (1)	Flowers in distinct axillary whorls	2. *H. phlomoides* Nees

10.1. ***Hygrophila auriculata*** (Sch.) Heine, Kew Bull. 16(2): 172.1962; Naskar, Aq. & Semi-Aq. Plants 202. 1990. *Barleria auriculata* Sch., Schumach. & Thom. Beskr. Guin. Pl. 285. 1827. *B. longifolius* L. Cent. Pl. 2: 22. 1756. *Ruellia longifolia* Roxb. Fl. Ind. 3: 50.1832. *Asteracantha longifolia* (L.) Nees, Wall. Pl. As. Rar. 3: 90.1832. *Hygrophila spinosa* T. And. Thw. Enum. Pl. Zeyl. 225. 1860. *H. longifolia* T. And, Thw. Enum. Pl. Zeyl. 225. 1860; Clarke, Hook.f. Fl. Brit. India 4: 408.1884. *(Vern. Name: Kanta Kalika* or *Kuliakhara).*

Erect marshy hispid undershrub; gregariously grow to the wet/ swampy areas. Naskar (1.c.) described this taxon from the lower Ganga delta after collecting from Baruipur *(K.N. - 92)* and Kumarzol *(K.N. - 307).* Root and leaves decoction of this taxon is diuretic in the treatment of dropsy, jaundice and urinogenital diseases. Seed juice after mixing with other medicine is useful to the mother after delivery.

10.2. ***Hygrophila phlomoides*** Nees, Wall. Pl. As. Rar. 3: 80. 1832; Naskar Aq. & Semi-Aq. Plants 203.1990.

Erect, unarmed herb; common on the road side marshy lands. Naskar (1.c.) described this taxon after collecting from Bhangankhali *(K.N. -207);* Garia *(K.N. - 361)* and Harinbari/Sagar Islands *(K.N. - 488).*

Leaves of this plant is used in poultice and headache.

10.3. ***Hygrophila polysperma*** (Roxb.) T. And., Journ. Linn. Soc. (Bot.) 9: 456.1867; Naskar, Aq. Semi-Aq. Plants 203. 1990. *Justicia polysperma* Roxb. Pl. 1: 119.1832. *Hygrophila phlomoides* Nees, Wall. Pl. Asiat. Rar. 3: 80. 1832. *Hemiadelphis polysperma* Nees, Wall. Pl. Asiat. Rar. 3: 80. 1832.

Small herb with creeping stems and sending up erect shoots. Common on the moist place. Naskar (1.c.) described this taxon from the Lower Ganga delta after collecting from Bhangankhali/opposite to Canning *(K.N.-156).*

11. *JUSTICIA* L.

Key to the Species

1a	Calyx 5 partite	(2)
1b	Calyx 4 partite	(3)

2a (1)	Diffused undershrub; rooting from lower nodes; leaves ovate-lanceolate; capsule clavate pubescent, stout	1. *J. betonica* L.
2b (1)	Glabrous shrub; not rooting at lower nodes; leaves linear-lanceolate; capsule clavate glabrous	3. *J. gendarussa* Burm.f.
3a (1)	Bracts and bracteoles as long or longer than the calyx in flowers	4. *J. simplex* D. Don
3b (1)	Bracts and bracteoles shorter than the calyx in flowers	2. *J. diffusa* Willd.

11.1. ***Justicia betonica*** L. Sp. Pl. 15. 1753; Clarke, Hook.f. Fl. Brit. India 4: 525. 1885; Prain Beng. Plants 2: 610.1905 (Rep. ed. 1963). *J. ramosissima* Roxb. Fl. Ind. 1: 129. 1832. *J. betonica* var. *ramosissima* Clarke, Hook.f. Fl. Brit. India 4: 525. 1884.

Erect diffusely branched, undershrub. Branches often rooting at the lower nodes, cylindric, often tinged with purple above the nodes. Leaves ovate or lanceolate, sub-obtuse, entire or slightly toothed. Flowers dull white, veined and spotted with pink, in simple or branched terminal spikes. Capsule stout, clavate pubescent. *Fls. & Frts.:* October-March.

Common on the hedge and village thickets.

Field Collection: Garia, *Dutta-101 (C.C.H.).*

Distributions: Almost throughout India, but often cultivated. Sri Lanka, Malaya and Tropical Africa.

Economic Importances: The plant is useful for swelling, also used in diarrhoea. In Sri Lanka, the leaves are used in poultice for boils.

11.2. ***Justicia diffusa*** Willd. Sp. Pl.1: 87.1797; Naskar, Aq. & Semi-Aq. Plants 204. 1990. *Rostellaris diffusa* Nees, Wall. Pl. As. Rar. 3: 100. 1832.

Erect, much branched herb. Stems 4-angled. Leaves opposite, ovate or elliptic-lanceolate, acute or axillary and terminal spikes; bracts linear-lanceolate, or linear, shorter than the calyx, sepals not hairy. Capsule oblong, glabrous. *Fls. & Frts.:* August-February.

Common weed on the paddy-fields and dampy/moist places.

Field Collection: Fort William, *Dutta-102* (C.C.H.).

Distributions: Almost throughout India. Rangoon and Siam.

11.3. ***Justicia gendarussa*** Burm.f. Fl. Ind. 10.1768; Clarke, Hook.f. Fl. Brit. India 4: 532. 1884; Watt Dict. Econ. Prod. 4: 557. 1890; Prain Beng. Plants 2: 610. 1903 (Rep. ed. 1963). *Gendarussa vulgarios* Nees, Wall. Pl. As. Rar. 3: 104. 1832. *(Vern. Name: Gandho-rusa* or *Jagatmadan).*

Glabrous, shrub; leaves lanceolate or linear-lanceolate, acute or sub-acute at the apex, narrowed down at the base. Flowers in terminal or upper axillary spikes; bracts upto 4 mm long, linear-capsule-clavate, glabrous, *Fls. & Frts.:* April-August.

Common in the hedges and thickets as an escape from cultivation.

Field Collection: Madhyamgram, *K.N. - 800.*

Distributions: Throughout India. Sri Lanka, Malacca, Malaya, China and Philippines.

Economic Importances: Leaves and tender shoots-diaphoretic, given in the form of decoction in chronic rheumatism. Infusion of leaves is given internally in cephalalgia, hemiplegia and facial paralysis.

11.4. ***Justicia simplex*** D. Don, Prodr. Fl. Nepal. 118. 1825; Naskar, Aq. & Semi-Aq. Plants 205. 1990.

Erect or prostrate herb. Common on the moist places. Naskar (1.c.) described this taxon from the lower Ganga delta after collecting from Baruipur, *K.N. - 25.*

Notes: Justicia procumbens L. also grow in the lower Ganga delta, but the present author have no collection of this species.

12. *NELSONIA* R. Brown

12.1. ***Nelsonia canescens*** (Lamk.) Spreng., Syst. Veg. 1: 42.1825; Naskar, Aq. & Semi-Aq. Plants 205. 1990. *Justicia canescens* Lamk., Tab. Encycl. Meth. Bot. 1: 41. 1791. *Nelsonia campestris* R. Brown, Prodr. Fl. Nov. Hall. 1: 481.1810. *(Vern. Name: Paramul).*

Diffused, shortly villous herb; frequently found in the moist waste lands. Naskar (1.c.) described this taxon from the lower Ganga delta.

13. *PERISTROPHE* Nees

13.1. ***Peristrophe bicalyculata*** (Retz.) Nees, Wall. Pl. As. Rar. 3: 113. 1832; Naskar Aq. & Semi-Aq. Plants 206.1990. *Dianthera bicalyculata* Retz. Vet. Akad, Handl. 297. t.9.1775. *Justicia bicalyculata* Vahl, Symb. Bot. 2: 13. 1791. *(Vern. Name: Nasa-Bhanga).*

Erect much branched herb or undershrub; common on the moist waste lands. Naskar (1.c.) described this taxon from the lower Ganga delta.

14. *PHAULOPSIS* Willd. *emend* Spreng. *nom. cons.*

14.1. ***Phaulopsis dorsiflora*** (Retz.) Santapau, Kew Bull. 1948 & Univ. Bomb. Bot. Mem. 2: 30. 1951. *Ruellia dorsiflora* Retz., Obs. 6: 31.1791. *P. parviflora* Willd. Sp. Pl. 3: 342. 1801; Clarke, Hook.f. Fl. Brit. India 4: 417.1884; Prain Beng. Plants 2: 601.1903 (Rep. ed. 1963).

Prostrate or decumbent, much branched, hairy herbs. Stem 20-80 cm long, suffruticose, slightly thickened at nodes. Leaves very unequal, ovate or elliptic, acute or sub-acute at apex; acute or acuminate and oblique at base. Flowers in terminal spikes or sometimes 1-3 clusters, in lower axils also. Capsule 5-6 mm long, glabrous, clavate, tip pointed. Seeds 4, brown. *Fls. & Frts.:* December-March.

Rarely grow on the waste lands and village shrubberies, prefer grassy lands.

Field Collection: Birati, *N.S. Chatterjee-46* (CAL).

Distributions: Throughout India. Sri Lanka, Madagascar and Tropical Africa.

15. *RHINACANTHUS* Nees

15.1. ***Rhinacanthus nasutus*** (L.) Kurz., As. Soc. Bengal 39: 79.1870. *Justicia nasuta* L. Sp. Pl. 16.1753. *Rhinacanthus communis* Nees, Wall. Pl. As. Rar.3: 109.1832; Clarke, Hook.f. Fl. Brit. India 4: 541.1884; Prain Beng. Plants 2: 611.1903 (Rep. ed. 1963). *(Vern. Name: Palak-jui).*

Undershrub, young pubescent. Stems angled or terete. Leaves opposite, elliptic or elliptic lanceolate, acute or sub-acute at apex, acute at base. Flowers sessile or shortly pedicelled in terminal pubescent panicles. Seeds black, glabrous, tuberculate. *Fls & Frts.:* December-April.

Rarely on the village thickets or on waste lands.

Field Collection: Satzelia, *K.N. - 2161.*

Distributions: Throughout India. Sri Lanka, Java and Madagascar.

Economic Importances: Roots, leaves and seeds are useful remedy for ringworm and other skin diseases. Roots after boiling with milk used as aphrodisiac; antidote to snake bite.

16. *RUELLIA* L. *emend.* Bremek.

16.1. ***Ruellia tuberosa*** L. Sp. Pl. 633.1753; Prain Beng. Plants 2: 599. 1903 (Rep. ed. 1963). *Dipteracanthus nudiflorus* Engelm. & Gray, Bost. J. Nat. Hist. 5: 229. 1845.

Much branched; undershrub; root-stock fleshy tuberous, leaves elliptic obovate, 5-10 cm long, open undulate crenate shortly petioled. Flowers deep blue, handsome in axillary and terminal cymes; calyx glandular. Capsule linear. *Fls. & Frts.:* June-October.

Frequently in dampy and shady areas.

Field Collection: Falta, *K.N.-2013.*

Distributions: Throughout India; native of Tropical America.

17. *RUNGIA* Nees

Key to the Species

1a	Leaves near the base petioled, all near glabrous; bracts dimorphic; barren bracts mucronate, often cuspidate, fertile ones hairy	1. *R. pectinata* (L.) Nees
1b	Leaves all sub-sessile, scabrid, puberulous; bracts round, apiculate	2. *R. repens* (L.) Nees

17.1. ***Rungia pectinata*** (L.) Nees, DC. Prodr. 11: 470. 1847; Naskar Aq. & Semi-Aq. Plants 207. 1990. *Justicia pectinata* L., Amoen. Acad. 4: 293. 1759. *Rungia parviflora* Nees var. *pectinata* (L.) Clarke, Hook.f. Fl. Brit. India 4: 550. 1885.

Much-branched, prostrate, annual herb on the grass lands and moist localities. Naskar (1.c.) described this taxon after collecting from Canning *(K.N. - 133)* and Saktinagar/Maslandapur *(K.N. - 550).*

17.2. ***Rungia repens*** (L.) Nees, Wall. Pl. As. Rar. 3: 110. 1832; Clarke, Hook.f. Fl. Brit. India 4: 549. 1885; Prain Beng. Plants 2: 613.1903 (Rep. ed. 1963). *Justicia repens* L. Sp. Pl. 15.1753. *(Vern. Name: Parpatha).*

Creeping, herb with quadrangular branches, rooting from the nodes. Leaves sub-sessile. Spikes often peduncled. Flowers violet or pinkish in colour, erect, terminal pubescent spikes; bracts broadly elliptic, cuspidate, margins ciliate. Capsule minutely, thinly hairy. *Fls. & Frts.:* October-April.

Widely distributed throughout the district 24-Parganas.

Field Collection: Durganagar/near Dum Dum Cant., *K.N. - 1718.*

Distributions: Throughout the warmer parts of India; Sri Lanka.

Economic Importances: Useful in fever, cough and expelling of worms. Fresh leaves pasted and mixed with castor oil are applied to scalp in case of skin disease from fungi.

18. *STAUROGYNE* Wall.

18.1. ***Staurogyne glutinosa*** (Wall. ex. Clarke) O. Kuntze, Rev. Gen. Pl. 1: 499.1891; Naskar, Aq. & Semi-Aq. Plants 208.1990. *Ebermaiera glutinosa* Wall. ex Clarke, Hook.f. Fl. Brit. India 4: 396.1884.

Diffuse trailing pubescent herb; occasionally in the moist lands. Naskar (1.c.) described this taxon from the lower Ganga delta.

19. *THUNBERGIA* Retz. *nom. cons.*

19.1. ***Thunbergia grandiflora*** (Roxb. ex Rottl.) Roxb. Fl. Ind. 3: 34. 1832; Clarke, Hook.f. Fl. Brit. India 4: 392.1884; Prain Beng. Plants 2: 593.1903 (Rep. ed. 1963). *Flemengia grandiflora* Roxb. ex Rottl. Gen. Naturf. Fr. Neue Schr. 4: 202.1803. *(Vern. Name: Nil lata).*

Large woody climber. Leaves thick, fleshy, large, deeply cordate, toothed or lobed. Flowers light or dark blue, large and showy, axillary from the uppermost leaves and in racemes. Capsules uncurved on the pendulous branches. *Fls. & Frts.:* September-December.

Commonly grow as an ornamental climber.

Field Collection: Dakshin Barasat, *K.N. - 1650.*

Distributions: Native of India. Old World Tropics.

Notes: Thunbergia Retz. is sometime treated under a separate and distinct family– THUNBERGIACEAE.

CIII. VERBENACEAE

Key to the Genera

1a	Inflorescence with lowest flowers opening first (centripetal)	(2)
1b	Inflorescence centrifugal	(5)
2a (1)	Flowers sessile; spikes simple	(3)
2b (1)	Flowers pedicelled; racemes simple or panicled	3. *DURANTA* L.
3a (2)	Plants herbaceous	7. *PHYLA* Lour.
3b (2)	Plants woody	(4)

4a (3)	Scandent spiny shrub	5. *LANTANA* L.
4b (3)	Branches slender erect or sub-erect	6. *LIPPIA* Houst.
5a (1)	Flowers distinctly zygomorphic; stamens didynamous	(6)
5b (1)	Flowers actinomorphic; stamens equal to the petals	(9)
6a (5)	Leaves digitate	10. *VITEX* L.
6b (5)	Leaves simple	(7)
7a (6)	Drupes with 1, 4-celled pyrene	(8)
7b (6)	Drupes with 4, 1-seeded pyrene or druplets; calyx not petaloid	2. *CLERODENDRUM* L.
8a (7)	Flowers greenish white	8. *PREMNA* L.
8b (7)	Flowers yellow or yellowish brown	4. *GMELINA* L.
9a (5)	Plants shrub; flowers pink	1. *CALLICARPA* L.
9b (5)	Large tree; flowers white	9. *TECTONA* Linn. f. *nom. cons.*

1. *CALLICARPA* L.

1.1. ***Callicarpa candicans*** (Burm.f.) Hochr., Candollea 5: 190.1934; Meeuse, Blumea 5: 71.1942. *Urtica candicans* Burm.f. Fl. Ind. 197.1768. *Callicarpa cana* L. Mant. 2: 198.1771; Clarke, Hook.f. Fl. Brit. India 4: 568. 1883; Prain Beng. Plants 2: 618.1903 (Rep. ed. 1963). *(Vern. Name: Arusha).*

Shrubs, leaves broadly elliptic, glabrous above, stellate-tomentose beneath, sub-acuminate at apex, cuneate at base, crenate-serrate; cymes 2-3 cm diam., peduncle 3-6 mm long. Fruit purple, finally black. *Fls. & Frts.:* March-August.

Rarely grow on the village thickets.

Field Collection: Hasnabad, *M.K. Ghosh-976* (CAL).

Distributions: Common in India, Philippines Islands and N. Australia; native of Malaya Peninsula.

2. *CLERODENDRUM* L.

Key to the Species

1a	Beautiful climbing shrub; flowers crimson, in many flowered terminal showy cymes; drupes black	3. *C. splandens* G. Don

1b	Plants not climbing, usually straggling or tall shrubs	(2)
2a (1)	Tall shrubs; leaves narrowly lanceolate, subsessile; corolla narrowly funnel-shaped; panicles terminal, elongated, lax	1. *C. indicum* (L.) O. Kuntze
2b (1)	Medium straggling shrub; leaves broad obovate or oblong; corolla irregularly salver-shaped	(3)
3a (2)	Panicles axillary or if terminal the panicles lax, leafy below; calyx in fruit closely applied to base of berry	2. *C. inerme* (L.) Gaertn.
3b (2)	Panicles terminal; calyx large, with broadly lanceolate, acute suberect segments panicles open, pyramidal	4. *C. viscosum* Vent.

2.1. ***Clerodendrum indicum*** (L.) O. Kuntze. Rev. Gen. 586. 1891. *Siphonanthus indica* L. Sp. Pl. 109. 1753. *Clerodendron siphonanthus* R. Br., Ait. f. Hort. Kew (ed. 2) 4: 65. 1812; Clarke, Hook.f. Fl. Brit. India 4: 595.1883; Prain Beng. Plants 2: 624.1903 (Rep. ed. 1963). *(Vern. Name: Bamanati).*

Tall perennial shrub. Stems grooved hollow. Leaves narrowly lanceolate, acuminate, tapering to the base. Flowers white in terminal lax panicle or cymes; calyx tube campanulate, green with lanceolate lobes. Drupes bluish green. *Fls. & Frts.:* July- January.

Common on the waste lands, roadsides and railway track sides.

Field Collection: Piyalinagar/Baruipur, *K.N. - 51.*

Distributions: Throughout India. Extensively cultivated in both hemispheres.

Economic Importances: This plant has great medicinal value; it is reported that its root is helpful to normal deliver if its smell is taken by the pregnant ladies. Occasionally planted as ornamentals.

2.2. ***Clerodendrum inerme*** (L.) Gaertn. Fruct. 1. 271. 1788; Clarke, Hook.f. Fl. Brit. India 4: 589.1883; Naskar & Guha Bakshi Mangr. Swamps Sundarbans 172. 1987. *(Vern. Name: Ban Jui* or *Batraj).*

Straggling much branched shrub up to 3 m high; branches very twiggy and long. No aerial roots. Leaves opposite, elliptic or obovate, obtuse, glabrous 2-3 cm long and 1.5 cm broad, apex acute. Flowers in axillary pedunculate cymes, generally 3- flowered, corolla white with long slender tube. Fruit pyriform drupe, woody 1.5-2.0 cm long. Germination hypogeal. Grow on the river banks and back-mangal localities, thickets; normally away from the tidal influences. *Fls. & Frts.:* July-February.

Field Collections: Surjyapur, *K.N. - 295;* Dhopa Gachi, *K.N. - 1157.*

Distributions: Throughout India; Sri Lanka.

Economic Importances: Checking soil erosion in the tidal river banks and add nutrients to the mangrove soil. Beside all these, this plant is also largely planted as dense hedge plant in the ornamental gardens throughout.

2.3. ***Clerodendrum splandens*** G. Don, Edinb. N. Phil. Journ. 11: 349.1824.

Climbing beautiful shrubs. Leaves oblong, orbicular to sub-orbicular, entire, coriaceous, dark green above. Cordate, acuminate. Flowers crimson, in dense, many flowered terminal, showy cymes. Drupes black. *Fls. & Frts.:* November-April.

Grow in the hedge as climber.

Field Collection: Dum Dum Cant., *N.S. Chatterjee-* 33 (CAL).

Distributions: Not so common in India.

Notes: D. Prain (1903a) did not mention this taxon in Bengal Plants.

2.4. ***Clerodendrum viscosum*** Vent. Gard. Malm. t.25. 1803. *Clerodendron infortunatum* Gaert. Fruct 1: 271.1788; Clarke, Hook.f. Fl. Brit. India 4: 594. 1883; Watt Dict. Econ. Prod. 2: 373. 1889; Prain Beng. Plants 2: 623. 1903 (Rep. ed. 1963). *(Vern. Name: Gnetu).*

Robust shrub. Branches clothed with yellowish-white tomentose. Leaves large, ovate and usually cordate. Flowers white, pink, in terminal panicles. Drupes seated on the enlarged pink calyx, sub- globose, black when ripe. *Fls. & Frts.:* August-July.

Gregariously grow on the waste places, roadsides and railway tracks.

Field Collection: Gosaba, *K.N. - 648;* Garia, *K.N. - 326.*

Distributions: Throughout India. Sri Lanka, Malaya and some other warmer countries.

Economic Importances: A bitter substances, *Clerodin* is found in this plant and it kills earthworms and small fish, within half an hour. It has been recommended as an effective vermifuse.

3. *DURANTA* L.

3.1. ***Duranta repens*** L. Sp. Pl. 637.1753. *D. plumieri* Jacq. Select. Am. 186, t.176. 1763; Clarke, Hook.f. Fl. Brit. India 4: 560. 1883; Prain Beng. Plants 2: 617. 1903 (Rep. ed. 1963). *(Vern. Name: Mudikanta).*

Erect, evergreen shrub with spinose branches. Leaves ovate, elliptic or oblong, lanceolate, entire or serrate. Flowers violet or blue in axillary and terminal racemose. Drupe globose yellow. *Fls. & Frts.:* April-December.

Planted as hedge plants for making boundary in the gardens.

Field Collection: Budge Budge, *K.N. - 733.*

Distributions: Throughout India. A native of S. America and West Indies; pantropic species.

Economic Importance: Effective hedge plants for the garden boundary.

4. *GMELINA* L.

4.1. ***Gmelina arborea*** Roxb. Fl. Cor. 3: 42,t. 246.1819; Clarke, Hook.f. Fl. Brit. India 4: 581.1883; Watt Dict. Econ. Prod. 3: 514.1890; Prain Beng. Plants 2: 619. 1903 (Rep. ed. 1963). *(Vern. Name: Gambar* or *Gamari).*

Moderate sized, unarmed, deciduous tree. Leaves broadly ovate, acuminate, entire, cordate, 10-23 cm long, cuneate at base; petioled glandular, at the top. Flowers yellowish brown in lateral or terminal panicles. Drupes pyriform or obovoid, smooth. *Fls. & Frts.:* February-July.

Rarely found in the hedges and village thickets, occasionally planted.

Field Collection: Specimen not collected or preserved.

Distributions: Indigenous in most parts of India; Burma.

Economic Importances: Fruits are edible by the tribals; roots are useful in abdominal pains, burning sensation and fevers. Flowers are useful in leprosy and blood diseases. Leaves juice is employed as a lotion for wounds and ulcers.

5. *LANTANA* L.

Key to the Species

1a	Flowers coloured	1. *L. camara* var. *aculeata* Mold.
1b	Flowers white	2. *L. indica* Roxb.

5.1. ***Lantana camara*** L. var. ***aculeata*** (L.) Moldenke, Torreya 34: 19. 1934. *L. aculeata* L. Sp. Pl. 627.1753. *L. camara* L. Sp. Pl. 627.1753; Clarke, Hook.f. Fl. Brit. India 4: 562. 1883; Prain Beng. Plants 2: 615. 1903 (Rep. ed. 1963). *(Vern. Name: Manindra Kanta).*

Profusely, much branched aromatic herb with small recurved prickles stem. Leaves ovate or ovate-oblong with cordate or sub-cordate base, rugose above. Flowers varied in colours; bracts distinct. Ovary 2-celled, 2 ovuled. Drupe black. *Fls. & Frts.:* Throughout the year.

Grow gregariously on the waste places, hedges, thickets, railways tracks and roadside jungles; very common everywhere.

Field Collections: Fulta, *K.N. - 43;* Canning/railway tracks, *K.N. - 136.*

Distributions: Throughout India; widely planted and common as weeds in forest clearings, cultivated in field and waste lands. Introduced in the Indian Botanic Garden in the year 1809. A native of America.

Economic Importances: Sometimes planted as ornamental hedge plants in the lawn or flower gardens. Leaves infusion is taken internally for bilious fever.

5.2. ***Lantana indica*** Roxb. Fl. Ind. 3: 89. 1832; Clarke, Hook.f. Fl. Brit. India 4: 562.1883; Prain Beng. Plants 2: 615. 1905 (Rep. ed. 1963). *L. alba* Schaucer, DC. Prodr. 11. 606. 1847. *Lantana trifoliata.* L.

Woody shrub. Branches rambling 4-angular, strigose. Leaves 3-7 cm long, ovate, coarsely crenate, ovate, acute or sub-obtuse, crenate, white-pubescent beneath. Flowers pure white, scentless, in axillary, peduncled, close heads of long spikes. Drupe purple, enclosed in thin transparent calyx. *Fls. & Frts.:* Throughout the year.

Frequently planted as a hedge plant in the garden, sometime as a garden escape.

Field Collection: Maslandapur, *K.N. - 514.*

Distributions: Almost throughout India, introduction in India at an early time. Widely distributed throughout the tropical America.

6. *LIPPIA* Houst.

6.1. ***Lippia javanica*** (Burm.f.) Spreng., Syst. 2: 752.1825; Naskar, Aq. & Semi. Aq. Plants 208.1990. *Verbena javanica* Burm.f., Fl. India 12: t. 6, f.2, 1766. *Lantana abla* Mill. Gand. Dict. Brit. (ed. 8) 8.1768. *Lippia geminata* H.B.& K., Nov. Gen. ed. Sp. 2: 266.1818. *L.alba* (Mill.) Br. ex Britton & Wilson, Sc. Surv. Porto Rico & Virgin Islands 6: 141.1925.

Small procumbent shrub; common on the margins of the canals and ponds. Naskar (l.c.) described this taxon from the lower Ganga delta (Mallickpore, K.N.-79). Considered stomachic and nervine in some parts of Brazil and Paraguay

7. *PHYLA* Lour.

7.1. ***Phyla nodiflora*** (L.) Greene, Bittonia 4: 46.1899. *Verbena nodiflora L.* Sp. Pl. 20.1753. *Lippia nodiflora* A. Rich. Michx. Pl. Bor-Amer. 2: 15. 1803. *(Vern. Name: Bhui-Okra).*

Annual creeping herb; gregariously grow on the moist waste places. Naskar, Aq. & Semi-Aq. Plants p. 204.1990, described this taxon after collecting from Gosaba (*K.N. - 659)* and Dakshineswar (*K.N. - 686).* This plant is useful in fever, as diuretic, pasted and applid as maturant for boils. Infusion of leaves and young stems are given to children in indigestion and to women after delivery.

8. *PREMNA* L.

Key to the Species

1a	Flowers foetid, greenish-yellow; fruits globose	1. *P. corymbosa* Rottl. & Willd.
1b	Flowers greensih-white; fruits globose	2. *P. mucronata* Roxb.

8.1. ***Premna corymbosa*** (Burm.f.) Rottl. & Willd., Gressla. Nat. Freunde Neue Schr. 4: 187.1803; Naskar & Guha Bakshi, Mangr. Swamps Sundarbans 173.1987. *Cornutia corymbosa* Burm.f., Ind. 133. t. 41.f.1.1768. *Premna integrifolia* L. Mant. 2:252.1771. *P. serratifolia* L. Mant. 2: 253.1771. *P. latifolia* Dalz. & Gibs. 200. (not of Roxb.) (*Vern. Name: Gamiari* or *Bhui Biravi).*

Shrub or trees. Occasionally grow on the scrub-jungles as back mangrove species. Naskar & Guha Bakshi (l.c.) described this taxon after collecting from Sagar Islands/Sundarbans. Leaves are said to be empoyed for curing piles and tumours.

8.2. ***Premna mucronata*** Roxb. Fl. Ind. 3: 80. 1832. *P. latifolia* var. *mucronata* Clarke, Hook.f. Fl. Brit. India 4: 578.1883; Prain Beng. Plants 2: 620.1903 (Rep. ed. 1963). *(Vern. Name: Bakar).*

Small to medium sized, deciduous tree. Leaves 7-15 cm long, broadly ovate, sharply acuminate, entire or undulate, deep green above, pale beneath, hairy especially on the midrib. Flowers greenish-white in broad corymbs, usually terminating short, leafy brances. Drupes globose. *Fls. & Frts.:* Almost throughout the year.

Rarely cultivated but mostly wild.

Field Collection: Canning/on the river sides, K.N.-190.

Distributions: Frequently found throughout India. Sri Lanka, Burma and other S.E. Asian countries.

9. *TECTONA* Linn. f. *nom. cons.*

9.1. ***Tectona grandis*** Linn.f.Suppl. 151. 1781; Clarke, Hook.f. Fl. Brit. India 4: 570.1885; Watt Dict. Econ. Prod. 6(4): 1.1893; Prain Beng. Plants 2: 618.1903 (Rep. ed. 1963). *(Vern. Name: Segun).*

Large, deciduous tree; branches 4-angular. Leaves large, elliptic or obovate, rough stellate tomentum beneath, acute or acuminate, entire, usually cuneate at the base. Flowers dull white. Drupes sub-globsoe, densely clothed with felted stellate hairs. *Fls. & Frts.:* July-January.

Planted as an avenue tree or as timber producing tree in the old orchards.

Field Collection: Specimens not preserved.

Distributions: Indigenous in most parts of the plains of Peninsula India. Burma, Malaya and the neighbouring islands.

Economic Importances: Yield a strong and valuable timber of high value; flowers are used in bronchitis and flower extracted oil promote the growth of hairs.

10. *VITEX* L.

10.1. ***Vitex negundo*** L. Sp. Pl. 638.1753; Clarke, Hook.f. Fl. Brit. India 4: 583.1883; Prain Beng. Plants 2: 622.1903 (Rep. ed. 1963). *(Vern. Name: Nishinda).*

Evergreen shrub or small tree. Branches 4-angled, densely white tomentose. Leaves 3-5 foliolate; leaflets stalked, lanceolate or narrowly lanceolate, acuminate, entire or rarely crenate. Flowers lavender blue, in loose clusters, arranged in a large terminal panicles. Drupes black, when ripe. *Fls. & Frts.:* Almost throughout the year.

As a hedge plant or planted in the roadsides and unused lands.

Field Collection: Champahati, *K.N. - 1248.*

Distributions: Throughout India. Japan to Southwards, throughout Malaysia to Western Polynesia, Tropical East Africa and Madagascar.

Economic Importances: Plants are said to be used to promote growth of hairs and in enlarged spleen and bronchitis; valuable insecticides.

CIV. AVICENNIACEAE

1. *AVICENNIA* L.

Key to the Species

1a	Leaves never glaucous, whitish underneath, coriaceous, elliptic-oblong or ovate-oblong, obtuse glabrous, smooth and shining above	3. *A. officinalis* L.
1b	Leaves glaucous, whitish, tomentose beneath; lanceolate, ovate-lanceolate to oblong lanceolate; glabrous above	(2)
2a (1)	Flowers orange in 10-32 flowered heads; ovary oblong glabrous below, hairy above; fruit ellipsoid, mucronate pilose, leaves ovate - lanceolate to oblong lanceolate	2. *A. marina* (Forssk.) Vierh.
2b (1)	Flowers yellow in long spike; ovary villous; hairy in the middle; fruits ovoid, smooth leaves lanceolate, acute	1. *A. alba Blume*

1.1. ***Avicennia alba*** Blume, Bijdr. Fl. Ned. Ind. 14: 821.1820; Naskar & Guha Bakshi Mangr. Swamps Sundarbans 94.1987. *(Vern. Name: Kal-Bain).*

Tall tree, upto 25 m high, bark dark greyish, lenticillate smooth, not fissured. Thin pencil like pneumatophores, upto 30 cm long, with little secondary thickening; occasionally or sporadically stilt roots can grow for additional support in the loosely consolidated soil and when grow on the river slopes. Leaves 12-15 cm long and 2.5-3.5 cm broad, usually pointed apically, pale green above but silvery grey or white beneath. Flowers 1.0-1.5 cm diam., tetramerous; corolla orange yellow, glabrous within. Fruit 4 cm long conical, each capsule containg 4 seeds; seeds flattened, 3-2 cm length within the testa. Seed germination epigeal or modified epigeal; crypto-viviparous or incipiant viviparous. *Fls. & Frts.:* November-June. (Fig.39).

Grow abundantly on the tidal river banks, and inter-tidal mangrove forests or sea-shores. This plant can grow in diverse saline condition; as

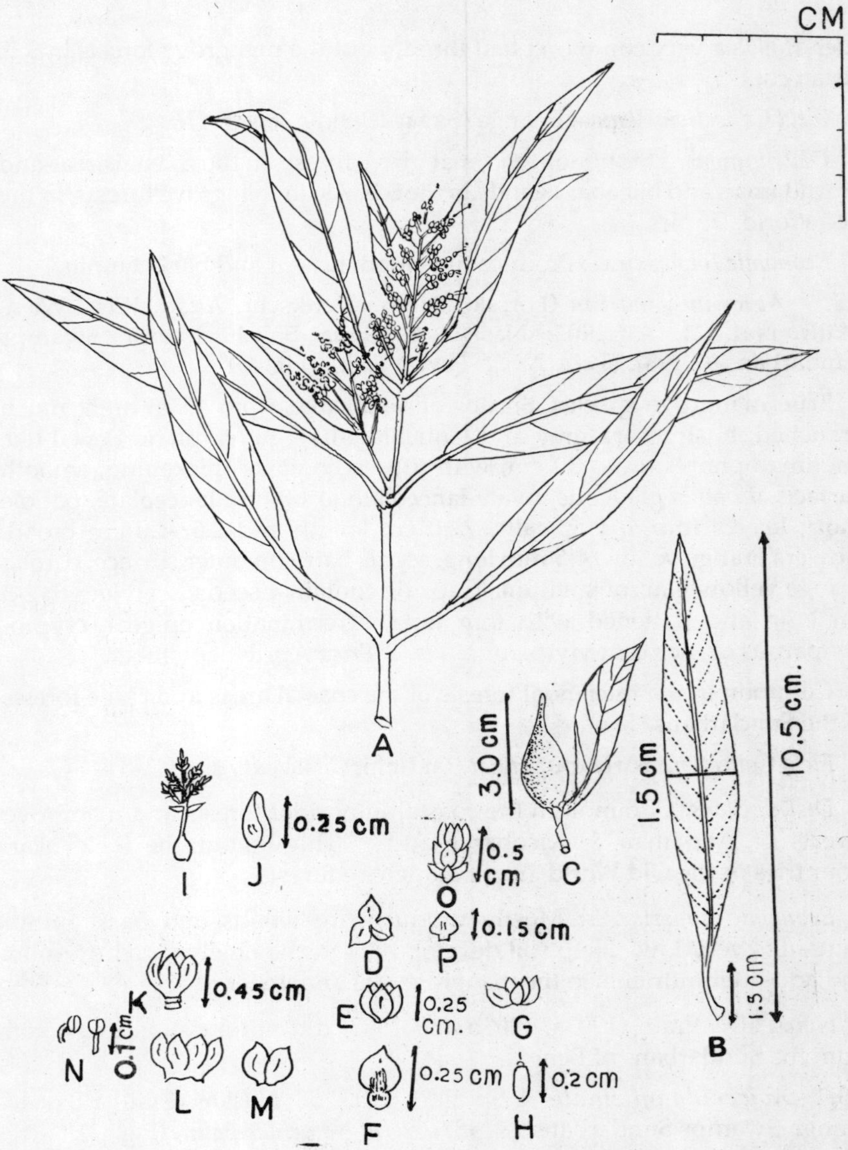

Fig.39. AVICENNIA ALBA Blume

A. Flowering twig, B. Single leaf, C. Single fruit with leaf, D. Single flower, E.–H. Parts of Flowers (sepals), I. Inflorescence J. Petals, K. Calyx, L. & M. Corolla with androecium, N. Anther lobes

such this is a very common plant throughout the mangrove forests in S.E. Asian countries.

Field Collection: Supari Bhanga/Sagar Islands, *K.N. - 416.*

Distributions: Throughout the mangrove forests in the coastal areas and in Andaman and Nicobar Islands.In the S.E. Asian mangrove forests; in the Old World Tropics.

Economic Importances: Produce fire wood, timber and bark- tannin.

1.2. ***Avicennia marina*** (Forssk.) Vierh., Denkschr. Akad. Wiss. Wien. Math. Nat. 71: 435.1907; Naskar & Guha Bakshi Mangr. Swamps Sundarbans 97.1987. *(Vern. Name: Peyara Bain).* (Fig.40).

True mangrove species. Shrubs or small tree, upto 15 m high; much branched, bushy, bark grey and lenticillate-like guava plant. Pencil-like pneumatophores, upto 30 cm, with little secondary thickening, smooth surface. Leaves glaucous, ovate lanceolate to oblong lanceolate, petiole short; leaves rounded apically, 7-10 cm length and 2.5 -4.0 cm broad. Flowers orange yellow, 4-5 mm long, sepals hairy on outer surface, corolla orange yellow glabrous within. Capsule contain 4 seeds, seed length 2-3 cm long and flattened with thin testa. Germination epigeal, crypto-viviparous or incipient viviparous. *Fls. & Frts.:* April- September.

Common in the inter-tidal forests of the coastal areas and ridge forests of the Sundarbans.

Field Collection: Surjyapur/near Raidighi, *G.G.Maity-s.n.* (CAL).

Distributions: Common in the coastal inter-tidal forests and mangrove forests of Andaman & Nicobar Islands. Throughout the S.E. Asian Countries or the Old World Tropic mangrove forests.

Economic Importances: Mostly produce fire woods and bark tannin contain is very low. Help soil development in the newly silted up soils; and add litter nutrients to the mangrove soil and water phases.

Notes: D. Prain (1903a, 1903b and 1905) did not mention this taxon from the Sundarbans of Bengal.

1.3. ***Avicennia officinalis*** L.Sp. Pl. 1: 110.1753; Naskar & Guha Bakshi Mangr. Swamps Sundarbans 97.1987. *(Vern. Name: Jat-Bain).* (Fig.41).

True mangrove species; medium tree, upto 20 m high with smooth bark. Branches both spreading and erect. Pencil like pneumatophores, occasionally stilt roots present; all these aerial roots with thick corky

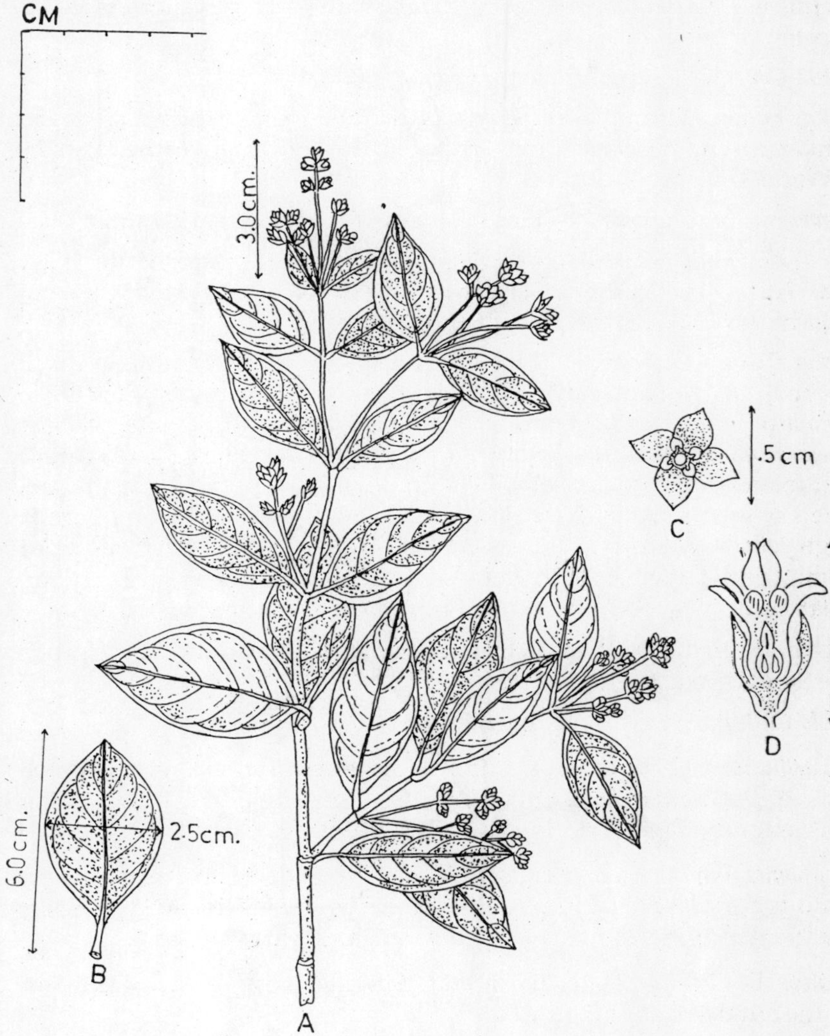

Fig.40. AVICENNIA MARINA (Forssk.) Vierh.
A. Flowering twig, B. Single leaf,
C. Single flower (Top view), D. T.S. of Flower.

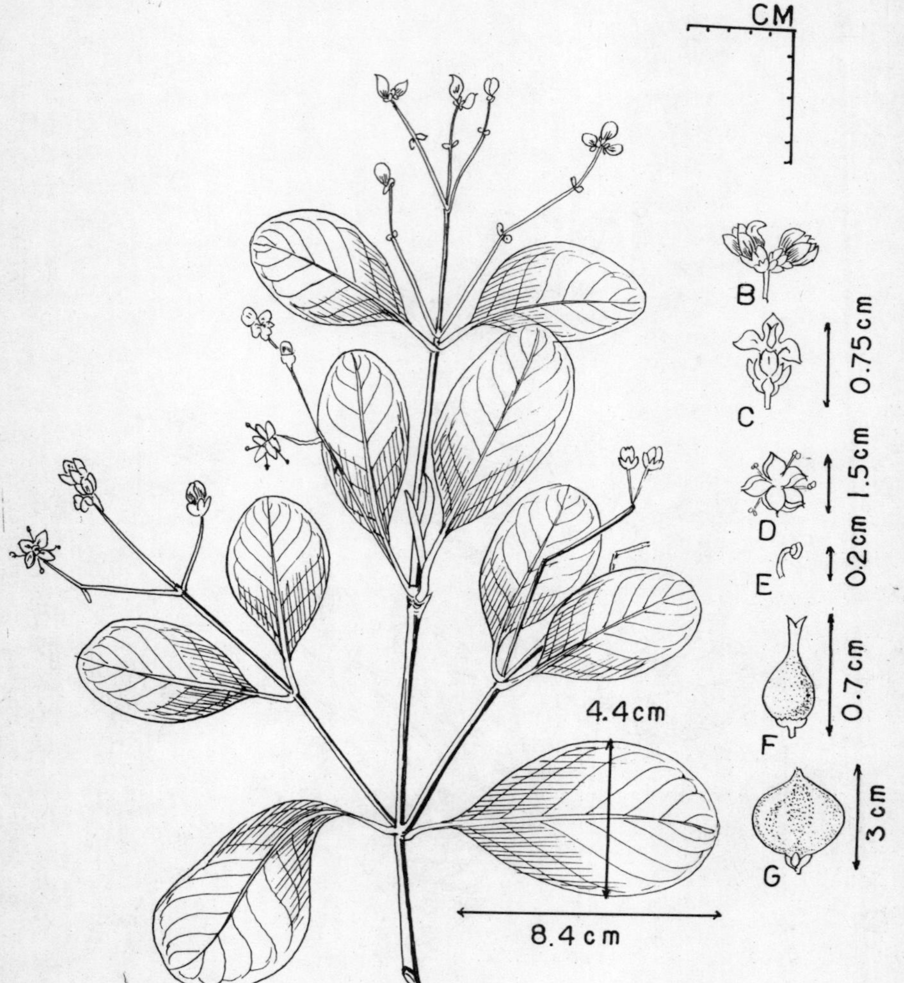

Fig.41. AVICENNIA OFFICINALIS L.

A. Flowering twig, B. Two flowers with peduncle, C. Single open flower, D. Top view of flower with androecium, E. Single androecium, F. Gynoecium, G. Single mature fruit.

arenchyma cells and numerous lenticels present on the cortex. Leaves coriaceous, elliptc, oblong or obovate oblong, very obtuse or rounded apically, glabrous, smooth and shining above, never glaucous whitish beneath, 10-14 cm long and 4-6 cm broad. Flowers zygomorphic, sessile in compact heads. 6-8 mm in diam., corolla orange yellow, glabrous within, tetramerous. Capsule contain 4 seeds. Seeds flattened with thin testa. Seed germination epigeal or modified epigeal, crypto-viviparous or incipient viviparous. *Fls. & Frts.:* April-August.

Very common on the inter-tidal mud-flats and ridge forests of the Sundarbans mangrove forests.

Field Collection: Bhangan Khali/Opposite to Canning, *K.N. - 298.*

Distributions: Throughout the mangrove forests and inter-tidal coast lines; Andaman & Nicobar Islands. In the Old World Mangrove areas.

Economic Importances: Beside its large conservation roles in this intertidal mangroves forests, this taxon produce fire wood, timber and bark tannins. Leaves litter enrich the soil and water with high organic manure.

Remarks: These *Avicennia* spp. are estimated about 40% in the Indian Sundarbans and play very dominant roles in protecting this coastal mangrove forests from several severe natural calamities. But the exploitation of these species are very high. Natural as well as the artificial regeneration power of these flora are very high; and these can grow in diverse climatic and edaphic conditions; as such these are the suitable species for afforestation programme.

CV. LAMIACEAE

(*nom. alt. LABIATAE*)

Key to the Genera

1a	Stamens declinate	(2)
1b	Stamens straight or ascending	(5)
2a (1)	Lower lip of corolla flat or nearly so	(3)
2b (1)	Lower lip of corolla conspicously concave or boat-shaped	(4)
3a (2)	Corolla-tube short; stigma 2-fid	9. *OCIMUM L.*
3b (2)	Corolla-tube usually long; stigma entire	10. *ORTHOSIPHON* Benth.

4a (2)	Corolla 5-lobed, obscurely 2-lipped, the lowest lobe shorter or equalling the other lobes	4. *HYPTIS* Jacq.
4b (2)	Corolla with upper lip very short, obtusely 3-4 toothed; filaments connate in a sheath	2. *COLEUS* Lour.
5a (1)	Perfect stamens 4	(6)
5b (1)	Perfect stamens 2	12. *SALVIA* L.
6a (5)	Anther 1 or 2 -celled, not separated; connective not filiform	(7)
6b (5)	Anther 2-celled, cells at length diverging	(9)
7a (6)	Anther 2-celled	8. *MENTHA* L.
7b (6)	Anther 1-celled	(8)
8a (7)	Corolla 2-lipped	11. *POGOSTEMON* Desf.
8b (7)	Corolla equally 4-fid	3. *DYSOPHYLLA* Blume
9a (6)	Upper lip of corolla flat, not hooked	1. *ANISOMELES* R. Brown
9b (6)	Upper lip of corolla hooked	(10)
10a (9)	Calyx 5-toothed or lobed	6. *LEONURUS* L.
10b (9)	Calyx 8-10 toothed	(11)
11a (10)	Upper lip of corolla shorter than the lower	7. *LEUCAS Brown*
11b (10)	Upper lip of corolla longer than lower	5. *LEONOTIS* R. Brown

1. *ANISOMELES* R. Brown

1.1. *Anisomeles indica* (L.) O. Kuntze., Rev. Gen. 512. 1891; Mukherjee, Rec. bot. Surv. India 14(1): 152.1940. *Nepeta indica* L. Sp. Pl. 571.1753. *Anisomeles ovata* R. Brown, Ait. Hort. Kew (ed.2) 3: 364.1811; Hook.f. Fl. Brit. India 4: 672.1885; Watt Dict. Econ. Prod. India 1: 255.1889; Prain Beng. Plants 2: 637. 1903 (Rep. ed. 1963). *Epimeredi indicus* (L.) Rothm., Fed. de. Report 53: 12. 1944. *(Vern. Name: Gobura).*

Densely pubescent or tomentose shrub. Leaves sub-coriaceous, ovate truncate or sub-cordate at base; acuminate or acute at the apex, crenate lateral nerves 5-8 pairs. Flowers in dense whorls in the axils of upper leaves, combined into terminal spikes, bracts linear. Nutlets broadly ovoid, black at maturity. *Fls. & Frts.:* October-January.

Profusely grow on the waste dry lands and hedges, roadsides and railway tracks.

Field Collections: Baruipur, *K.N.* - *23;* Garfa, *B. Safui-7* (CAL).

Distributions: Common throughout India. A native of temperate Asia, Sri Lanka, Malaya Archipelago, China and Philippines.

Economic Importances: This plant is used as carminative astringent and also used for flavouring foods.

Notes: Prain (1903a) reported this taxon from Chota Nagpur and Chittagong.

2. *COLEUS* Lour.

2.1. ***Coleus amboinicus*** Lour. Fl. Cochinch. 2: 372.1790. *C. aromaticus* Benth., Wall. Pl. As. Rar. 2: 15. 1831; Watt Dict. Econ. Prod. 2: 503. 1889; Hook.f. Fl. Brit. India 4: 625.1885; Prain Beng. Plants 2: 692. 1903 (Rep. ed. 1963). *Plectranthus aromaticus* Roxb. Fl. Ind. 3: 22.1832. *(Vern. Name: Pathor-chur).*

Medium herb. Leaves fleshy, aromatic. Flowers in lax or close panicles; calyx-throat glabrous within; corolla pale purple; stamens shortly exserted. Nutlets 4 or orbicular or ovoid, smooth, dry. *Fls. & Frts.:* August-November

Cultivated garden herb.

Field Collections: Specimen not preserved.

Distributions: Planted throughout the ornamental gardens; indigenous to Bombay.

Economic Importances: This plant is used for flavouring 'Cooling cups' with thick succulent aromatic leaves.

3. *DYSOPHYLLA* Blume

3.1. ***Dysophylla verticillata*** Benth., Wall. Pl. As. Rar. 1: 30. 1830; Hook.f. Fl. Brit. India 4: 639.1885; Prain Beng. Plants 2: 635.1903 (Rep. ed. 1963). *Mentha verticillata* Roxb. Fl.Ind.3: 5.1832. *(Vern. Name: Pani-kula).*

Much-branched, decumbent herb. Stems hollow. Leaves linear or linear lanceolate, 4-5 or upto 10 cm in a whorl, entire or crenate serrate. Flowers in spikes, the widely open fruiting calyx looking like large holes. Nutlets pale brown. *Fls. & Frts.:* October-March.

Occasionally found on the paddy fields.

Field Collection: Bhojerhat/near Salt Lake, *M.K. Ghosh- 2678* (CAL).

Distributions: Eastern part of India. Bangladesh and Burma.

4. *HYPTIS* Jacq.

4.1. ***Hyptis suaveolens*** (L.) Poit, Ann. Mus. Par. 7: 472, t. 29, f.2, 1806; Hook.f. Fl. Brit. India 4: 630.1885; Prain Beng. Plants 2: 633. 1903 (Rep. ed. 1963). *Ballota suaveolens* L. Syst. (ed.10) 1100. 1759. *(Vern. Name: Ganga-Tulsi or Bilati-Tulsi).*

Tall, branched, scented herb; stem quadrangulate; leaves ovate, minute, sinuate and crenate denticulate, slightly cordate, hairy. Flowers small, blue when young often capitate. Nutlets compressed ovoid, oblong. *Fls. & Frts.:* November-March.

Grow on the moist waste lands, on the roadsides and railway tracks.

Field Collection: Dum Dum Cant., *N.S. Chatterjee-2* (CAL).

Distributions: As a weed plant, almost throughout; introduced into India during 1872-1897.

Economic Importance: Sometime planted as an ornamental plant.

5. *LEONOTIS* R. Brown

5.1. ***Leonotis nepetaefolia*** (L.) R. Brown. Prodr. 504.1810; Hook.f. Fl. Brit. India 4: 691.1885; Watt Dict. Econ. Prod. 4: 625. 1890; Prain Beng. Plants 2: 640.1903 (Rep. ed. 1963). *(Vern. Name: Hejurchi).*

Erect annual herb. Leaves ovate, acute at apex, coarsely crenate, serrate. Flowers in axillary dense globose, many flowered verticillasters; bracts slender, linear, spine-tipped. Nutlets linear, oblong, widening upwards, truncate with a deep triangular pit at the apex. *Fls. & Frts.:* October-February.

Waste places, road-sides and thickets.

Field Collection: Purandarpur, *K.N. - 1148.*

Distributions: Throughout the hotter parts of India.A native of Tropical America.

Economic Importances: Ash of the flower-heads after mixing with curds is applied to ringworms; seeds are said to have anti-malarial properties.

6. *LEONURUS* L.

6.1. ***Leonurus sibiricus*** L. Sp. Pl. 584.1753; Hook.f. Fl. Brit. India 4: 678.1885; Watt Dict. Econ. Prod. 4: 626.1890; Prain Beng. Plants 2: 638.1903 (Rep. ed. 1963). *(Vern. Name: Rakta-drone).*

Annual, branched, erect, pubescent herb. Leaves long, petioles deeply palmate or pinnate-partite with linear insisted segments. Flowers in axillary verticillasters, often collected into dense terminal spikes. Nutlets dark-brown, glabrous. *Fls. & Frts.:* March-August.

Abundantly grow on the roadsides, railway tracks, hedges and waste places, also self-grown on the vegetable garden as weed.

Field Collection: Magrahat, *K.N. - 211.*

Distributions: Almost throughout India. Widely distributed in all the warmer countries, a native of Asia.

Economic Importances: The entire plant is medicinal; the dried plant is credited with tonic, alterative vulnerary and emmeragogic properties; also used in puerperal and menstrual disorder; the leaves extract has been found to be effective towards uterus contraction.

7. *LEUCAS* R. Brown

Key to the Species

1a	Calyx mouth straight or nearly so	(2)
1b	Calyx mouth oblique	(3)
2a (1)	Whorls 6-many, 1-flowered	4. *L. lavendulaefolia* Sm.
2b (1)	Whorls 1-3 flowered	2. *L.biflora* var. *procumbens* Gamble
3a (1)	Calyx mouth ribbed below and hispid above	1. *L. aspera* (Willd.) Spreng.
3b (1)	Calyx strongly ribbed throughout	5. L. nutans Spreng.
3c (1)	Calyx mouth villous within	3. *L. cephalotes* (Roxb.) Spreng.

7.1. ***Leucas aspera*** (Willd.) Spreng. Syst. 2: 743.1823; Hook.f. Fl. Brit. India 4: 690.1885; Watt Dict. Econ. Prod. 4: 632.1889; Prain Beng. Plants 2: 639.1903 (Rep. ed. 1963). *Phlomis aspera* Willd. Link. Enum. Pl. Hort. Berol. 2: 621.1922. *P. esculenta* Roxb. Fl. Ind. 3: 10.1832. *Leucas mollisima* Wall., Pl. As. Rar. 1: 62.1830. *(Vern. Name: Chhota-halkusa).*

Erect or diffused, much branched, annual herb. Stems hispid or scabrid, branched 4-angled. Leaves elliptic-linear, narrowly oblong, lanceolate-oblong, entire or crenate. Flowers whitish in terminal and axillary whorls; bracts linear, acute and tipped with a bristle, as long as the calyx; calyx tube curved, mouth very oblique. Nutlets oblong, smooth, brown. *Fls. & Frts.:* Major parts of the year.

Common weed on waste places and cultivated fields; roadsides and railway tracks.

Field Collection: Bagha Jatin, *K.N. - 1052.*

Distributions: Almost throughout India. China, Mauritius, Java and Philippines.

Economic Importances: Flower juice after mixing with honey is useful for cough and cold. Alcoholic extract of leaves has anti-bacterial properties.

7.2. ***Leucas biflora*** (Vahl) R. Brown, Prodr. 504.1810 var. ***procumbens*** *(Desf.)* Gamble, Fl. Mad. 1151.1924. *Phlomis biflora* Vahl, Symb. 3: 77. 1794. *Leucas procumbens* Desf., Mem. Par. 11: 7. t. 3. f. 2.1824; Hook.f. Fl. Brit. India 4: 683.1885; Prain Beng. Plants 2: 640. 1903 (Rep. ed. 1963).

Procumbent herb. Stems and branches 4-angled. Leaves stalked, lanceolate, acute, serrate. Flowers white in whorls of 1-3 only, pedicelled, calyx 10 nerved. Nutlets obovoid, tracts at the apex. *Fls. & Frts.:* August-March.

Rarely seen in the moist waste places, near the paddy fields and road-side thickets.

Field Collection: Bamanghata, *K.N. - 2311.*

Distributions: Bengal and W. Peninsula. Bangladesh.

Notes: Prain (1903a) did not mention this taxon from the lower Bengal.

7.3. ***Leucas cephalotes*** (Roxb.) Spreng., Syst. 2: 743.1825; Hook.f. Fl. Brit. India 4: 689.1885; Watt Dict. Econ. Prod. 4: 633.1890; Prain Beng. Plants 2: 639.1903 (Rep. ed. 1963). *Phlomis cephalotus* Roth, Nov. Sp. 262.1821. *(Vern. Name: Bara-halkusa).*

Annual, stout herb. Stem and branches quadrangular with hairs. Leaves elliptic-lanceolate or ovate lanceolate, membranous, pubescent, crenate-serrate, tapering at the base. Flowers sessile, in large globose dense terminal whorls, bracts prominently nerved, foliaceous, green lanceolate, acute, awned. Nutlets obovoid-oblong, smooth brown. *Fls. & Frts.:* September-December.

Common on the roadside waste low lands, hedges and railway tracks.

Field Collection: Hasnabad, *G.Sen Gupta-123* (CAL).

Distributions: India, mainly Assam and Bengal. Afghanistan and in the Tropics.

Economic Importances: Leaves used in urinary discharges and a good therapeutic value in stomach troubles. Juice of the leaf is applied in the nostrils for cold and applied on forehead for headache.

7.4. ***Leucas lavandulaefolia*** Sm. Rees Cycl. 20 (2): 53.1819; Wlh. India 6: 80.1962. *L. linifolia* Spreng. Syst. 2: 743.1825; Hook.f. Fl. Brit. India 4: 690.1885; Watt Dict. Econ. Prod. 4: 633.1890; Prain Beng. Plants 2: 639.1903 (Rep. ed. 1963). *(Vern. Name: Halkussa).*

Annual, erect, branched herb. Leaves linear-lanceolate or linear-oblong, obtuse at apex, margin distinctly orenculate or sub-entire. Flowers white in axillary and terminal verticillasters. Nutlets dark-brown, oblong, rounded at apex. *Fls. & Frts.:* February-July.

Common on the paddy fields, it is common in the dry season on the road-side thickets and hedges or on the waste places.

Field Collection: Baruipur, *K.N. - 13.*

Distributions: Throughout India. South China and Malaya.

Economic Importances: Leaves used in bronchitis, dyspepsia and leucoma.

7.5. ***Leucas nutans*** Spreng. Syst. 2: 743.1825; Hook.f. Fl. Brit. India 4: 688.1885; Prain Beng. Plants 2: 639.1903 (Rep. ed. 1963). *Leucas vestita* Woodr., Jorun. Bomb. Nat. 12: 362.1899 (not of Benth.).

Dwarf annual herb; stem and branches quadrangular, hirsute with spreading hairs. Leaves oblong or obong-lanceolate, obtude, sub-serrate with gland-dotted. Flowers sessile or sub-sessile in terminal and axillary dense many flowered whorls. Furiting calyx strongly ribbed. Nutlets narrowly oblong-ellipsoid, rounded at apex. *Fls. & Frts.:* July-November.

Occasionally on the waste lands,

Field Collection: Kamdohri/Garia, *M.K.Ghosh- 2456* (CAL).

Field Collections: North and North West Povinces, Bihar and Western Peninsula; Burma.

Notes: D. Prain (1903a) did not mention this taxon from the Bengal.

8. *MENTHA* L.

8.1. ***Mentha spicata*** L. Sp. Pl. 576.1753; Bailey Man. Cult. Plants 863.1949. *M.viridis* L. Sp. Pl. 804.1753; Hook.f. Fl. Brit. India 4: 647.1885; Watt Dict. Econ. Prod. 5: 231.1891. *(Vern. Name: Pudina).*

Aromatic, perennial, glabrous, ascending herb. Leaves 4-5 cm long, sessile, ovate-lanceolate, sharply serrate. *Fls. & Frts.:* Not observed.

Rarely cultivated in the vegetable gardens.

Field Collection: Rahara/Danga Para, *K.N. - 2193.*

Distributions: Cultivated almost throughout India,

Economic Importances: Leaves are edible for its scented properties; and said to be very good for diarrhoea and dysentery.

Notes: Prain (1903a) did not mention this taxon from Bengal.

9. *OCIMUM* L.

Key to the Species

1a	Pedicels as long as or longer than the calyx	4. *O. sanctum* L.
1b	Pedicels shorter than the calyx	(2)
2a (1)	Calyx glabrous within	3. *O. gratissimum* L.
2b (1)	Calyx hairy within	(3)
3a (2)	Fruiting calyx 3 mm long and corolla 6 mm	1. *O. americanum* L.
3b (2)	Fruiting calyx 6mm long and corolla 10-12 cm	2. *O. basilicum* L.

9.1. ***Ocimum americanum*** L. Sp. Pl. 833.1753 (non Jacq.) *O. canum* Sims, Bot. Mag., t. 2452.1824; Hook.f. Fl. Brit. India 4: 607.1885; Watt Dict. Econ. Prod. 5: 442.1891; Prain Beng. Plants 2: 629.1903 (Rep. ed. 1963). *(Vern. Name: Bharbari).*

Much-branched, erect-pubescent herb. Leaves elliptic lanceolate or rhomboid, entire or shallowly serrate. Flowers white, tinged violet. Nutlets ellipsoid, black. *Fls. & Frts.:* August- February.

Gregariously grow as a weed plant in the waste lands and cultivated fields and road-sides.

Field Collection: Falta Fort, *M.K. Ghosh-2635* (CAL).

Distributions: Plains and low-hills of India. Sri Lanks, Java, W. Asia, Tropcial Africa, Madagascar and cultivated in America.

Economic Importances: Leaves are used for flavouring sauces, soups and salads. Seeds are diuretic, tonic and are used in the preparation of cooling drinks. Useful for cough, dysentery and useful for toothache.

9.2. ***Ocimum basilicum*** L. Sp. Pl. 597.1753; Hook.f. Fl. Brit. India 4: 608.1885; Watt Dict. Econ. Prod. 5: 440.1891; Prain Beng. Plants 2: 629.1903 (Rep. ed. 1963). *(Vern. Name: Babui-Tulsi).*

Erect, much-branched, strongly scented herb. Stems pubescent, hairy at the nodes, often purple or violet. Leaves ovate, acute entire or toothed. Flowers whitish pink or purplish in whorls on simple or branched racemes. Nutlets ellipsoid, black. *Fls. & Frts.:* Almost throughout the year.

Commonly found on the cultivated ground, on the houses and sometime on the waste lands.

Field Collection: Dhapdhapi/Public House, *K.N. - 39.*

Distributions: Almost throughout India. Tropical Asia and Africa.

Economic Importances: This plant has a strong aromatic flavour and is employed as a seasoning for culinary purposes; the seeds are also medicinally used.

9.3. ***Ocimum gratissimum*** L. Sp. Pl. 1197.1753; Hook.f. Fl. Brit. India 4: 608.1885; Watt Dict. Econ. Prod. 5: 443.1891; Prain Beng. Plants 2: 629.1903 (Rep. ed. 1963). *(Vern. Name: Ram-Tulsi.).*

Erect, much branched, shrub, woody at the base. Young parts pubescent. Leaves elliptic-lanceolate, acute. Flowers greenish-white, small, clustered in short, simple or branched racemes, bracts broadly ovate. Nutlets sub-globose, brown. *Fls. & Frts.:* October-April.

Rarely seen in the waste places along the road-sides, hedges and the railway tracks.

Field Collection: Ghoramara/adjacent to Sagar Islands, *K.N. - 402.*

Distributions: Often cultivated throughout India. Sri Lanka and Java.

Economic Importance: It has the mosquito-repellent properties.

9.4. ***Ocimum sanctum*** L. Mant. 1: 85.1767; Hook.f. Fl. Brit. India 4: 609.1885; Watt Dict. Econ. Prod . 5: 445.1891; Prain Beng. Plants 2: 629.1903 (Rep. ed. 1963). *(Vern. Name: Tulsi).*

Much-branched, woody undershrub, the whole plant often purple coloured. Stems and branches clothed with soft spreading hairs. Leaves ovate, ovate oblong or elliptic, oblong, hairy or pubescent both sides. Nutlets reddish or yellow. *Fls. & Frts.:* Almost throughout the year.

It is a holy herb to the Hindus; very common to every Hindu's house; also as an escape in the waste places.

Field Collection: Atghora/Baruipur, *K.N. - 61.*

Distributions: Throughout India; Malay Archipelago, Australia, W. Asia and Arabia.

Economic Importances: Leaves juices are much used with honey in cough and cold, especially to the children.

10. *ORTHOSIPHON* Benth.

10.1. ***Orthosiphon rubicundus*** Benth., Wall. Pl. As. Rar. 2: 14.1831; Hook.f. Fl. Brit. India 4: 614.1885; Prain Beng. Plants 2: 630.1903 (Rep. ed. 1963). *O. tuberosum* Roxb. Fl. Ind. 3: 18.1832.

Undershrub. Leaves sessile. Flowers medium, corolla tube longer than the calyx; stamens included. Fruits dry, ovoid or orbicular nutlets. Smooth or nearly so; basal scar small. *Fls. & Frts.:* April-September.

Rarely on the village thickets.

Field Collection: Gobardanga, *K.N. - 535.*

Distributions: North Bengal Bihar and Chhota Nagpur; Sri Lanka.

Economic Importances: Tubers edible; largely used as remedy against colic.

Notes: D. Prain (1903a) did not mention this taxon from Bengal Proper.

11. *POGOSTEMON* Desf.

11.1. ***Pogostemon benghalense*** (Burm.) O. Kuntze, Rev. Gen. Pl. 529.1891. *Origanum benghalense* Burm., Pl. Ind. 128. t. 38, f.3.1768. *Pogostemon plectranthoides* Desf., Ann. Mus. Paris 2: 155.1803; Hook.f. Fl. Brit. India 4: 632.1885; Prain Beng. Plants 2: 634.1903 (Rep. ed. 1963). *(Vern. Name: Jui-Lata).*

Large, aromatic herb. Leaves ovate, rarely lanceolate, acute serrate. Flowers white or pale but appearing pink at maturity. Nutlets broadly ellipsoid, shining dark brown. *Fls. & Frts.:* January-April.

Gregariously grow in the shady places and waste lands.

Field Collection: Barrackpore, *K.N. - 2393.*

Distributions: W. Himalaya, Lower Bengal, Bihar and W. Peninsula.

12. *SALVIA* L.

12.1. ***Salvia plebeja*** R. Brown., Prodr. 501.1810; Hook.f. Fl. Brit. India 4: 655.1885; Watt Dict. Econ. Prod. 6(2): 451.1893; Prain Beng. Plants 2: 642.1903 (Rep. ed. 1963). *(Vern. Name: Bhui-Tulsi).*

Erect, deep-rooted, roughly pubescent annual herb. Stems branched, quadrangular and grooved. Leaves oblong, lanceolate, crenate obtuse or sub-acute. Flowers bluish white, small in panicled, bracts reflexed. Nutlets minute, ovoid, brown. *Fls. & Frts.:* January-July.

Gregariously grow as weed in the moist and shady areas, near the canals, dry ditches and jhils.

Field Collection: Malatipur, *K.N. - 1587.*

Distributions: Throughout India. China, Malaya and Australia.

CVI. PLANTAGINACEAE

1. *PLANTAGO* L.

1.1. ***Plantago ovata*** Forssk., Fl. Aegypt.-Arab. 31.1776; Hook.f. Fl. Brit. India 4: 707.1885; Watt Dict. Econ. Prod. 6(1): 285.1892. *P. ispaghula* Roxb. Fl. Ind. 1: 404.1832. *(Vern. Name: Ispghul).*

Erect herb. Leaves radical, spreading in rosette, 15-30 cm long, linear, acuminate. Scapes straight, erect. Flowers in cylindric or ovoid spikes, bracts winged. Capsules 2-seeded. *Fls. & Frts.:* November-March.

Very rarely planted for its medicinal value.

Field Collection: Nurpur/public garden, *K.N. - 1938.*

Distributions: Plain and Low-hills of India, throughout. Doubtfully, indigenous in Sind; Spain and Canary Islands.

Economic Importances: Seeds are said to be used in dysentery and diarrhoea.

CVII. NYCTAGINACEAE

Key to the Genera

1a	Climbing shrub; bracts coloured	1. *BOUGAINVILLEA* Comm. ex Juss.

1b	Erect, straggling or sub-scandent; bracts not petioled	(2)
2a (1)	Flowers showy, in a calyx like involucre	3. *MIRABILIS* L.
2b (1)	Flowers small, in heads or umbles	1. *BOERHAVIA* L.

1. *BOERHAVIA* L.

1.1. ***Boerhavia diffusa*** L. Sp. Pl. 3. 1753; Watt Dict. Econ. Prod. 1: 485.1889; Stemmerick, Fl. Males. 6: 454.1967. *B. repens* L. Sp. Pl.3.1753; Hook.f. Fl. Brit. India 4: 709.1885; Prain Beng. Plants 2: 645.1903 (Rep. ed. 1963). *B. procumbens* Roxb. Fl. Ind. 1: 146.1832. *(Vern. Name: Punarnava).*

Diffusely branched herb, more robust branches long, stout glabrous pubescent or viscous. Leaves ovate obtuse or acute usually cordate green beneath. Inflorescence axillary. Flowers minute capitate. Fruits bluntly 5-ribbed, very glandular. *Fls. & Frts.:* Almost throughout the year.

Very common on the waste places, railway tracks, road-sides and sometime on the old walls.

Field Collection: Nurpur/on the roadside, *K.N. - 1932.*

Distributions: Throughout India. Asia, Africa and America.

Economic Importances: Roots are reported to be beneficial in the diuretic cases, leaves are edible.

Remarks: Generally two forms are collected one with white flowers and the other with deep violet flowers.

2. *BOUGAINVILLEA* Comm. ex Juss. Corr. Spach. *nom. coms.*

Key to the Species

1a	Leaves and branches glabrous	1. *B. glabra* Choisy
1b	Leaves and branches softy pubescent	2. *B. spectabilis* Willd.

2.1. ***Bougainvillea glabra*** Choisy, DC. Prodr. 13(2): 437.1849; Prain Beng. Plants 2: 645.1903 (Rep. ed. 1963).

Scandent or straggling herb. Branches and leaves are glabrous or sub-glabrous. Leaves lanceolate, broad-ovate, ovate-oblong, bracts colured. Perianth tubular with 5-lobed limb. *Fls. & Frts.:* Almost thoughout the year, profusely in the winter season.

Cultivated or planted as an ornamental plants in gardens and walls or gate of the houses or lawns.

Field Collections: Specimen not preserved.

Distributions: Commonly planted throughout India as an ornamental, both in the public and private gardens; introduced in India during 1884-1894. A native of Brazil.

Economic Importance: Ornamental, very beautiful garden plants.

2.2. ***Bougainvillea spectabilis*** Willd. Sp. Pl. 2: 348.1799; Prain Beng. Plants 2: 645.1903 (Rep. ed. 1963).

Woody, scandent, thorny, extensively climbing shrub. Branches and leaves are softy pubescent. Leaves ovate to orbicular-ovate. Perianth greenish-yellow, clustered at the ends of the branches; bracts 3, pink, red or violet large, showy. *Fls. & Frts.:* Almost throughout the year, specially in winter.

Planted in the gardens as an ornamental climbers.

Field Collections: Specimen not preserved.

Distributions: Introduced in India from England, during the year 1860; now commonly planted throughout India. A native of Brazil.

Remarks: A large number of horticultural varieties are available from this area.

3. *MIRABILIS* L.

3.1. ***Mirabilis jalapa*** L. Sp. Pl. 177.1753; Watt Dict. Econ. Prod. 5: 255.1891; Prain Beng. Plants 2: 644.1903 (Rep. ed. 1963). *(Vern. Name: Krishna-Kali).*

Erect, much branched, undershrub with large perennial tuberous. Leaves 5-10 cm long, ovate, truncate or cordate. Flowers white, red or yellow, showy, in a calyx like involucre. Fruits often ribbed or rugose. *Fls. & Frts.:* June-March.

Commonly planted in the gardens as an ornamental, rarely wild in the waste lands.

Field Collection: Taki, *K.N. - 1030.*

Distributions: Throughout India. A native of Tropical America.

Economic Importances: Ornamental herb to the village gardens.

CVIII. AMARANTHACEAE

Key to the Genera

1a	Anthers 1-celled	(2)
1b	Anthers 2-celled	(3)
2a (1)	Staminal tube short, stigma capitate	4. *ALTERNANTHERA* Forssk.
2b (1)	Staminal tube long; stigma 2-fid on long style	8. *GOMPHRENA* L.
3a (1)	Leaves alternate	(4)
3b (1)	Leaves opposite; stamens with interposed staminodes	1. *ACHYRANTHES* L.
3c (1)	Leaves alternate or fascicled, or with opposite and alternate mixed	2. *AERVA* Forssk.
4a (3)	Seeds numerous	6. *CELOSIA* L.
4b (3)	Seeds single	(5)
5a (4)	Flowers unisexual	5. *AMARANTHUS* L.
5b (4)	Flowers hermaphrodite	(6)
6a (5)	Flowers capitate; fruits a circumscissile capsule, seeds articulate	3. *ALLMANIA* R. Brown
6b (5)	Flowers spicate; fruit a crustaceous nut; seeds not arillate	7. *DIGERA* Forssk.

1.1. *Achyranthes aspera* L. Sp. Pl. 204.1753; Hook.f. Fl. Brit. India 4: 730. 1885; Watt Dict. Econ. Prod. 1: 81.1889; Prain Beng. Plants 2: 654. 1903 (Rep. ed. 1963). *A. indica* Mill. *Aerva aspera* (L.) Spreng.; *(Vern. Name: Apang).*

Erect or sub-erect, scandent annual herb. Leaves variable, elliptic obovate or sub-orbicular. Flowers greenish white, in long terminal spikes, bracts persistent reflexd. Utricle oblong, enclosed in the hardened perianth; seeds brown. *Fls. & Frts.:* Almost throughout the year, specially March-October.

Very common on the waste lands, road-side hedges, railway tracks, old wall crevises and unused lands.

Field Collection: Dum Dum Cantt., *K.N. - 1731.*

Distributions Common, throughout India; Sri Lanka, Tropical Asia, Africa, Australia and America.

Economic Importances: Used in piles, boils, skin eruptions and in colic pain; acts as purgative and diuretic. Roots are astringent; seed emetic. The whole plant is much used in Ayurvedic Medicine.

2. *AERVA (AERUA)* Forssk. *nom. cons*

Key to the Species

1a	Climbing undershrub; leaves alternate, spikes all axillary	1. *A. lanata* (L.) Juss.
1b	Erect undershrub; leaves both alternate and opposite; spikes also axillary and terminal	2. *A. sanguinolenta* (L.) Blume

2.1. ***Aerva lanata*** (L.) Juss., Ann. Mus. Par. 2: 131.1803; Naskar & Guha Bakshi Mangrs. Swamps Sundarbans 83.1987. *Achyranthes lanata* L. Sp. Pl. 204.1753. *(Vern. Name: Chhaya).*

Erect, diffused or prostrate herb. Naskar & Guha Bakshi (1987) described this taxon after collecting from Mollakhali *K.N. - 1853.*

2.2. ***Aerva sanguinolenta*** (L.) Blume, Bijdr. 547.1826; Baker, Fl. Males. I, 4(2): 85.1949. *Achyranthes sanguinolenta* L. Sp. Pl. 259.1762. *A. scandens* Roxb. Fl. Ind. 1: 1676.1820. *Aerva scandens* Wall. ex Moq., DC. Prodr. 13(2): 302.1849; Hook.f. Fl. Brit. India 4: 727.1885; Watt Dict. Econ. Prod. 1: 125.1889; Prain Beng. Plants 2: 653.1903 (Rep. ed. 1963). *(Vern. Name: Nuriya).*

Erect or straggling, scarcely scandent herb. Leaves 5-15 cm long, elliptic-lancelolate or ovate, obtuse, acute or mucronate; softly hairy base tapering into a short petiole. Flowers whitish, in ovoid or cylindric axillary and terminal, sub-sessile terminal panicles. Utricle broadly ovate, acute, seeds lenticular, black and shining. *Fls. & Frts.:* August-April.

Common on the waste places and hedges.

Field Collections: Madarhat, *K.N. - 63;* Baruipur, *K.N. - 109.*

Distributions: Common in most parts of India, China, Malaya, Philippines and Tropical Africa.

3. *ALLMANIA* R. Brown

3.1. ***Allmania nodiflora*** R. Brown, Wall. Cat. Hort. Monspel. 77. 1813; Naskar Aq. & Semi-Aq. Plants 210.1990. *Achyranthes nodiflora* Roxb. Fl. Ind. 1: 678.1820.

Diffused, dichotomously branched annual herb. Common on the canal sides. Naskar (l.c.) described this taxon after collecting from Jadavpur *K.N. - 1283.*

4. *ALTERNANTHERA* Forssk.

Key to the Species

1a	Leaves in equal pairs	(2)
1b	Leaves in unequal pairs	(3)
2a (1)	Flower-heads sessile, stamens 3	4. *A. sessilis* (L.) R. Brown
2b (1)	Flower-heads peduncle, stamens 5	2. *A. philoxeroides* (Mart.) Griseb.
3a (1)	Bracts and tepals spine tipped	3. *A. pungens* Kunth.
3b (1)	Bracts and tepals not spine tipped	1. *A. paronychioides* St. Hill.

4.1. ***Alternanthera paronychioides*** St. Hill. Voy. Brazil 2: 439.1833; Naskar Aq.& Semi-Aq.Plants 211.1990. *A. ficoides* Grisb., Fl. Brit. W. India 67.1959 *Telanthera polygonoides* Moq., DC. Prodr. 13: 363.1949 (ed.2) (in part). *Gomprena polygonoides* L. Sp. Pl. 225.1753 (in part).

Prostrate herb with cylindric much branched stem. Very common along the side of the roads; prefer moist places but also can grow in the xerophytic conditions. Naskar (l.c.) described this taxon after collecting from Subuddhipur/Baruipur, *K.N. - 3*

4.2. ***Alternanthera philoxeroides*** (Mart.) Griseb., Abh. Ges. Eiss. Goett. 24: 36.1879; Naskar Aq.& Semi-Aq. Plants 212.1990.

Perennial much-branched herb. Common on the moist, waste lands. Naskar (l.c.) described this taxon from Sitakundu/Baruipur *K.N. - 58.*

4.3. ***Alternanthera pungens*** Kunth. H.B. & K., Nov. Gen. & Sp. 2: 206.1818; Naskar Aq. & Semi-Aq. Plants 213.1990. *Achyranthes repens* L. Sp. Pl 205. 1753. *Alternanthera echinata* Sm., Rees, Cyclop. 39.1819. *A repens* (L.) Link. Enum. Pl. Hort. Berol. 1: 154.1821.

Prostrate or decumbent, spreading herb. Common on road-sides. Naskar (l.c.) described this taxon after collecting from Nurpur, *K.N. - 1933.*

4.4. ***Alternanthera sessilis*** (L.) R. Brown Prodr. 417.1810; Naskar Aq. & Semi-Aq. Plants 213.1990. *Gomphrena sessilis* L. Sp. Pl. 225.1753. *Alternanthera traindra* Lamk. Encycl. 1: 95.1783. *A. repens* J.F. Gmel. Syst.

Nat. 2: 106.1791. *A. nodiflora* R. Brown Prodr. 1: 417.1810. *(Vern. Name: Sanchi* or *Chanchi).*

Decumbent herb with purple stem. Very common weed in the cultivated land and other moist waste places. Naskar (l.c.) described this taxon from the lower Ganga after collecting from Sashan Rd./Baruipur *K.N. - 30; Canning K.N. - 138* and Gobardanga *K.N. - 533.*

Economically, this plant is useful as galactic cholagugue and febrifuge; stems and leaves are used as poultice in snake bite, as a stimulant to increase the breast milk of the milking mother and also used in eye troubles.

5. *AMARANTHUS* L.

Key to the Species

1a	Plants spinous	2. *A. spinosus L.*
1b	Plants not spinous	(2)
2a (1)	Tepals 5; stamens 5	1. *A. caudatus* L.
2b (1)	Tepals 2; stamens 2	3. *A. tenufolius* Willd.
2c (1)	Tepals 3; stamens 3	(3)
3a (2)	Bracts longer than tepals	4. *A. tricolor* L.
3b (2)	Bracts shorter than tepals	5. *A. viridis* L.

5.1. ***Amaranthus caudatus*** L. Sp. Pl. 990. 1753; Hook.f. Fl. Brit. India 4: 719.1885: Watt Dict. Econ. Prod. 1: 211.1889; Prain Beng. Plants 2: 650.1903 (Rep. ed. 1963).

Tall, 25-60 cm high robust, erect plant. Stems softly pubescent. Leaves long-petioled, ovate to ovate-oblong or ovate-lanceolate, 5-25 cm long, alternate at base. Flowers in panicle of numerous or few spikes. Utricle circumscissile, smooth. Seeds black to yellowish-white. *Fls. & Frts.:* July-April.

Rarely planted or as an garden escape.

Field Collection: Gosaba, *K.N. - 651.*

Distributions: Planted almost throughout India.

Economic Importances: Ornamenal garden plants.

5.2. ***Amranthus spinosus*** L. Sp. Pl. 991.1753; Hook.f. Fl. Brit. India 4: 718.1885; Watt Dict. Econ. Prod. 1: 215.1889; Prain Beng. Plants 2: 660.1903 (Rep. ed. 1963). *(Vern. Name: Kanta Note).*

Erect, spinose herb. Leaves alternate, green, 5-10 cm long, petiolate, mucronate, ovate, lanceolate or rhomboid, cuneate. Flowers unisexual, greenish in terminal or axillary clusters and also in long terminal and axillary more or less spreading spikes. Perianth 5, dissimilar in male and female flowers. Stamens equal to the number of perianth. Ovary with one erect ovule. Fruits ovoid, seeds dark brown or black, shining. *Fls. & Frts.:* Almost throughout the year.

Very common to the moist and waste lands, roadside hedges, thickets railway tracks and often in the cultivated fields.

Field Collection: Sarberia, *K.N. - 782.*

Distributions: Throughout India. Common in all the Tropical countries of the world.

Economic Importance: During scarcity this plant is eaten as vegetable.

5.3. ***Amaranthus tenuifolius*** Willd. Sp. Pl. 1805; Hook.f. Fl. Brit. India 4: 722.1885; Watt Dict. Econ. Prod. 1: 216.1889; Prain Beng. Plants 2: 651.1903 (Rep. ed. 1963). *Mengea tenuifolia* Moq., DC. Prodr. 13 (2) 271.

Glabrous, annual herb; much branched, leaves linear or oblanceolate-oblong, rounded and usually notched at the apex. Flowers minute in axillary clusters. Utricle indehiscent, orbicular, compressed; seeds lenticular, brown and shining. *Fls. & Frts.:* March-October.

Rarely found in the moist waste lands.

Field Collection: Durgapur, *K.N. - 1719.*

Distributions: Throughout India. Tropical countries of the world.

5.4. ***Amaranthus tricolor*** L. Sp. Pl. 989. 1753; *A. gangeticus* L. Syst. Nat. (ed. 10) 1268. 1759; Hook.f. Fl. Brit. India 4: 719.1885; Watt Dict. Econ. Prod. 1: 212. 1889; Prain Beng. Plants 2: 651.1903 (Rep. ed. 1963). *A. polygamous* L. Amoen. Acad. 4: 294. 1759. *A. melancholicus* L. Sp. Pl. 989.1753. *A. oleraceus* Willd. Sp. Pl. 4: 386.1805.

Erect diffuse, glabrous, annual herb. Stems often tinged with purple, branched. Leaves variable, rhomboid or deltoid ovate, quite variable. Flowers in axillary, dense clusters, forming long, distantly interrupted spikes. Capsules ovoid; seeds black, biconvex. *Fls. & Frts.:* July-March.

Commonly found in the moist and waste places, on the bank of canals or nalas.

Field Collection: Sagar Islands, *A.K. Mukherjee-7046* (CAL).

Distributions: Throughout India. Sri Lanka, Tropical Asia, Africa and America.

Economic Importance: Leaves are edible. Plants is astringent, used in haemorrhage and also used for the wash of ulcers of throat and mouth.

5.5. ***Amaranthus viridis*** L. (ex parte) Sp. Pl. (ed.2) 1405.1763. *A. gracilis* Desf., Tabl. Ec. Bot. 43.1804; Baker, Fl. Males, 1, 4: 76.1949: Naskar, Aq. & Semi-Aq. Plants 214. 1990. *(Vern. Name: Ban-nate).*

Erect, glabrous, branching, annual herb. Common on moist and waste lands, along the water canals and in cultivated fields. Naskar (1.c.) described this taxon from the lower Ganga delta after collecting from Maslandapur *(K.N.-501)* and Madhyamgram *(K.N.-803).*

6. *CELOSIA* L.

Key to the Species

1a	Spikes simple	1. C. *argentea* L.
1b	Spikes usually fasciated	2. C. *cristata* L.

6.1. ***Celosia argentea*** L. Sp. Pl. 205.1753; Roxb. Fl. Ind. 1: 678.1814; Hook.f. Fl. Brit. India 4: 714. 1885; Watt Dict. Econ. Prod. 2: 240. 1889; Prain Beng. Plants 2: 648. 1903 (Rep. ed. 1963). *(Vern. Name: Swet-Murga* or *Morag-Phul).*

Annual, terrestrial, erect glabrous herb. Leaves linear lanceolate, alternate. Inflorescence terminal or axillary on ovoid spikes. Flowers bisexual glistering white or pink. Perianth 5, scarious. Stamens 5 united. Utricles membranous. Capsule ellipsoid; seeds black or brown, polished. *Fls. & Frts.:* April-March.

Commonly found as a weed in the moist waste lands.

Field Collection: Duttapukur, *K.N. - 795.*

Distributions: Common weed in marshy places throughout India. Sri Lanka, introduced in most of the Tropical countries.

Economic Importances: Seeds are used in diarrhoea, aphroidis, useful in blood diseases and mouth sores, for clearing the vision and diseases of the eyes.

6.2. ***Celosia cristata*** L. Sp. Pl. 205. 1753; Hook.f. Fl. Brit. India 4: 715. 1885; Watt Dict. Econ. Prod. 2: 241. 1889; Prain Beng. Plants 2: 649. 1903 (Rep. ed. 1963). *(Vern. Name: Marag Jata).*

Erect, glabrous, tall and branched, annual herb. Leaves linear to ovate-lanceolate. Flowers pink crimson or yellow on fasciated and crest-like divisions of branches, spikes. *Fls. & Frts.:* November-April.

Often cultivated in the garden, but sometime grows as an escape in the waste lands.

Field Collection: Santoshpur/near Budge Budge, *K.N. - 495.*

Distributions: Almost throughout India. As ornamental garden plant.

7. *DIGERA* Forssk.

7.1. ***Digera muricata*** (L.) Mart., Teitr. Amar. 2: 77. 1825; Backer, Fl. Mals, I, 4: 80. 1949. *Achyranthes muricata* L. Sp. Pl. (ed.2) 295. 1762. *A. alternifolia* L. Mant. 50.1767. *Digera arvensis* Forssk., Fl. Aegypt. Arb. 65.1775; Hook.f. Fl. Brit. India 4: 717.1885; Watt Dict. Econ. Prod. 3: 112. 1890; Prain Beng. Plants 2: 649.1903 (Rep. ed. 1963). *Desmochaeta muricata* DC. Cat.Hort. Monsp. 103.1813. *Digera alternifolia* (L.) Aschera, Schwfth. Baetr. Fl. Aethiopiens 180. 1867.*(Vern. Name: Luta-mahauria).*

Erect or prostrate annual herb. Leaves alternate, variable, glabrous, ovate or elliptic or lanceolate obtuse or acute. Spike axillary, elongated. Flowers about 3 mm long, pink; perianth 5, outer larger, purplish; stamens 5 free, anther 2-celled. Fruits slightly compressed and margined, muricated. Seeds small, yellowish brown, *Fls. & Frts.:* February-March.

As a weed of the cultivated lands and also in moist waste places.

Field Collections: Gocharan, *K.N. - 243.* and D. Chatra, *K.N. - 520.*

Distributions: Major parts of India. Sri Lanka and North Africa.

8. *GOMPHRENA* L.

Key to the Species

1a	Flower in globose heads	2. *G. globosa* L.
1b	Flowers in spikes	1. *G. celosioides* Mart.

8.1. ***Gomphrena celosioides*** Mart., Bietr. Amar. 193. 1825; Backer, Fl. Males. I, 4: 96.1949.

Low spreading, much branched herb with stout tap root; branches clothed with white hairs. Leaves 2.5-6.0 cm long, spathula-shaped or oblong elliptic, mucronate, cuneate at base, with white hairs beneath. Flowers white, compressed in dense, cylindrical or ovoid, terminal spikes, subtended by two small leaves. Fruits enclosed by the hardened perianth, seeds by two small leaves. Seeds brown. *Fls. & Frts.:* Almost throughout the year.

Common on the waste ground along the road sides and railway tracks.

Field Collection: Bantala, *K.N. - 1746.*

Distributions: Cultivated or planted almost throughout India. A native of South America. Oramental garden plants.

8.2. ***Gomphrena globosa*** L. Sp. Pl. 224. 1753; Hook.f. Fl. Brit. India 4: 732. 1885; Watt Dict. Econ. Prod. 3: 208.1890; Prain Beng. Plants 2: 655.1903 (Rep. ed. 1963). *(Vern. Name: Gomphorna).*

Erect, branched, annual herb. Leaves elliptic, obovate-oblong, opposite, margins entire and ciliate. Flower heads globular, purple, red or variously coloured, long, peduncled. Fruit mostly indehiscent; 1-seeded. *Fls. & Frts.:* September-March.

Commonly planted in the gardens and often found as an escape.

Field Collections: Belekhali/Salt Lake, *M.K.Ghosh-61* (CAL).

Distributions: Planted in the ornamental gardens throughout India. Probably a native of America, but was originally described from India. Naturalized in the Tropical Regions.

Economic Importance: Ornamental Garden plants.

CIX. CHENOPODIACEAE

Key to the Genera

1a	Stem leafy, not jointed	(2)
1b	Stem without leaves, jointed	(5)
2a (1)	Leaves membranous, flat; embryo annular	(3)
2b (1)	Leaves fleshy; semi-terete, embryo spiral	6. *SUAEDA* Forssk.
3a (2)	Flowers usually bisexual; perianth all alike	(4)
3b (2)	Flowers unisexual; perianth of 2 kinds	5. *SPINACIA* L.
4a (3)	Root tuberous	2. BETA L.
4b (3)	Root not tuberous	3. *CHENOPODIUM* L.

5a (1)	Flowers 2-6 together in the axils of the scales of sessile, cone-like spikes; perianth ovoid or pyramidal; mouth narrowed, 3-4 fid	1. *ARTHROCNEMUM* Moq.
5b (1)	Flowers 3-rate, sunk in superposed cavities in successive joints, perianth obpyramidal, mouth wide	4. *SALICORNIA* L.

1. *ARTHROCNEMUM* Moq.

1.1. ***Arthrocnemum indicum*** (Willd.) Moq. Chenopod. Enum. 113. 1840; Naskar Aq. & Semi-Aq. Plants 216. 1990. *Salicornia indica* Willd., Ges. Naturf. Fr. Neue Schr. 2: 111, t.4, f.2.1799. *Arthrocnemum glaucum* Woodr., Journ. Bomb. Nat. 12: 365.1899. *(Vern. Name: Jadu Palang).*

Stem fleshy, usually jointed. Common on the river sides of Sundarbans. Naskar (1.c.) reported this taxon after collecting it from Suparibhanga/Sagar Islands, *(K.N.-421).*

2. *BETA* L.

2.1. ***Beta vulgaris*** L. Sp. Pl. 222. 1753;Hook.f. Fl. Brit. India 5: 5. 1886; Prain Beng. Plants 2: 658.1903 (Rep. ed. 1963). *Beta maritima* Watt Dict. Econ. Prod. 1: 488. 1889. *B. benghalensis* Roxb. Fl. Ind. 2: 59. 1820. *(Vern. Name: Beet).*

Succulent, tuberous herb. Leaves mostly basal, ovate to oblong-ovate. Flowers greenish, many in panicles. *Fls. & Frts.:* November-April.

Commonly cultivated in the vast area of the Sundarbans for its Sugar extraction and for their vegetable values.

Field Collection: Garia, *K.N. - 345.*

Distributions: Cultivated almost throughout India. A native of Europe.

Economic Importances: The tubers/ roots are cooked as vegetable or eaten as raw; it has high starch content and sugar obtained from roots.

3. *CHENOPODIUM* L.

Key to the Species

1a	Strongly foetid-smelling undershrub	2. *C. ambrosioides* L.
1b	Scentless herbs	(2)
2a (1)	Seeds smooth, shining	1. *C. album* L.
2b (1)	Seeds rugose, dull black	3. *C. murale* L.

3.1. ***Chenopodium album*** L. Sp. Pl. 219.1753; Naskar Aq. & Semi-Aq. Plants 213. 1990. *(Vern. Name: Bethua-sak).*

Erect or ascending herb; commonly found on the cultivated ground as weed, on the river banks and canal sides. Naskar (1.c.) described this taxon after collecting from Diamond Harbour, *(K.N.-400).* Its leaves have immense medicinal values.

3.2. ***Chenopodium ambrosioides*** L. Sp. Pl. 219.1753; Naskar Aq. & Semi-Aq. Plants 217. 1990.

Erect, strongly foetid smelling, much branched shrub. Common on the moist wastelands on the roadsides and canal sides. Naskar (1.c.) described this taxon after collecting from Baruipur *(K.N. - 94);* Sarberia *(K.N.-785);* Malancha *(K.N.-776);* Budge Budge *(K.N.- 739);* Madhyampur *(K.N.- 951)* and Sankchura/Taki *(K.N. - 993).*

3.3. ***Chenopodium murale*** L. Sp. Pl. 219. 1753; Naskar Aq. & Semi Aq. Plants 218.1990.

Glabrous, erect herb. Commonly found on moist waste lands. Naskar (1.c.) described this taxon from the lower Ganga delta based on the collection, *G.Das-505* (CAL) from Sodepur.

4. *SALICORNIA* L.

4.1. ***Salicornia brachiata*** Roxb. Fl. Ind. (ed. Carey & Wall.) 1: 82.1820; Naskar & Guha Bakshi Mangrs. Swamps Sundarbans 103. 1987; Naskar Aq. & Semi-Aq. Plants 219. 1990.

Stem fleshy, usually jointed. Occasionally found on the waste lands and river beds. Naskar & Guha Bakshi (1.c.) and Naskar (1.c.) described this taxon after collecting from Lakshmikantapur *(K.N. - 1784).* The young shoots sometime planted as pot-herb.

5. *SPINACIA* L.

5.1. ***Spinacia oleracea*** L. Sp. Pl. 1027. 1753; Hook.f. Fl. Brit. India 5: 6. 1886; Watt Dict. Econ. Prod. 6(3): 330. 1893; Prain Beng. Plants 2: 658.1903 (Rep. ed. 1963). *(Vern. Name: Palang-Sak).*

Erect, annual, smooth herb. Radical leaves large, cauline leaves smaller, becoming lanceolate in the inflorescence. Flowers unisexual. Utricle enclosed in a spine. *Fls. & Frts.:* Almost throughout the year.

Commonly cultivated as vegetable, rarely self-grown as garden escape.

Field Collection: Madhyamgram, *K.N. - 810.*

Distributions: Cultivated almost throughout India; native country unknown.

Economic Importances: Commonly cooked as very nutritious vegetable; vitamins and food values are very high.

6. *SUAEDA* Forssk.

Key to the Species

1a	Styles 2; seeds usually horizontal	1. *S. maritima* Dumort.
1b	Styles 3; seeds erect	2. *S. nudiflora* Moq.

6.1. ***Suaeda maritima*** Dumort. Fl. Belg. 22. 1827; Naskar & Guha Baksh Mangrs. Swamps Sundarbans 103.1987. *(Vern. Name: Gire sak).*

Erect, reddish-green glabrous herb; leaves linear, filiform, semi terete, succulent or fleshy. Commonly grow on the river flats. Naskar & Guha Bakshi (1.c.) described this taxon after collecting from Hingalgunj *(K.N.-1489).* It is sometime eaten as vegetables.

6.2. ***Suaeda nudiflora*** Moq., Ann. Sc. Nat. 23.1831; Naskar & Guha Bakshi Mangr. Swamps Sundarbans 104.1987.

Diffused branched woody perennial undershrub. Common on the inter-tidal river slopes of the Sundarbans. Naskar & Guha Bakshi (1.c.) described this taxon after collecting from Bhangankhali *(K.N. - 209).*

CX. BASELLACEAE

(This family has been separated out from Chenopodiaceae)

1. *BASELLA* L.

1.1. ***Basella alba*** L. Sp. Pl. 272. 1753; Watt Dict. Econ. Prod. 1: 170. 1889. *Basella rubra* L. Sp. Pl. 272.1753; Hook.f. Fl. Brit. India 5: 20. 1886; Watt Dict. Econ. Prod. 1: 404. 1889; Prain Beng. Plants 2: 659. 1903 (Rep. ed. 1963). *(Vern. Name: Pui-Sak).*

Glabrous, succulent twiner, stem green, rarely reddish. Leaves cordate-ovate. spikes axillary, fleshy with ovoid white buds. Fruit reddish black with purple juice. *Fls. & Frts.:* August-January.

Commonly planted/cultivated in the kitchen gardens in every house of the villages; often found as an escape, climbing the bushes or shrubs. The wild form is very thin in comparison to the cultivated form.

Field Collection: Piyalinagar/Baruipur, *K.N. - 108.*

Distributions: Almost throughout India. In the Tropics.

Economic Importance: Common vegetable, has some purgative properties.

Remarks: Bailey (Man. Cult. Plants 367.1949) distinguished *B. alba* L. & *B. rubra* L. as distinct species, while van Steenis in Fl. Males. 1 (5): 300.1957, considered both as a single species – only the colour varies among the two forms. Here the author followed van Steenis (1.c.)

CXI. PHYTOLACCACEAE

Key to the Genera

1a	Stamens 4; fruit globose, glabrous	2. *RIVINA* L.
1b	Stamens 8; fruit elongate, spinous	1. *PETIVERIA* L.

1. *PETIVERIA* L.

1.1 ***Petiveria alliacea*** L. Sp. Pl. 342. 1753. *P. follis* Browne, Hist. Jam. 274.1789. *P. foetida* Salisb. Prod. 214. 1796.

Erect, deeply rooted herb. Stem puberolous in upper region. Leaves alternate, simple, wavy, elliptic, acuminate or acute, stipules minute. Flowers in long terminal spikes, rachis 10-40 cm long; bracts ovate, acuminate; bisexual, haplochlamydous, regular, hypogynous. Perianth 4-merous, persistent, petaloid, white, imbricate and connate in rotate or semi-funnel shaped perigonium. Ovary monocarpellary, ellipsoid or semi-globose oblong, style absent. Fruit semi-globose to elongate cuneate, covered at base by 4 persistent perianth leaves, tip with curved spines. Seeds erect, linear, solitary, adnate to the pericarp, cotyledons leafy, unequal. *Fls. & Frts.:* October-February.

Rarely grow on the dry waste land and roadside thickets.

Field Collections: Sodepur, *K.N. - 1487.*

Distributions: Throughout India. Tropical America, West Indies, Jamaica and Java.

Notes: Prain (1903a) did not mention this taxon from Bengal.

2. *RIVINA* L.

2.1. ***Rivina humilis*** L. Sp. Pl. 121. 1753; Hook.f. Fl. Brit. India 5: 21. 1886; Prain Beng. Plants 2: 660. 1903. (Rep. ed. 1963).

Herb, woody below, with sulcate branches. Leaves alternate, petioled, caducous. Flowers hermaphrodite, in axillary, rarely terminal slender, many flowered racemes. Perianth corolline, 4-partite. Fruits compressed, subglobose. Seeds lenticular. *Fls. & Frts.:* July-November.

Rarely found in the village shrubberies.

Field Collection: Ballygunge, *K.N.* - 122.

Distributions: Very rare in India. A native of America.

Economic Importances: Ornamental plant as well as produce red-dye.

CXII. POLYGONACEAE

Key to the Genera

1a	Climbing shrub	1. *ANTIGONON* Endl.
1b	Prostrate or erect herbs or shrubs	(2)
2a (1)	Perianth 3-5 cleft; stigmas capitate	2. *POLYGONUM* L.
2b (1)	Perinath 6-cleft, stigmas fimbriate	3 *RUMEX* L.

1. *ANTIGONON* Endl.

1.1. ***Antigonon leptopus*** Hook. & Arn., Bot. Beech. Voy. 308,t. 69.1841; Prain Beng. Plants 2: 665. 1903 (Rep. ed. 1963).

Climbing shrub. Leaves ovate, cordate or deltoid, acute to acuminate, glabrous. Flowers showy, white, pink or red, in panicled racemes. *Fls. & Frts.:* Almost throughout the year.

Generally planted along the trellises of the ornamental gardens; and rarely found as wild-climbing on thickets.

Field Collection: Bhebe Choumatha/near Basirhat, K.N.-1253.

Distributions: Almost throughout India in the ornamental gardens and wild forms. A native of South America.

Economic Importances: It's red or pink flowers are very showy and ornamental.

2. *POLYGONUM* L.

Key to the Species

1a	Flowers in clusters, axillary	5. *P. plebejum* R. Brown
1b	Flowers in racemes, terminal	(2)
2a (1)	Nuts biconvex, not triquetrous	(3)
2b (1)	Nuts trigonous or triquetrous	(4)
3a (2)	Ochrea ciliate	4. *P. orientale* L.
3b (2)	Ochrea not ciliate	2. *P. glabrum* Willd.

4a (2)	Perianth and leaves without gland	1. *P. barbatum* L.
4b (2)	Perianth and leaves with glands	(5)
5a (4)	Perianth greenish	3. *P. hydropiper* L.
5b (4)	Perianth reddish	6. *P. pubescens* Blume

2.1. ***Polygonum barbatum*** L. Sp. Pl. 362.1753; Naskar Aq. & Semi-Aq. Plants 220. 1990. *P. rivulare* Roxb. Fl. Ind. 2: 290. 1824. *(Vern. Name: Bekh-Unjubaz).*

Undershrub, quite glabrous stem; common on the moist waste places. Naskar (1.c.) described this taxon from the lower Ganga delta after collecting from Magrahat, *(K.N. - 253)*. Its root is used as an astringent and as a cooling agent. Seeds are useful to relieve gripping pains of colic.

2.1./1. ***Polygonum barbatum*** L. var. ***gracile*** (Denser) Steward, Contr. Gray. Herb. 5(88): 55. 1830. *P. barbatum* L. sub. sp. *gracile* Denser, Bull. Jard. Bot. Buitz. Ser. 3(8): 146. 1927. *P. flaccidum* Roxb. Fl. Ind. (ed. Carey) 2: 291. 1832. *P. serrulatum* sensu Hook.f. Fl. Brit. India 5: 38.1886.

Erect or creeping below, slightly pubescent. Generally in the waste marshy places, on the side of the ponds, hills or drains. Naskar (1.c.) described this taxon after collecting from Sashan Road/near Baruipur; *(K.N. -1171).*

2.2. ***Polygonum glabrum*** Willd. Sp. Pl. 2: 447. 1799; Naskar Aq. & Semi-Aq. Plants 221. 1990.

Erect, glabrous, shrub, often red in colour, stems slightly branched; common on the ditches and other water courses. Naskar (1.c.) described this taxon from the lower Ganga delta. This plant is useful in dispelling fever; infusion of leaves used in colic pain.

2.3. ***Polygonum hydropiper*** L. Sp. Pl. 361. 1753; Naskar Aq. & Semi-Aq. Plants 222. 1990. *(Vern. Name: Pakurmul).*

Annual herb, glabrous, rather robust; generally grow on the road-side ditches and margins of the ponds, rivers, drainage, jhills. Naskar (1.c.) described this taxon after collecting from Dandihat *(K.N. - 972)*; Purandarpur *(K.N. - 1147)* and Ultadanga *(K.N. - 1460)*.

This plant has acrid juice, used in diuretic cases, uterine disorder and as immersion in causing precacious maturity of the young and also used as fish poison.

2.4. ***Polygonum orientale*** Sp. Pl. 362.1753; Naskar Aq. & Semi-Aq. Plants 223.1990. *(Vern. Name: Bara-Panimirich).*

Annual tall herb; branches softly pubescent, leaves long-petioled. Generally in the waste places. Naskar (1.c.) described this taxon from the lower Ganga delta after collecting from Dakshineswar, *(K.N. - 700;* Bongaon *(K.N. - 1316).*

2.5. ***Polygonum plebejum*** R. Brown Prodr. 420. 1810; Naskar Aq. & Semi-Aq. Plants 224. 1990.

Diffusely branched, prostrate, smooth or minutely scaberulous, branched terete grooved; commonly found on the side of ditches and roadsides. Naskar (1.c.) described this taxon after collecting from Saktinagar *(K.N. - 548)* and Dhanyakuria *(K.N. - 561).* Dried and powdered plants are taken internally in pneumonia; root is given in bowel complaints.

2.5./1. ***Polygonum plebejum*** R. Brown var. ***effusa*** R. Brown Prodr. 420. 1810; Hook.f. Fl. Brit. India 5: 28. 1886; Prain Beng. Plants 2: 662. 1903 (Rep. ed. 1963). *P. effusum* Meissn., DC. Prodr. 14. 95. 1856.

Lax, slender, flaccid, internodes longer; leaves narrowly linear or dilated upward from a very narrow base, tip rounded; flowers almost pedicelled, sepals short-broad. *Fls. & Frts.:* June-October.

Rarely found in the shady waste places and rice-fields.

Field Collection: Behala, *Dutta 145* (C.C.H.).

Distributions: Almost throughout India. Distributed in the Tropics.

2.5./2. ***Polygonum plebejum*** R. Brown var. ***elegans*** Hook.f. Fl. Brit. India 5: 29. 1886; Prain Beng. Plants 2: 663. 1903 (Rep. ed. 1963). *P. elegans* Roxb. Fl. Ind. 2: 291. 1832.

Black when dry, branches stout-grooved smooth or scaberulous. Stipules brownish fimbriate, leaves upper as long as the lower, linear-lanceolate acute or acuminate rarely obtuse, margins flat or revolute. Flower sessile or shortly pedicelled, sepals narrow, 2 outer acute. *Fls. & Frts.:* November-May.

Rarely on the roadside hedges.

Field Collection: Ghatakpukur, *K.N. - 1361.*

Distribution: Throughout India. In the Tropics.

2.6. ***Polygonum pubescens*** Blume; Majumdar, Ind. Agr. 6: 115.1962. *P. flaccidum* Meisan., DC. Prodr. 14(1): 107.1856; Naskar Aq. & Semi-Aq. Plants 224. 1990.

Shrub, erect, grow on the marshy places. Naskar (1.c.) described this taxon from the lower Ganga delta.

3. *RUMEX* L.

Key to the Species

1a	Leaves oblong obtuse	1. *R. dentatus* L.
1b	*Leaves lanceolate-oblong*	1. *R. maritimus* L.

3.1. ***Rumex dentatus*** L. Sp. Pl. 2: 226. 1771; Naskar Aq. & Semi-Aq. Plants 225. 1990. *(Vern. Name: Ambavati).*

Annual herb. Commonly found near the marshy places, road-side canals and ditches. Naskar (1.c.) described this taxon after collecting from Garia *(K.N. - 338)* and Sarbedia *K.N. - 781).*

Roots of this taxon are used as astringent and applied in cutaneous disorders.

3.2. ***Rumex maritimus*** L. Mant. 2: 226. 1771; Hook.f. Fl. Brit. India 5: 59. 1886; Prain Beng. Plants 2: 665. 1903 (Rep. ed. 1963). *(Vern. Name: Ban-Palang).*

Terrestrial, erect herb. Leaves lanceolate-oblong. Flowers in axillary or terminal clusters, bracteate, bi-sexual, perigynous, perianth 6, outer 3 unchanged, inner 3 enlarging in fruit-with hardly toothed margins. Stamens 6, styled 3. Nut enclosed in inner, sepals with smooth tubercles. *Fls. & Frts.:* December-March.

Growing on the waste lands.

Field Collection: Nalban, *M.K. Ghosh-1315* (CAL).

Distributions: Almost throughout India. In Europe, Caucasus and Africa.

CXIII. ARISTOLOCHIACEAE

1. *ARISTOLOCHIA* L.

1.1. ***Aristolochia indica*** L. Sp. Pl. 360. 1753; Hook.f. Fl. Brit. India 5: 75. 1886; Watt Dict. Econ. Prod. 1: 315.1889; Prain Beng. Plants 2: 666. 1903 (Rep. ed. 1963). *(Vern. Name: Ishermul).*

Climbing perennial herb. Stem long, slender, woody at the base, glabrous. Leaves highly variable, ovate, acuminate at apex, 3- nerved from the base. Flowers 2-5, in axillary clusters. Capsule broadly oblong, from below upwards, 6-valved. Seeds upto 6 mm, ovate triangular, acute, winged. *Fls. & Frts.:* October-January.

Rarely grow on the hedges and waste lands, along the railway tracks.

Field Collection: Malatipur, *K.N. - 1581.*

Distributions: Common in Nepal and Sri Lanka; and greater parts of India; said to be as the native of India.

Economic Importances: Root is given in bowel complaints of children and in intermittent fever. Popular belief about its use against poisonous bite.

CXIV. PIPERACEAE

Key to the Genera

1a	Erect; flowers bisexual	1. *PEPEROMIA* Ruiz. & Pav.
1b	Creeping; flowers unisexual	2. *PIPER* L.

1. *PEPEROMIA* Ruiz. & Pav.

1. ***Peperomia pellucida*** (L.) H.B. & K., Nov. Gen. et. sp. 1: 64. 1815; Prain Beng. Plants 2: 669. 1903 (Rep. ed. 1963); Naskar Aq. & Semi-Aq. Plants 225. 1990. *(Vern. Name: Luchipata).*

Very succulent herb. Very common in the marshy waste lands, old walls during rainy season. Naskar (1.c.) described this taxon after collecting from Atghora/Madarhat *(K.N. - 60)*, Brace Bridge *(K.N. - 758.)* and Sankchura/near Taki *(K.N. - 974).*

2. *PIPER* L.

Key to the Species

1a	Purely cultivated	1. *P. betle* L.
1a	Mostly wild	2. *P. longum* L.

2.1. ***Piper betle*** L. Sp. Pl. 28. 1753: Hook.f. Fl. Brit. India 5: 85. 1886; Watt Dict. Econ. Prod. 6(1): 247. 1892; Prain Beng. Plants 2: 668. 1903 (Rep. ed. 1963). *Chavica betle* Miq. Syst. Pip. 28. *(Vern. Name: Pan).*

Climbing or twining shrub. Rooting from the nodes; roots fibrous. Leaves alternate or opposite; glabrous, dark green cordate, base notched, acuminate, entire. *Fls. & Frts.:* Very occasional.

Commonly cultivated, there are two types of cultivation (i) under complete shading and (ii) allow to grow on the wall or mango tree or on other tree trunks.

Field Collection: Subhashgram, *K.N. - 1348.*

Distributions: Eastern and South India. South East Asia.

Economic Importances: Betal leaf is aromatic stimulocurminative, astringent and aphrodisiac. There are many application of leaves as house remedy like promotion of milk secretion when applied on the breast of lactating women, pulmonary affections of children, cough and dysponex. The habit of chewing 'pan' leaves is a part of social life in India. While moderate dose is not injurious but heavy doses have adverse effect in causing excessive salivary secretion, which is thrown away, reduces digestive function.

2.2. ***Piper longum*** L. Sp. Pl. 29. 1753; Hook.f. Fl. Brit. India 5: 83. 1886; Watt Dict. Econ. Prod. 6(1): 258. 1893; Prain Beng. Plants 2: 668.1903 (Rep. ed. 1963). *(Vern. Name: Pipla).*

Stem prostrate or ascending. Leaves cordate, glarbous, lower with long petioles, the upper sessile and stem-clasping. Fruits small, blackish green, sunk in the fleshy spikes. *Fls. & Frts.:* June-September.

Rarely found on the dampy soils. Occasionally cultivated here.

Field Collection: Kazipara/Barasat, *K.N. - 931.*

Distributions: Native of the hotter parts of India. Tropical flora.

Economic Importances: The dried unripe fruits and roots are much used in indigenous medicine.

CXV. LAURACEAE

Key to the Genera

1a	Parasitic, leafless, twining herb; inflorescene compact	1. *CASSYTHA* L.	
1b	Large shrubs or trees; inflorescence lax		(2)
2a (1)	Anthers of the two outermost rows opening inwards, of the third row opening outwards by valves; flowers hermaphrodite	2. *CINNAMOMUM* Blume	
2b (1)	Anthers of all the rows opening inwards by valves, flowers dioecious	3. *LITSAEA Lamk. nom. cons.*	

1. *CASSYTHA* L.

1.1. ***Cassytha filiformis*** L. Sp. Pl. 35.1753; Hook.f. Fl. Brit. India 5: 188.1886; Naskar & Guha Bakshi Mangr. Swamps Sundarbans 128.1987. *(Vern. Name: Akash bel).*

Parasitic leafless twining herb. Naskar and Guha Bakshi (l.c.) described this taxon after collecting from Diamond Harbour, *K.N.-392.* This plant is useful in bilious affections, urethritics, chronic dysentery and skin diseases.

2. *CINNAMOMUM* Blume

Key to the Species

1a	Buds enclosed in large naked scales; leaves alternate, penninerved; perianth lobes wholly deciduous	1. *C. camphora* T. Nees & Eberm.
1b	Buds naked or with very small scales; leaves opposite, 3-nerved	(2)
2a (1)	Perianth lobes deciduous in fruits; stamens and ovary villous	3. *C. tamala* T. Nees & Eberm.
2b (1)	Perianth lobes rounded, persistent in fruits	(3)
3a (2)	Leaves 20-30 cm long, secondary nerves very faint	2. *C. obtusifolium* Ness & Eberm.
3b (2)	Leaves 12.5-15.0 cm long, nerves distinctly reticulate beneath	4. *C. zeylanicum* Blume

2.1. ***Cinnamomum camphora*** T. Nees & Eberm., Handb. Med. Pharm. Bot. 2: 430.1831; Hook.f. Fl. Brit. India 5: 134.1886; Watt Dict. Econ. Prod. 2: 317.1889; Prain Beng. Plants 2: 673. 1903 (Rep. ed. 1963). *(Vern. Name: Karpur).*

Small to medium tree, much branched. Leaves dark green, shining, lanceolate, entire, acute or acuminate. Flowers bisexual; perianth lobes deciduous. *Fls. & Frts.:* February-April.

Very occasionally planted in the ornamental gardens.

Field Collection: Panihati, *K.N. - 1603.*

Distributions: Planted almost throughout India. A native of China, Japan and Formosa.

Economic Importances: Camphor is produced from this plant. Wood oil contain camphor and safrole used for perfumes. Camphor is also used for inflammations, rheumatic pain, sprain and internally used in diarrhoea and as heart stimulant.

2.2. ***Cinnamomum obtusifolium*** T. Nees & Eberm., Handb. Med., Pharm.. Bot. 2: 426.1831; Hook.f. Fl. Brit. India 5: 128.1886; Prain Beng.

Plants 2: 673.1903 (Rep. ed. 1963). *Laurus obtusifolia* Roxb. Fl. Ind. 2: 302. 1824. *(Vern. Name: Ram- Tejpataor Kinton).*

Large evergreen tree. Leaves 20-30 cm long, secondary nerve very faint. Buds almost naked; perianth lobes rounded, persistent in fruit; stamens and ovary sparsely hairy or glabrous. Fuit a berry. *Fls. & Frts.:* December - March.

Occasionally planted on the village area or grow wild sometimes.

Field Collections: Bhebe, *K.N. - 709;* Birati, *K.N. - 819.*

Distributions: On the foothills of Himalayas; Bangladesh and Sri Lanka.

Notes: Prain (1903a) has not reported this taxon from Bengal.

2.3. ***Cinnamomum tamala*** T. Nees & Eberm. Handb. Med. Pharm. Bot. 2: 426.1831; Hook.f. Fl. Brit. India 5: 128.1886: Watt Dict. Econ. Prod. 2: 319.1889 Prain Beng. Plants 2: 673.1903 (Rep. ed. 1963). *Cinnamomum albiflorum* T. Nees, Wight Ic. t. 140. *Laurus cassia* Roxb. Fl. Ind. 2: 297. 1824. *(Vern. Name: Tejpata).*

Moderate sized tree. Leaves very aromatic, glabrous, opposite or nearly so, often alternate on the same branch. Flowers 0.5 -0.6 mm, perianth silky. Fruit black when ripe, succulent. *Fls. & Frts.:* February-May.

Rarely planted in the orchards or ornamental gardens.

Field Collection: Kadampur/Nabgram, *K.N. - 627.*

Distributions: Himalaya, Jumna eastwards, chiefly in damp ravines. Khasi Hills and Tripura. Bangladesh, Upper Burma and China.

Economic Importances: Leaves are extensively used as spices, condiment and for flavouring tea. Leaves are carminative and are used in colic and diarrhoea.

3.4. ***Cinnamomum zeylanicum*** Blume, Bijdr. 568.1825. Hook.f. Fl. Brit. India 5: 131.1886; Watt Dict. Econ. Prod. 2: 324. 1889; Prain Beng. Plants 2: 675.1903 (Rep. ed. 1963). *(Vern. Name: Dalchini* or *daruchini).*

Small much-branched tree. Leaves distinctly reticulate beneath. Perianth lobes distictly persistent in fruit. Stamens and ovary globose. *Fls. & Frts.:* November-March.

Rarely cultivated/planted in the orchards or ornamental gardens.

Field Collection: Harinbari/Sagar Islands, *K.N. - 485.*

Distributions: Planted almost throughout India. A native of Sri Lanka.

Economic Importance: Stem-bark is a valuable and highly scented spice. Medicinally it is astringent, stimulats and carminative and also used for checking nausea and vomiting. Bark oil is useful for perfumery and pharmaceutical industries.

3. *LITSEA* Lamk. *nom.cons.*

Key to the Species

1a	Perianth - segments incomplete or absent; fruits globose and small, thickened perianth-tube	1. *L. glutinosa* (Lour.) Rob.
1b	Perianth segments distinct, usually 6; perianth tube not enlarging in fruit	2. *L. monopetala* (Roxb.) Pers.

3.1. ***Litsea glutinosa*** (Lour.) Rob., Philip. J.Sc., Bot. 6: 321.1911; Wlh. India 6: 153.1962. *Sebifera glutinosa* Lour. Fl.Cochinch. 638.1790. *Litsea sebifera* Pers. Syn.2: 4.1807; Hook.f. Fl. Brit. India 5: 157.1886; Prain Beng. Plants 2: 675.1903 (Rep. ed. 1963).*(Vern. Name: Kukur-Chita).*

Medium tree, upto 3 m tall. Leaves variable both in shape and size, glabrous above and pubescent beneath, oblong to elliptic-oblong, apex acute or acuminate, base acute to rounded. Flowers greenish yellow, globose, pubescent. Perianth tube long, pubescent out-side. Stamens many. Fruits 6-7 mm diam., sub- globose, black or dark-brown. *Fls. & Frts.:* May-November

Common on the village shrubberies and thickets, everywhere.

Field Collections: Surjyapur, *K.N. - 284;* Naihati, *K.N. - 1996.*

Distributions: Punjab Salt range, Assam, Khasi Hills, Bengal, Central and South India, Burma.

Economic Importances: Leaves are used in infusion or as poultice for bruises and wounds. Fruits are edible, the roots are sweetish bitter, astringent and tonic.

3.2. ***Litsea monopetala*** (Roxb.) Pers. Syn. 2: 4.1807; Hook.f. Fl. Brit. India 5: 162.1886; Watt Dict. Econ. Prod. 5: 82.1891. *L. polyantha* Juss., Ann. Mus. Par. 6: 211.1805; Prain Beng. Plants 2: 676.1903 (Rep. ed. 1963). *Tetranthera monopetala* Roxb. Cor. Pl. t. 148. 1820. *(Vern. Name: Bara-kukur-chita).*

Medium tree. Leaves elliptic-oblong. Inflorescence with short brown or rust coloured pubescence. Fruits 0.5-0.6 cm long, seated on the persistent base of perianth. *Fls. & Frts.:* March-May.

Occasionally grow on the village thickets and shrubberies.

Field Collections: Bhebe, *K.N. - 713;* Dattapukur/Barasat, *K.N. - 794.*

Distributions: Punjab, Assam, Khashia, Naga Hills, Cachar and Bengal Plains. Bangadesh, Burma and Malaya.

Economic Importances: Bark is mildly astringent and is reported to be used for diarrhoea; powder barks and roots are used in external application for pains, bruises and combusions; also used for fractures in animals.

CXVI. PROTEACEAE

1. *GREVILLEA* R. Brown

1.1. *Grevillea robusta* A. Cunn., R. Brown Prot. Nov. 24.1820; Brandis Ind. Trees 544. 1978.

Robust tree. Leaves fern like, bipinnate or rarely tripinnate, entire or deeply pinnatifid, lanceolate, lower surface whitish. Flowers orange, showy racemes solitary or several together on short leafless branches; style long, lateral. Fruits oblique, coriaceous follicle, brown; seeds 1or 2 winged. *Fls. & Frts.:* March-July.

Planted as ornamental plant in the gardens.

Field Collections: Specimnes not preserved.

Distributions: Planted almost throughout India. A native of Queensland and New South Wales.

Economic Importances: Planted as avenue tree. Produce cellulose raw materials; bark contain lignin and produce lignin compounds. Leaves are valuable green manure; wood is hard but light and good for cabinet works.

CXVII. LORANTHACEAE

Key to the Genera

1a	Flowers bisexual	(2)
1b	Flowers unisexual	3.*VISCUM* L.
2a (1)	Corolla mostly 5-cleft; style not articulated	1. *DENDROPHTHOE* Mart.
2b (1)	Corolla mostly 6-cleft; style articulated above	2. *MACROSOLEN* (Bl.) Reichenb.

1. *DENDROPHTHOE* Mart.

1.1. ***Dendrophthoe falcata*** (Linn.f.) Etting., Denkschr. Akad. Wissen. Math. Naturw. Cl. 32: 52,53,58 t. 13,f. 14,1872. *Loranthus falcatus* Linn.f. Suppl. Pl. 211.1781. *L. longiflorus* Desr., Lamk.Encycl. Meth.3: 598. 1789; Hook.f. Fl. Brit. India 5: 214.1886; Watt Dict. Econ. Prod. 5: 92.1891; Prain Beng. Plants 2: 682.1903 (Rep. ed. 1963). *(Vern. Name: Baramanda).*

Semi-parasitic herb.Leaves simple, coriaceous. Flowers bi-sexual, red. Perianth segment present. Fruits one seeded, globose berry with copious mucilage. *Fls. & Frts.:* November- February.

Parasitic on the *Ficus benghalensis, F. religiosa, Mangifera indica* and *Melia azedarach.*

Field Collections: Amtala, *K.N. - 373;* Rangabelia, *K.N. - 1511.*

Distributions: Throughout India.Sri Lanka, Australia and Malaya

2. *MACROSOLEN* (Bl.) Reichenb.

2.1. ***Macrosolen cochinchinensis*** (Lour.) Van. Teigh. Bull. Soc. Bot. Fr. 41: 122.1894. *Loranthus cochinchinensis* Lour. Fl. Coch. 1: 195.1790. *L. ampullaceus* Roxb.Fl.Ind.2: 209.1824; Hook.f. Fl. Brit. India 5: 220.1886; Prain Beng. Plants 2: 683.1903 (Rep. ed. 1963). *(Vern. Name: Chhota-Manda).*

Much branced, bushy shrub; semi-parasitic on the old trees.

Naskar & Guha Bakshi (Mangr. Swamps. Sundarbans 130.1987) described this taxon after collecting from Maliddhar/Canning (*K.N. - 126);* Diamond Harbour *(K.N. - 395).* Leaves are said to be useful for headache as poultice. (Fig.42).

3. *VISCUM* L.

Key to the Species

1a	Leaves obliquely ovate or falcate, acute.	1.*V. monoicum* Roxb.
1b	Leaves equally elliptic or elliptic lanceolate, obtuse	2.*V. orientale* Willd.

3.1. ***Viscum monoicum*** Roxb. Fl.Ind. 3: 763.1832; Naskar & Guha Bakshi Mangr. Swamps Sundarbsns 130.1987. *(Vern. Name: Manda).*

Large shrub; parasitic on the trees of Sundarbans mangrove forests. Naskar & Guha Bakshi (l.c.) described this taxon after collecting from Harinbari/Sagar Islands (*K.N. - 480).*

The plant is a powerful narcotic and is poisonous to animals.

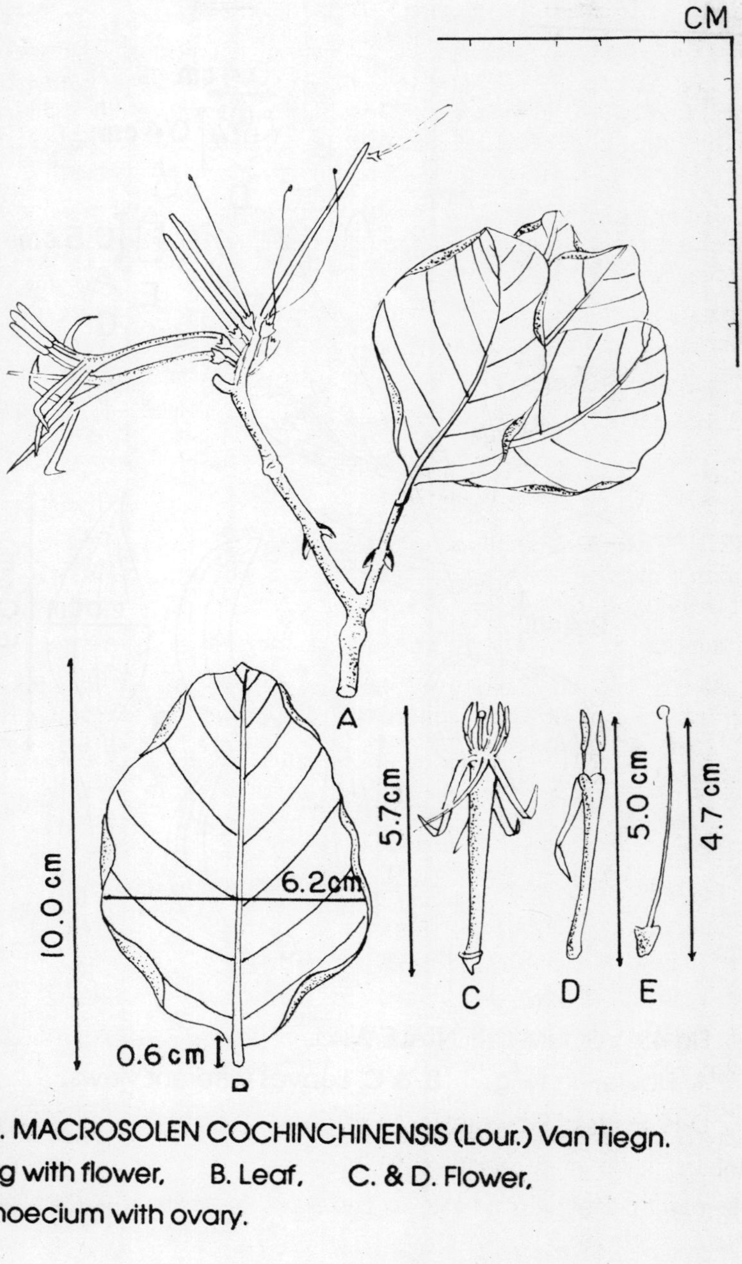

Fig.42. MACROSOLEN COCHINCHINENSIS (Lour.) Van Tiegn.

A. Twig with flower, B. Leaf, C. & D. Flower,

E. Gynoecium with ovary.

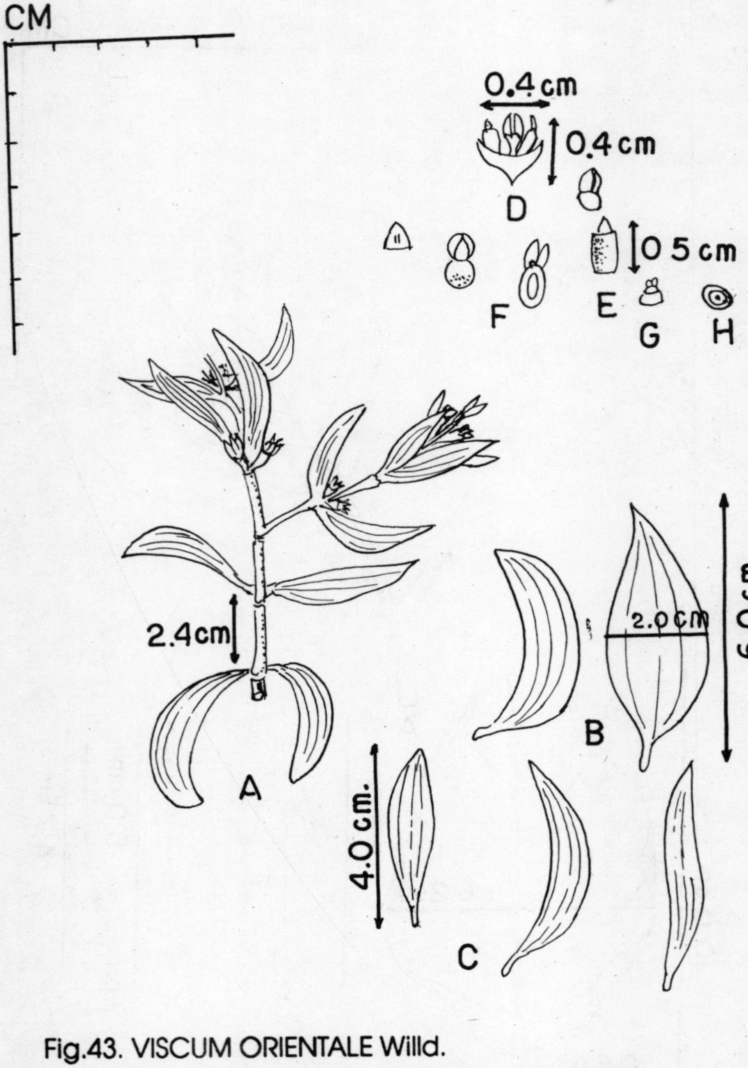

Fig.43. VISCUM ORIENTALE Willd.

A. Flowering twig, B. & C. Leaves different views,

D-H. Parts of Flowers.

3.2. ***Viscum orientale*** Willd. Sp.Pl.4 (2): 737.1806; Naskar & Guha Mangr. Swamps Sundarbans 131.1987. *(Vern. Name: Banda)*. (Fig.43).

Parasitic menoecious herb; commonly found on *Excoecaria agallocha*. Naskar & Guha Bakshi (l.c.) described this taxon from the lower Ganga delta after collecting from Kulti *(K.N. - 1430)* and Naihati *(K.N. - 1537)*. This taxon is said to be poisonous and is said to posses medicinal properties and more or less similar of those of the tree on which it grows.

CXVIII. EUPHORBIACEAE

Key to the Genera

1a	Flowers aggregate-monoecious	(2)
1b	Flowers generally dioecious or separate-monoecious	(3)
2a (1)	Involucre regular, campanulate or turbinate or slightly irregular	10. *EUPHORBIA* L.
2b (1)	Involucre irregular, oblique, slipper shaped declinate	15. *PEDILANTHUS* Poit. *nom. cons.*
3a (1)	Stamens of the outer or only series opposite the sepals, or if stamens all in centre of flowers, the cells of the 2-many-carpelled ovary each 2-ovulate; petals when present minute; sepals 1-2 seriate, inflorescence lateral or axillary	(4)
3b (1)	Stamens of the outer or only series alternate with the sepals, or if stamens all central, the cells of the 2-3 rarely 4-carpelled ovary each 1-ovuled; petals if present, often exceeding the sepals	(10)
4a (3)	Leaves tri-foliotate	4. *BISCHOFIA* Blume
4b (3)	Leaves generally simple, entire	(5)
5a (4)	Male flowers in axillary fascicles or solitary axillary	(6)
5b (4)	Male flowers in axillary spikes, racemes panicles stamens free round a rudimentary pistillode; anther didynamous; fruit indehiscent, seeds without an arillus	2. *ANTIDESMA* L.

6a (5)	Corolla of 5 (rarely 6), scale like petals; calyx lobes or segments 5, valvate	6. *BRIDELIA* Willd.
6b (5)	Corolla absent in both sexes; sepals imbricate	(7)
7a (6)	Styles much dilated; fruit indehiscent	17. *PUTRANJIVA* Wall.
7b (6)	Styles or style arms slender	(8)
8a (7)	Disk none in either sex, or if present then only in the male flowers; stamens 3; filaments united in a column	(9)
8b (7)	Disk present in both sexes, or if absent then only so form male flowers; fruits a dry capsule	16. *PHYLLANTHUS* L.
9a (8)	Fruit a capsule with three 2-valved carpels; anther sessile at apex of column; disk present in male flowers	21. *SYNOSTEMON* Mueller
9b (8)	Fruit of 3-6 indehiscent cocci; disk absent from flowers of both sexes	5. *BREYNIA* J. R. & G. Forsst.
10a (3)	Flowers in terminal spikes, racemes or panicles	(11)
10b (3)	Flowers in axillary clusters, spikes, racemes or panicles	(12)
11a (10)	Flowers in 2-3 chotomously cymose panicles; stamens erect; leaves always digitately lobed	12. *JATROPHA* L.
11b (10)	Flowers in androgynous spikes or racemes; stamens inflexed in bud with anthers reversed, many, inserted on the hairy receptacle; fruit dry, capsule	9. *CROTON* L.
12a (10)	Petals present in male flowers; filaments straight; ovary 3-celled, fruit a capsule	(13)
12b (10)	Petals absent from both sexes	(14)
13a (12)	Sepals imbricate; disc developed in both sexes; stamens 15-30, free	8. *CODIAEUM* Juss.

13b (12)	Sepals valvate; disk in male obscure; stamens 5-15, connate below	7. *CHROZOPHORA* Juss. Corr.
14a (12)	Sepals imbricate, 4-6, free	(15)
14b (12)	Sepals or calyx-segments valvate or occasionally, calyx obsolate; flowers in axillary, rarely terminal spikes, racemes, or panicles	(16)
15a (14)	Flowers in both sexes in axillary, clusters; dioecious	20. *SUREGADA* Roxb. ex Rottl.
15b (14)	Flowers in both sexes in clusters on numerous axillary racemes or panicles; monoecious or dioecious	3. *BALIOSPERMUM* Blume
16a (14)	Calyx of male flower closed in bud, usually membranous, oblong, ovoid or globose	(17)
16b (14)	Calyx of male flower open in bud; styles entire, free or shortly connate at the base; stamens always 1-seriate; disk absent	(22)
17a (16)	Styles distinct, usually long, entire, 2-fid, multifid or papillose-fimbriate	(18)
17b (16)	Styles connate; twining herbs with alternate leaves; racemes always axillary; androgynous; stamens always 1-seriate; disk absent or obscure; fruit always capsule	22. *TRAGIA* L.
18a (17)	Filaments free	(19)
18b (17)	Filaments connate in bundles or branched; disk absent; fruit capsule	18. *RICINUS* L.
19a (18)	Anther cells 2, united only by their bases; stamens many or few	(20)
19b (18)	Anther cells 2, oblong-parallel or globose, laterally attached to the connective; stamens many	(21)

20a (19)	Styles undivided, spreading, fringed, short, flowers dioecious; disk of 3 hypogynous scales	14. *MICROCOCCA* Benth.
20b (19)	Styles filiform, long, laciniate or fimbriate; flowers monoecious, males minute, ebracteate, female at base of male spikes or in separate spikes, usually with a large bracts; disk absent	1. *ACALYPHA* L.
21a (19)	Leaves opposite; flowers rather large, dioecious; males in lax lateral racemes	23. *TREWIA* L.
21b (19)	Leaves alternate or opposite; flowers small or minute, monoecious or dioecious, in axillary or terminal, simple or panicled spikes or racemes	13. *MALLOTUS* Lour.
22a (16)	Calyx 2-3 lobed; flowers monoecious, in terminal, simple or panicled spikes or racemes	19. *SAPIUM* R. Browr
22b (16)	Calyx deeply 3-partite; flowers in lateral axillary or terminal, spiciform, unisexual (dioecious or monoecious) or androgynous racmes or spikes	11. *EXCOECARIA* L.

1. *ACALYPHA* L.

Key to the Species

1a	Purely cultivated garden plants	1. *A. hispida* Burm.f.
1b	Annual weed plants	(2)
2a (1)	Bracts rather shroter than capsules	3. *A. lanceolata* Willd.
2b (1)	Bracts longer than capsules	2. *A. indica L.*

1.1. ***Acalypha hispida*** Burm. f. Fl. Ind. 303.t.61, f.1; Muell, DC. Prodr. 15(2): 815. 1866; Hook.f. Fl. Brit. India 5: 417.1887.

Shrubs, leaves broadly ovate. Female racemes drooping, scarlet, 30-45 cm long. Spikes long; bracts minute; style deeply pinnatifid, segments linear subulate. *Fls. & Frts.:* Throughout the year.

Cultivated garden plants.

Field Collections: Amtala, *K.N. - 367*; Basanti, *K.N. - 856*.

Distributions: Cultivated throughout India. A native of East Indies; planted in Malay Archipelago.

Notes: D. Prain (1903a) did not mention this plant from Bengal.

1.2. ***Acalypha indica*** L. Sp. Pl. 1003. 1753; Hook.f. Fl. Brit. India 5: 416.1887; Prain Beng. Plants 2: 710.1903 (Rep. ed. 1963). *(Vern. Name: Mukta-jhuri).*

Erect, annual herb. Leaves thin, glabrous, ovate or rhomboid-ovate, obtuse or sub-acute, crenate serrate; cuneate at base, 3-nerved. Flowers minute, green in axillary androgynous spikes. Capsules concealed by persistent bracts, seeds ovoid, plae brown. *Fls. & Frts.:* June-December.

Common weed on the moist waste lands.

Field Collection: Maslandapur, *K.N. - 513.*

Distributions: Almost throughout India. Tropical Africa and Asia throughout.

Economic Importances: The fresh plant is an excellent remedy for severe cough, associated with bleeding from lungs, haemoptysis and incipient phthisis.

1.3. ***Acalypha lanceolata*** Willd. Sp. Pl. 4: 524. 1805: *A. fallax* Muell.-Arg. Linn. 34: 43.1865; Hook.f. Fl. Brit. India 5: 416.1887; Prain Beng. Plants 2: 710.1903 (Rep. ed. 1963).

Terrestrial, erect herb. Leaves alternate, palminerved, cordate lanceolate. Flowers inconspicous, male flowers without bracts; perianth leaves 4, sepaloid; stamens 8 or many female flowers situated on a leafy bracts, solitary, perianth 3-4, ovary 3 celled, each cell with one ovules. *Fls. & Frts.:* June- October.

Common on the waste places, and near the garbage-heaps of the villages.

Field Collections: Dewanganj, *K.N. - 625;* Brace Bridge, *K.N. - 727.*

Distributions: More or less throughout India. In the Tropical Asia.

2. *ANTIDESMA* L.

2.1 ***Antidesma ghaesembilla (ghesaembilla*** Gaertn., De Fruct. 1: 189. 1788; Hook.f. Fl. Brit. India 5: 357. 1887; Prain Beng. Plants 2: 703. 1903 (Rep. ed. 1963). *A. pubescence* Roxb. Pl. Cor. 2: 35, t. 157.1768. *A. paniculata* Roxb. Fl. Ind. (ed.2) 3: 770. 1832. *(Vern. Name: Khundi-Jam).*

Large shrub, branches pubescent or fulvuous-tomentose. Leaves 5-15 cm long, broadly elliptic, orbicular or elliptic oblong. Flowers greenish-yellow, in densely tomentose panicled racemes, minutely pedicelled. Drupe sub-globose, reddish or black when ripe. *Fls. & Frts.:* July December.

Commonly grow on the waste places and thickets.

Field Collections: Maslandapur, *K.N. - 528.*

Distributions: Throughout India. Sri Lanka, Tropical Africa, South China, through Malaya to Tropical Australia.

Economic Importances: The acrid fruits are said to be edible and the leaves are occasionally used for giving a flavour to food; leaves also used in various ailments, *e.g.*, headache, abdominal swelling and scruff.

3. *BALIOSPERMUM* Blume

3.1. ***Baliospermum montanum*** (Willd.) Muelll.-Arg., DC. Prod. 15(2):1125.1866. *Jatropha montana* Willd. Sp. Pl. 4: 563.1805. *Baliospermum axillare* Blume, Bijdr. 604.1826; Hook.f. Fl. Brit. India 5: 461. 1887; Prain Beng. Plants 2: 709.1903 (Rep. ed. 1963). *(Vern. Name: Danti* or *Hakun).*

Stout undershrub. Leaves very variable both in size and shape, alternate, lanceolate, sinuate or serrate. Flowers green, fascicled axillary and in the axils of bracts. Capsules 3-lobed, pubescent. *Fls. & Frts.:* November-April.

Rarely found in marshy, shady and waste places.

Field Collections: Nurpur, *K.N. - 1934;* Budge Budge, S.K. Mukherjee-s.n. (CAL).

Distributions: A native of India. In Malaya.

Economic Importances: Roots are useful for dropsy and jaundice. Seed-oil is also used as hydragogue cathartic.

4. *BISCHOFIA* Blume

4.1. ***Bischofia javanica*** Blume Bijdr. 1168.1827; Hook.f. Fl. Brit. India 5: 345.1887; Prain Beng. Plants 2: 693.1903 (Rep. ed. 1963). *Micraelus roeperianus* W. & A., Edinburgh New Philos. J. 14: 298.1833. *(Vern. Name: Kain).*

Medium, handsome, deciduous tree. Leaves alternate, long petioled, 3-foliolate; leaflets elliptic-oblong to obovate, acuminate. Flowers greenish yellow. Fruits reddish-brown; seeds brown, shining. *Fls. & Frts.:* March-July.Grow abundantly in the thickets sometime planted.

Field Collection: Malatipur, *K.N. - 1585.*

Distributions: Almost throughout India. A native of Malaya, Pacific Islands, Burma and Bangladesh.

Economic Importances: Wood is hard and durable, used to build bridges and also boat-building. The fruits are edible by the villagers. Leaves-juice is considered a cure for sores.

5. *BREYNIA* J.R. & G. Forsst. *nom. cons.*

5.1. ***Breynia rhamnoides*** (Retz.) Muell.-Arg., DC. Prodr. 15: 440.1866; Hook. f. Fl. Brit. India 5: 330. 1887; Prain Beng. Plants 2: 698.1903 (Rep. ed. 1963). *Phyllanthus rhamnoides* Retz.. Obs. 30. 1791. *Melanthesa rhamnoides* Retz. Blume Bijdr. 591.1825. *(Vern. Name: Kali-Sitki).*

Much-branched, dense glabrous shrubs. Branches spreading; branchlets angular. Leaves subulate from deltoid or broad-base, persistent. Flowers minute. Male flowers in small fascicles; female flowers mostly solitary; pedicels; filiform-variable in length. Berry globose, Seated on slightly or hardly enlarged calyx, red when ripe. *Fls. & Frts.:* November-May.

Common on the village shrubberies, hedges and river bank.

Field Collections: Joynagar, *K.N. - 862;* Garia, *G. Sen Gupta-41* (CAL) .

Distributions: Almost throughout India.

6. *BRIDELIA* Willd.

6.1. ***Bridelia stipularia*** (L.) Blume, Bijdr. 597.1825; Hook.f. Fl. Brit. India 5: 270.1887;Prain Beng. Plants 2: 694.1903 (Rep. ed. 1963). *Cluytia stipularia* L. Mant.1: 127.1767. *(Vern. Name: Harinhara).*

Large woody evergreen climber with pendent branches. Leaves broadly elliptic-oblong or obovate, obtuse or sub-acute, rounded or cordate at the base. Flowers monoecious, greenish-yellow, in small, dense, axillary clusters or spikes. Drupes oblong, obtuse, smooth. *Fls. & Frts.:* March-August.

Rarely found in the waste lands.

Field Collection: Diana/Salt Lake, *M.K.Ghosh-4050* (CAL).

Distributions: More or less throughout India. In the Tropics.

7. *CHROZOPHORA* Juss. Corr., Benth & Hook.f. *nom.cons.*

7.1. ***Chrozophora rottleri*** (Geis.) Klotzsch. ex Juss., Tent. Euph. 28. 1824; Prain Kew Bull. 1918:25.1918. *Croton rottleri* Geis., Crot. Monogr. 54.1807. *Chrozophora plicata* A. Juss., Tent. Euph. 28.1824; Hook.f. Fl. Brit. India 5: 409.1887; Prain Beng. Plants 2: 708.1903 (Rep. ed. 1963). *(Vern. Name: Khudi-Okra).*

Erect hoary annual herb. Leaves 5-15cm long ovate to orbicular with 3-rounded lobes. Flowers monoecious, yellow in racemes. Stamens 15, stigma red. Capsules densely stellate tomentose. *Fls. & Frts.:* Almost throughout the year.

Commonly grow on the sandy soil near the canals, river banks and roadside hedges.

Field Collections: Garia, *K.N. - 327;* Ghoramar/Sagar Island, *K.N. - 415.*

Distributions: Throughout India, said to be a native of India. Asia, Europe and Australia.

8. *CODIAEUM* Juss.

8.1. ***Codiaeum variegatum*** Blume, Bijd. 606. 1826; Hook.f. Fl. Brit. India 5: 399.1887;Prain Beng. Plants 2: 707.1903 (Rep. ed. 1963). *(Vern. Name: Patabahar.).*

An evergreen shrub. Leaves very variable from oblong to narrowly linear, often variegated green and yellow. Racemes long axillary. Flowers in unisexual racemes, ovary 3-celled, style long, slender recurved, entire. *Fls. & Frts.:* Throughout the year.

Often planted as ornamental plants for its beautiful variegated foliages.

Field Collection: Amtala, *K.N. - 369.*

Distributions: Planted almost throughout in India. In the Tropics; A native of Molucca Islands.

Economic Importance: Ornamental garden plants.

9. *CROTON* L.

Key to the Species

1a	Much-branched herb or undershrub	1. *C. bonplandianum* Baill.
1b	Medium-sized, deciduous tree	2. *C. oblongifolius* Roxb.

9.1. ***Croton bonplandianum*** Baill., Adans. 4: 339.1864; Prain, Rec. Bot. Surv. India 2: 337. 1903; Muell.-Arg., DC. Prodr. 15(2): 671.1866. *C. sparsiflorum* Morong. Ann. NY. Acad. Sci. 7: 221. 1893. *(Vern. Name: Jhunjhuni* or *Banlanka).*

Erect, much branched herb or undershrub. Stem stellate hairy. Leaves simple, lanceolate, ovate-lanceolate, warty and toothed, sparsely stellately hairy beneath, base 3-nerved. Flowers pale white in terminal erect androgynous spike. Capsule 3-angled, stellate hairy. *Fls. & Frts.:* Almost throughout the year.

Gregariously grow along the roadsides, railway tracks and waste lands everywhere, a common weed of the cultivated land.

Field Collections: Fultala/Baruipur, *K.N. - 47;* Canning, *K.N. - 171.*

Distributions: Throughout India. First introduced into Chittagong (Bangladesh) in the year 1897 with ballast of mud. A native of South America.

Economic Importances: Juice of the young stem is very good for checking bleeding in the fresh cut.

9.2. ***Croton oblongifolius*** Roxb. Fl. Ind. 3: 685.1832; Hook.f. Fl. Brit. India 5: 386.1887; Prain Beng. Plants 2: 705.1903 (Rep. ed. 1963). *(Vern. Name: Chucka).*

Medium sized deciduous tree. Leaves 15-25 cm long, oblong lanceolate, narrowly elliptic, sub-acute, crowded, towards the ends of the branches. Flowers monoecious, yellowish-green. Capsule sub-globose, covered with flat scales, splitting into 2- valves cocci. *Fls. & Frts.:* January-May.

Rarely found on the waste lands.

Field Collection: Nabagram, *K.N. - 640.*

Distributions: Drier parts of India. In the Tropics; indigenous in Burma and Sri Lanka.

Economic Importances: Fresh bark is pounded and applied to swellings, bruises and sprains; root bark is used as a remedy for enlarged liver.

10. *EUPHORBIA* L.

Key to the Species

1a Perennial shrubs or small tree (2)

1b	Annual herbs with slender, erect ascending or prostrate stems and branches, leafy throughout	(6)
2a (1)	Stems slender, woody, unarmed	11. *E. pulcherrima* Willd. ex Klotz.
2b (1)	Stem thick and fleshy, armed	(3)
3a (2)	Branches ascending, armed with paired, short, stipulary thorns	(4)
3b (2)	Branches spreading, unarmed, cylindric, scattered or whorled or clustered	13. *E. tirucalli* L.
4a (3)	Branches prominently 3, sometime 4-6 angled, the angles fleshy, sinuately repand; styles 2-cleft	1. *E. antiquorum* L.
4b (3)	Branches terete or slightly 5-angled; styles simple, thickened at the base	(5)
5a (4)	Branches with stipular thorns, arising from thick, sub-confluent tubercles arranged in 5-more or loss regular	7. *E. neriifolia* L.
5b (4)	Branches with stipular thorns, arising from low, distant tubercles, spirally arranged on the cylindric branches	8. *E. nivulia* Buch. Ham.
6a (1)	Plants armed	6. *E. milii* Desmoul
6b (1)	Plants unarmed	(7)
7a (6)	Involucral glands 4, winged	(8)
7b (6)	Involucral with a single large glands on one side; floral bract green	4. *E. geniculata* Orteg.
8a (7)	Lower leaves alternate	3. *E. dracunculoides* Lamk.
8b (7)	Leaves all opposite	(9)
9a (8)	Leaves 1-4 cm long	(10)
9b (8)	Leaves generally absent, if present 1 cm long	(11)

10a (9)	Gland with a distinct limb	10. *E. parviflora* L.
10b (9)	Glands with an obscure-limb	5. *E. hirta* L.
11a (9)	Capsule glabrous	9. *E. orbiculata* H.B. & K.
11b (9)	Capsule pubescence or glabrous	(12)
12a (11)	Leaves crenulate; stems usually hispidly hairy	12. *E. thymifolia* L.
12b (11)	Leaves sometime serrate at the wide end; stems usually glabrous	2. *E. chamaesyce* L.

10.1. ***Euphorbia antiquorum*** L. Sp. Pl. 450. 1753; Hook.f. Fl. Brit. India 5: 271.1887; Prain Beng. Plants 2:691.1903 (Rep. ed. 1963). *(Vern. Name: Bajvaran).*

Erect fleshy glabrous shrub or small tree; branches jointed, curving upwards. Leaves obovate-oblong or spathulate, rounded at apex, entire, cuneate at base; stipular spines divaricate, short. Involucres 3-nate, in short peduncled simple or less often compound cymes. Light yellow, lateral ones of the cymes short stout-pedicelled, the central sessile. Capsule; cocci compressed, glabrous. *Fls. & Frts.:* December-April.

Commonly cultivated as a hedge plant or naturalized in the village shrubberies; prefer dry soil as a xerophytic plant.

Field Collection: Sagar Islands, *K.N. - 492.*

Distributions: Throughout the drier and hotter parts of India. Sri Lanka and Sind.

Economic Importances: Latex is poisonous to the fishes; many valuable medicinal properties are attributed to the plants; the bark, juice and roots all used to cure a wide variety of diseases, *e.g.*, rheumatism, toothache, warts and nervous diseases.

10.2. ***Euphorbia chamaesyce*** L. Sp. Pl. 455.1753. *E. prostrata* Ait., Hort. Kew. 2: 139. 1789; Hook.f. Fl. Brit. India 5: 266. 1887.

Annual herb, prostrate. Leaves opposite, oblong or oblong-obovate, rounded at apex, margin pinkish, entire to faintly dentate obtuse or rounded sometimes oblique at base; stipules free or united. Involucre in short axillary leafy racemes like branches, in only one axil of a node, campanulate, glabrous or sub-glabrous. Capsule globose trigonous, angles pubescent. Seeds pale reddish, oblong, 4-angled, transversely wrinkled. *Fls. & Frts.:* February-May.

Rarely in the waste places.

Field Collections: Taki, *K.N. - 1014;* Kalyanpur, *K.N. - 1165.*

Distributions: Introduced into India, early in the 19th Century; common throughout as a weed in gardens and waste lands. A native of Tropical America.

10.3. ***Euphorbia dracunculoides*** Lamk., Encycl. 2: 428.1786; Hook.f. Fl. Brit. India 5: 262. 1887; Prain Beng. Plants 2: 691. 1903 (Rep. ed. 1963). *(Vern. Name: Chhagal Puputi).*

Glabrous, dichotomously, much branched, annual herb. Leaves sessile, linear or linear-lanceolate. Flowers pale yellow; involucre sub-sessile, solitary; glands transversely oblong 2- cornuate; anther cells globose. Capsule smooth; seeds tubercled, depressed. *Fls. & Frts.:* February-August.

A common weed in the cultivated lands.

Field Collection: Tapsia, *K.N. - 1756.*

Distributions: Widely distributed in the tropics.

10.4. ***Euphorbia geniculata*** Orteg., Nov. Rar. Pl. Matr. Dec. 18.1797; Hook.f. Fl. Brit. India 5: 262. 1887; Prain Beng. Plants 2: 691.1903 (Rep. ed. 1963).

Erect or decumbent, annual herb. Leaves variable in shape and size, ovate-elliptic, elliptic oblong to elliptic, long petioled. Involucres small, campanulate in dense corymbose cymes. Capsule glabrous. *Fls. & Frts.:* November-April.

Commonly grow on the moist waste places, along the road-sides and thickets.

Field Collection: Naihati, *K.N. - 1597.*

Distributions: Introduced in India before 1872; A native of Tropical America.

10.5. ***Euphorbia hirta*** L. Sp. Pl. 454. 1753. *E. pilulifera* L. Sp. Pl. 454.1753; Hook.f. Fl. Brit. India 5: 250.1887; Prain Beng. Plants 2: 692. 1903 (Rep. ed. 1963). *(Vern. Name: Kerui).*

Erect, decumbent, prostrate, ascending, hispid herb, branching from the rootstock. Leaves elliptic-oblong, obovate or oblong-lanceolate, opposite, unequal-sided, serrulate, acute or sub-acute often dark green or reddish above. Involucre stalked, many, cyothin axillary and terminal, clustered in dense crowded cymes. Capsules hairy, seeds reddish brown, trigonous. *Fls. & Frts.:* Almost throughout the year.

Gregariously grow in the cultivated lands as weed, in waste lands.

Field Collections: Baruipur, *K.N. - 20;* Maslandapur, *K.N. - 913.*

Distributions: Throughout India. Pantropic.

Economic Importances: It is useful for bronchial affections, cough, asthma and worms. Latex is used for warts.

10.6. ***Euphorbia milii*** Desmoul, Bull. Hist. Nat. Soc. Linn. Bordeaux 1: 27-30, Pl. 1. 1826. *E. splendens* Boj. ex Hook., Bot. Mag. t. 2909. 1929. *E. bojeri* Hook. Bot. Mag. t. 3527.1836.

Small, much-branched, perennial, spiny shrub. Stem angled. Leaves compound, fascicled, spathulate or obovate, obtuse or retuse, mostly apiculate. Flowers crimson in long peduncled, dichotomous cymes; each cyathium closely sub-tendered by 2-broadly ovate, bright red bracts. *Fls. & Frts.:* Almost throughout the year.

Slightly xerophytic habitat, grow on the railwy tracks, road-sides and river banks.

Field Collection: Panihati, *K.N. - 1602.*

Distributions: Cultivated almost throughout India. A native of Madagascar.

Economic Importance: Ornamental plants.

10.7. ***Euphorbia neriifolia*** L. Hort. Cliff. 196. (in part); Hook.f. Fl. Brit. India 5: 255.1887; Prain Beng. Plants 2: 691.1903 (Rep. ed. 1963). *E. ligularis* Roxb. Fl. Ind. 2: 465.1824. *(Vern. Name: Mansa Sij).*

Erect, fleshy glabrous shrub; branches jointed, cylindric, the nodes arranged in 5, more or less spirally twisted ribs, branches 5 angled. Leaves few, cuneate or oblanceolate. Involucres in small compact shortly pedunculate dichotomous cymes. *Fls. & Frts.:* April-June.

Sometime planted in the public gardens.

Field Collections: Specimen not preserved.

Distributions: Throughout India. Burma, Malay Islands and Baluchistan.

Economic Importances: Ornamental garden Plants. Milky latex is used as purgative, rubefacient and expectorant.

10.8 ***Euphorbia nivulia*** Buch. Ham., Trans. Linn. Soc. 14: 286.1825; Hook. f. Fl. Brit. India 5: 255.1887; Prain Beng. Plants 2: 691.1903 (Rep. ed. 1963). *E. neriifolia* Roxb. Fl. Ind. 2: 467.1824. *(Vern. Name: Sij).*

Branches spreading, often whorled, jointed, terete often armed with pairs spines inserted on flat corky-areas. Leaves fleshy, sessile, obtuse, cuneate or obvate 10-30 cm long, deciduous. Involucres yellow usually 3 in short cymes near the ends of the branches; the lateral 2 females, the central 1 male; anther purple; capsule prominently 3-lobed: lobes compressed. *Fls. & Frts.:* Very occasionally flowering occur.

Often grow as hedge plant in the village shrubberies.

Field Collection: Rudranagar/Sagar Island, *K.N. - 498.*

Distributions: Native of the dry and rocky parts of India.

Economic Importances: Latex of the leaves are used as a purgative, and as a cure of rheumatism and ear ache. The pulp of the stem, mixed with green ginger is given to people bitten by mad dogs.

10.9. ***Euphorbia orbiculata*** H.B. & K., Nov Gen Sp. 2: 52.1817. *E. microphylla* Heyne ex Ròth, Nov. Pl. Sp. 229.1821; Hook.f. Fl. Brit. India 5: 252. 1887; Prain Beng. Plants 2: 692.1903 (Rep. ed. 1963). *E. bombaiensis* Santapau, Bull. bot. Soc. Bengal 8: 17.1954. *(Vern. Name: Chhota Kerui).*

Sub-glarbous, prostrate herb. Stamens many, spreading from the roots. Leaves opposite, obliquely oblong. Involucres axillary, lobes triangular, apiculate; capsules glabrous; seeds obtusely 4-angled, blunt at the apex. *Fls. & Frts.:* Almost throughout the year.

Commonly found on waste lands; as a weed of cultivated fields.

Field Collections: Ghoramara/Sagar, K.N.- 412; Bamanpukur, K.N. - 1976.

Distributions: Almost throughout India, Java.

10.10. ***Euphorbia parviflora*** L. Syst. (ed. 10) 2: 1047.1759. *E. hypercifolia* L. Sp. Pl. 454.1753; Hook.f. Fl. Brit. India 5: 249.1887; Prain Beng. Plants 2: 692. 1903 (Rep. ed. 1963). *E. hypercifolia* var. *parviflora* Prain Beng. Plants 2: 691.1903 (Rep. ed. 1963). *(Vern. Name: Dudhi).*

Branched, erect or decumbent, glabrous, annual herb. Leaves variable linear oblong to elliptic ovate. Involucres in numerous small cymes, axillary or terminating, short axillary branches; lobes usually with a minute white or pink petaloid limb; glands petaloid, pink. Capsules pubescent, seeds 4 angled, ellipsoid. *Fls. & Frts.:* Almost throughout the year.

Common weed of the waste lands.

Field Collections: Fultala/Baruipur, *K.N. - 41;* Canning, *K.N. - 135.*

Distributions: Common throughout the hotter parts of India. Sri Lanka, also Tropics of both hemispheres, except Australia and the Pacific Islands.

Economic Importances: Infusion of dried leaves are considered as astringent, useful in dysentery, diarrhoea, irregularity and excessive menstrual flow.

10.11. ***Euphorbia pulcherrima*** Willd. ex Klotz., Otto & Dietr. Allgen. Gartens. 2: 27.1834. *Poinsettia pulcherrima* R. Grali., Edinb. N. Phil. Jorurn. 20: 412.1836; Hook.f. Fl. Brit. India 5: 239.1887. *(Vern. Name: Lalpata).*

Ornamental erect shrub with spreading branches from the base. Leaves scattered, ovate-elliptic to lanceolate, entire or sinuate-toothed or lobed. Cyathia in cymes, involucres greenish, baring a large yellow gland on one side; flowers small, insignificant yellowish. *Fls. & Frts.:* October-March.

Rarely cultivated as an ornamental plant.

Field Collections: Specimen not collected.

Distributions: Almost throughout India. A native of Mexico and Central America.

Economic Importance: Ornamental garden plants.

10.12. ***Euphorbia thymifolia*** L. Sp. Pl. 454.1753; Hook.f. Fl. Brit. India 5: 252.1887; Prain Beng. Plants 2:692.1903 (Rep. ed. 1963). *E. prostrata* Grahm, Cat. 179.1839 *(nom.Ait.) (Vern. Name: Swet-kerui).*

Small, prostrate, hispidly pubescent, annual herb. Leaves small, obliquely oblong, obtuse, crenulate. Involucre minute, axillary, solitary. Capsules erect, pubescent, obtusely keeled, hairy; seed quadrangular, transversely furrowed. *Fls. & Frts.:* Almost throughout the year.

Weed of waste lands and banks of the tanks or hills.

Field Collections: Falta Fort, *MK. Ghosh-898* (CAL).

Distributions: Throughout the hotter parts of India. Pantropic.

10.13. ***Euphorbia tirucalli*** L. Sp. Pl. 452. 1753; Hook.f. Fl. Brit. India 5: 254.1887; Prain Beng. Plants 2: 691.1903 (Rep. ed. 1963). *(Vern. Name: Lanka-sij).*

Fleshy shrub or small tree, branches cylindric, spreading, branchlets smooth, polished, whorled or clustered, erect cylindric. Leaves of the branchlets alternate, linear or linear-oblong, obtuse or rounded, entire. Bracteoles numerous, lacerate. Capsule obtuse, 3-lobed, minutely puberulous. Seed ellipsoid, smooth, glabrous. *Fls. & Frts.:* July-November.

Commonly planted as hedge plant.

Field Collection: Budge Budge, *S.K. Mukherjee-6013* (CAL).

Distributions: Throughout the hotter parts of India. A native of Africa, now planted in most of the Tropical countries.

Economic Importances: Milky latex is used for rheumatism, neuralgia, toothache, cough and asthma.

11. *EXCOECARIA* L.

Key to the Species

1a	Handsome halophytic tree	1. *E. agallocha* L.
1b	Halophytic shrub	2. *E. bicolar* Hassk.

11.1. ***Excoecaria agallocha*** L. Syst. Nat. (ed.10) 1288. 1759; Naskar & Guha Bakshi Mangr. Swamps Sundarbans 112.1987. *(Vern. Name: Gneo).*

Medium tree, upto 15 m high, partially deciduous, with poisonous milky latex, bark-grey. Without any aerial roots, but tidal water current washed away the bottom soil and horizontal underground roots are exposed, look like snake-cable roots. Occasionally large gall like structures are developed on trunk base; these are laterally perforated and help the tree excrete the salt content. Leaves spirally arranged, coriaceous or fleshy, simple, terete, petiole 1-2 cm long, 9 cm long and 6 cm braod, ovate elliptic to obovate, bluntly acuminate. Unisexual flowers in axillary inflorescences. Fruit 3 lobed, schizocarp 1.5-2.0 cm diam., seeds length 0.3 cm; epigeal or modified epigeal germination. *Fls. & Frts.:* June-October.

Mangrove associate; grow both in halophytic intertidal and in saline environment, on silted up ridge forests and tidal river slopes. (Fig. 44 &45).

Field Collections: Surjyapur, *K.N. - 293;* Naraharipur, *K.N. - 467.*

Distributions: Throughout the coastal inter-tidal zones and in Andaman and Nicobar Islands. Mangrove forests and coastal areas of South East Asia and Australia.

Economic Importances: Milky latex is very poisonous; wood is soft and largely used for packing boxes; paper pulp, match-box and splints of matches.

11.2. ***Excoecaria bicolar*** Hassk. Retz. 154. 1855; Naskar & Guha Bakshi Mangr. Swamps Sundarbans 113. 1987. *(Vern. Name: Lata Gneo).*

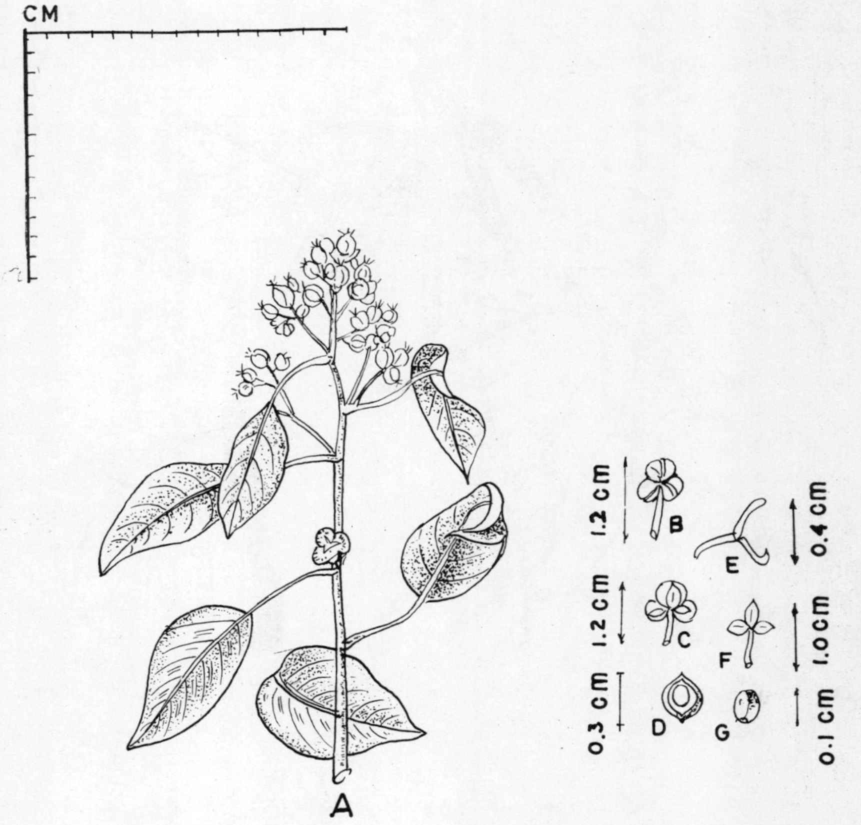

Fig.44. EXCOECARIA AGALLOCHA L. (Female plant)

A. ♀ Flowering twig, with female inflorescence, B. & C. Single fruit, D. T.S. of fruit, E. Stigma head, F. & G. Parts of Flowers.

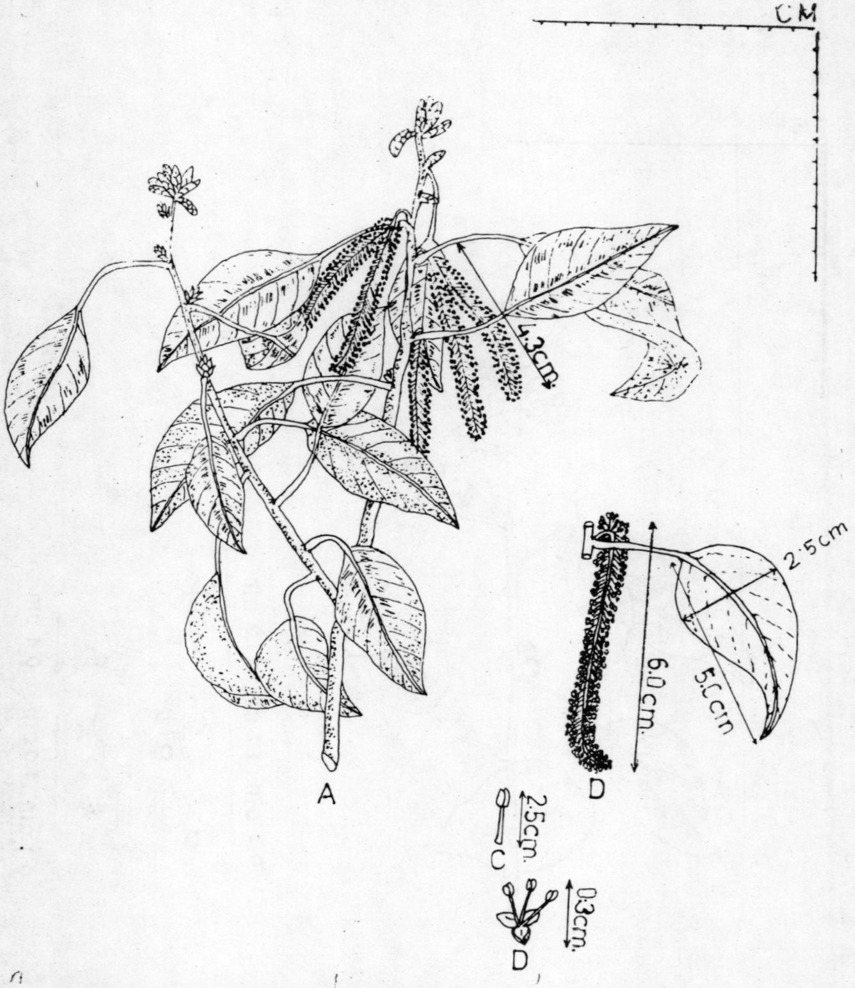

Fig.45. EXCOECARIA AGALLOCHA L. (Male plant)
A. O Flowering twig, with male inflorescence,
B. Single leaf with axillary male inflorescence,
C. Anther lobe—androecium, D. Section of flower

Handsome shrub, upto 1.5 m tall; no aerial roots. Leaves ovate lanceolate, opposite, olive green above and crimson beneath. Flowers very small. *Fls. & Frts.:* November-March.

Occasionally found in the river banks and mangrove forests of the inter-tidal areas of the Sundarbans. It is an associate flora of the Mangrove forest.

Field Collection: Sajnakhali, *K.N. - 2371.*

Distributions: Very rare in the other mangrove forests in India.

Remarks: D. Prain (1903a, 1903b and 1905) did not mention this taxon from the Sundarbans.

12. *JATROPHA* L.

Key to the Species

1a	Flowers greenish yellow	1. *J. curcas* L.
1b	Flowers red	2. *J. gossypifolia* L.

12.1. ***Jatropha curcas*** L. Sp. Pl. 1006. 1753; Hook.f. Fl. Brit. India 5: 383. 1887; Watt Dict. Econ. Prod. 4: 545. 1890; Prain Beng. Plants 2: 705. 1903 (Rep. ed. 1963). *(Vern. Name: Bagh-Bherenda).*

Deciduous shrub, latex watery. Leaves 3-5 lobed. Flowers monoecious, pale yellowish-green with a disagreeable odours in a panicle of cymes. Capsules woody, sub-globose to ellipsoid, black. *Fls. & Frts.:* June-January.

Frequently planted as hedge plant; sometime naturalised in shrubberies.

Field Collections: Canning/Bisonar, *K.N. - 168;* Nabagram. *K.N. - 641.*

Distributions: Introduced into India by the Portuguese, now almost naturalised. A native of Tropical America.

Economic Importances: Seeds are said to be used as a purgative and an illuminating oil.

12.2. ***Jatropha gossypifolia*** L. Sp. Pl. 1006. 1753; Hook.f. Fl. Brit. India 5: 383. 1887; Prain Beng. Plants 2: 705. 1903 (Rep. ed. 1963). *(Vern. Name: Lal-Bherenda).*

Much branched shrub. Leaves 3-5 lobed, lobes elliptic acute, petioles, stipules and leaf margins covered with glandular bristles. Flowers purplish-red, monoecious, in terminal cymes. Capsules 3-lobed, truncate at both ends; seeds greyish-red. *Fls. & Frts.:* June-January.

Grow everywhere on waste lands or village shrubberies and road-sides.

Field Collections: Bhangankhali, *K.N. - 152;* Ultadanga, *Dutta-134* (C.C.H.).

Distributions: Hotter parts of India. A native of Brazil, Tropical America, now a pantropic weed.

Economic Importances: Aquous extract of the plant possesses insecticidal properties. The roots are used in leprosy, also reported to be used as an antidote for snake-bite.

13. *MALLOTUS* Lour.

Key to the Species

1a	Capsule 3-coccous, densely covered with red, resinous glands	1. *M. philippensis* Muell.-Arg.
1b	Capsule 2-coccous, velvety tomentose	2. *M. rependus* Muell. Arg.

13.1. ***Mallotus philippensis*** (Lamk.) Muell.-Arg., Linnaen 34: 196. 1865; Hook.f. Fl. Brit. India 5: 442. 1887; Watt Dict. Econ. Prod. 5: 114. 1891; Prain Beng. Plants 2: 713.1903 (Rep. ed. 1963). *Croton philippense* Lamk. Encycl. 2: 206. 1976. *Rottlera lincotoria* Roxb. Pl. Cor. 2: 36. t. 167. 1798. *(Vern. Name: Tunj).*

Evergreen, much-branched, small to medium sized tree. Leaves alternate, variable in size and shape, ovate, ovate-oblong or lanceolate, acuminate, with a greenish-yellow, glandular pubescent. Flowers small; male flowers clustered in long racemes. Capules 3-lobed; densely covered with red glands. *Fls. & Frts.:* September-April.

Grow abundantly along road-sides and waste lands.

Field Collection: Ramnagar-Jugdia, *K.N. - 218.*

Distributions: Throughout the Tropical India. Sri Lanka, Malay Islands, Australia and China.

Economic Importances: Fruit powder mixing with milk, curd or honey is effective against ascaris and tape-worms for its mild action; it is beneficial to the children; red powder of the seeds is a valuable dye, employed in dyeing silk a bright flame colour.

13.2. ***Mallotus rependus*** (Willd.) Muell.-Arg., Linnaea 34: 197. 1865; Hook.f . Fl. Brit. India 5: 442. 1887; Prain Beng. Plants 2: 713.1903 (Rep. ed. 1963). *Croton rependus* Willd., Neus Schift. Natur.f., Freunde Berlin 4: 206. 1803. *Rottlera dicocca* Roxb. Fl. Ind. 3: 829.1832. *(Vern. Name: Akus).*

Large scandent or sub-scandent dioecious shrub; stem-base throny, young shoots tomentose. Leaves ovate, sometimes sub-peltate, acute truncate or rounded at base, alternate 3-nerved, brown downy tomentose. Male flowers shortly pedicelled, in clusters, in brownish downy-tomentose on terminal or upper axillary paniculate racemes. Capsule didynamous, stellately brownish tomentose. *Fls. & Frts.:* December-April.

Common on the hedges and village thickets.

Field Collection: Garia, *K.N. - 316.*

Distributions: Throughout India. Sri Lanka, Malaya, South China and Formosa.

14. *MICROCOCCA* Benth.

14.1. *Micrococca mercurialis* (L.) Benth. Hook. Niger. Fl. 503.1849; Prain Ann. Bot. 25: 631. 1911. *Tragia mercurialis* L. Sp. Pl. 980.1753. *Claoxylon mercurialis* (L.) Thw. Enum. Pl. Zeyl. 142.1859.

Annual herb, glabrous. Leaves alternate, ovate or elliptic-ovate or ovate-lanceolate, acute or shortly acuminate at apex, margin crenate, acute or truncate at base; slender. Flowers solitary or sometimes 2 on slender 4-6 cm long, racemes. Seeds globose, dark-brown, pitted. *Fls. & Frts.:* April-August.

Rarely on the railway tracks and road-sides.

Field Collection: Ichhapur, *K.N. - 1825.*

Distributions: Throughout India. Sri Lanka, Burma, Arabia and Tropical Africa.

Notes: Prain (1903a) did not report this taxon from Bengal proper.

15. *PEDILANTHUS* Poit. *nom. cons.*

15.1. *Pedilanthus tithymaloides* (L.) Poit., Ann. Mus. Hist. nat. Paris 19: 390, t.19.1812; Hook.f. Fl. Brit. India 5: 239.1887; Prain Beng. Plants 2: 693. 1903 (Rep. ed. 1963). *Euphorbia tithymaloides* L. Sp. Pl. 453. 1753. *(Vern. Name: Rangchita).*

Succulent shrub, stem dark green. Leaves petioled, dark green, fleshy, stipule dark-brown. Cyathium; involucre tube glabrous without pubescent or sub-glabrous within, greenish yellow at base, red above. Capsule grey to deep brown, ovoid or sub-globose, within an acute apex truncate at base. *Fls. & Frts.:* June-October.

Very common as hedge plant in the village shrubberies.

Field Collection: Arbelia, *K.N. - 568.*

Distributions: Almost throughout India. A native of Mexico; introduced and naturalised elsewhere.

Economic Importances: Largely used as garden hedge and the root is powerful emetic.

16. *PHYLLANTHUS* L.

Key to the Species

1a	Sepals 4 (rarely 5-6) in both sexes; stamens 4, free; anthers oblong and slits vertical; styles 4, free; fruit fleshy	1. *P. acidus* (L.) Skeels
1b	Sepals 5-6 in both sexes; stamens either 5 or 3; styles 3	(2)
2a (1)	Stamens 5 in 2 series, the inner filaments connate; anthers erect, slits vertical; styles very minute; stigmas 2-lobed; fruit 4-8 celled berry	4. *P. reticulatus* Poir.
2b (1)	Stamens 3; filaments united in a short or long column	(3)
3a (2)	Anthers erect. slits verticle; connective usually produced	(4)
3b (2)	Anthers didynamous or reniform	(5)
4a (3)	Fruit large, fleshy, with 3 bony, 2 valved cocci	3. *P. emblica* L.
4b (3)	Fruit a capsule of 3 crustaceous, 2-valved cocci	5. *P. urinaria* L.
5a (3)	Stipules peltate, subsagittate	6. *P. virgatus* Forsst.
5b (3)	Stipules simple, not peltate	2. *P. asperulatus* Hutuch.

16.1. ***Phylanthus acidus*** (L.) Skeels USDA Bureau. Fl. Ind. Bull. 148: 17.1909. *Averrhoa acida* L. Sp. Pl. 428. 1753. *Cicca disticha* L. Mant. 1: 124. 1767. *Phyllanthus distichus* (L.) Muell.-Arg. DC. Prodr. 15(2): 413. 1866; Hook.f. Fl. Brit. India 5: 304. 1887; Prain Beng. Plants 2: 701. 1903 (Rep. ed. 1963). *Cicca acida* (L.) Merrill, Interpr. Herb. Amb. 314. 1917; Benthall Trees Calc. 383. 1946. *(Vern. Name: Nore* or *Note).*

Small to medium tree. Leaves distichous, ovate, acute. Flowers reddish, minute, in many flowered panicles. Berries depressed globose, angled, greenish yellow. *Fls. & Frts.:* February-October.

Rarely planted in the gardens, not very common.

Field Collection: Shyamnagar, *K.N. - 1816.*

Distributions: Almost throughout India. A native of Malaya Islands and Madagascar.

Economic Importances: Fruits are edible, acidic in taste; leaves medicinally used in small pox; roots and seeds are cathartic used in Aurveda as a remedy of jaundice.

16.2. ***Phyllanthus asperulatus*** Hutuch., Kew. Bull. 1920: 27-28. 1920. *P. fraternum* Webster, Contr. Gray. Herb. No. 176: 53. 1955. *P. nirui* auct. plun. (non.L.); Hook.f. Fl. Brit . India 5: 298.1887; Prain Beng. Plants 2: 701. 1903 (Rep. ed. 1963). *(Vern. Name: Bhui amla).*

Erect, glabrous, annual herb. Leaves simple, but looks like compound, oblong, distichous, often overlapping. Flowers yellowish, axillary. Capsules depressed-globose, smooth, seeds trigonous. *Fls. & Frts.:* June-November.

Commonly found as an weed of cultivated fields and waste lands.

Field Collection: Dapdhapi, *K.N. - 2302.*

Distributions: Almost throughout India. Pantropic, but probably introduced in the New World.

Economic Importances: Leaves are used in the treatment of jaundice.

16.3. ***Phyllanthus emblica*** L. Sp. Pl. 982. 1753; Hook.f. Fl. Brit. India 5: 289. 1887; Prain Beng. Plants 2: 700. 1903 (Rep. ed. 1963). *(Vern. Name: Amlaki).*

Small to moderate sized deciduous tree. Leaves simple, distichous, resembling pinnate leaves, narrowly linear, obtuse, pale-green, stipule minute. Flowers greenish yellow. Fruits fleshy, globose, pale-yellow, tinged with pink. *Fls. & Frts.:* Almost throughout the year.

Rarely planted in the gardens.

Field Collections: Purandarpur/near Baruipur, *K.N. - 1235.*

Distributions: Throughout India; a native of the Tropical and sub-Tropical India. Sri Lanka and Malaya.

Economic Importances: Fruits are edible; used as medicine; bark and leaves are used in dyeing and tanning, the myrobolan is an important article in Indian medicine–liver tonic, as an aperient.

16.4. ***Phyllanthus reticulatus*** Poir., Ency. Meth. 5: 298.1804; Hook.f. Fl. Brit. India 5: 288. 1887; Watt Dict. Econ. Prod. 6(1): 223. 1892; Prain Beng. Plants 2: 700.1903 (Rep. ed. 1963). *P. multiflorus* Willd. Sp. Pl. 4:581.1805. *Anisonema multiflora* Wight., Icon. t. 1899. *Kirganelia reticulata* Baill. Etud. Gen. Euph. 613.1885. *(Vern. Name: Panjuli).*

Much-branched glabrous shrubs; branchlets somewhat angled. Cataphylls dark brown, triangular, acuminate at apex, margin ciliolate, truncate at base. Leaves chartaceous, elliptic, obtuse or rounded at both ends; stipules dark-brown, lanceolate, acuminate at apex, margin ciliolate, base truncate. Flowers solitary, male-in fascicles of 2-6, female solitary. Seeds finely granulate. *Fls. & Frts.:* March-September.

Common on the thickets and hedges.

Field Collections: Jugdia-Benipur, *K.N. - 227*; Sankchura, *K.N. - 981.*

Distributions: Throughout India. South China and Formosa, southwards through Malaya to Timor and Celebes.

16.5. ***Phyllanthus urinaria*** L. Sp. Pl. 982.1753; Hook.f. Fl. Brit. India 5: 293. 1887; Watt Dict. Econ. Prod. 6(1): 224.1892; Prain Beng. Plants 2: 701.1903 (Rep. ed. 1963). *P. leprocarpus* Wight, Icon.t.1895,f. 4,1852. *(Vern. Name: Hazarmani).*

Annual herb with erect stem. Leaves sessile, distichously imbricate, oblong or linear oblong, tip rounded or apiculate, pale beneath. Flowers monoecious, axillary. Capsule depressed globose, lobed, verrucose, seeds brown, transversely furrowed. *Fls. & Frts.:* June-December.

A common weed in the waste places and even on the cultivated fields.

Field Collections: Jugdia-Benipur, *K.N. - 226;* Dakshineswar. *K.N. - 696.*

Distributions: Almost throughout India. Sri Lanka, Pantropic, probably introduced in the Tropical America.

Economic Importances: Used as diuretic in dropssy and also useful in urino-genital troubles. Act as fish poison; root is useful for the sleepless children.

16.6. ***Phyllanthus virgatus*** Forsst.f. Fl. Ins. Austr. Prodr. 65.1786. *P. simplex* Retz. Obs. Bot. fasc. 5: 29. 1789; Hook.f. Fl. Brit. India 5: 295. 1887;

Watt Dict. Econ. Prod. 6(1): 223. 1892; Prain Beng. Plants 2: 701. 1903 (Rep. ed. 1963).

Erect, ascending deep-rooted, perennial herb. Leaves many, linear-oblong distichous, obtuse and apiculate, sub-sessile, stipules peltate. Flowers whitish, minute from an axillary cluster or minute bracts. Capsule depressed, globose, seeds trigonous. *Fls. & Frts.:* August-January.

Rare weed on the cultivated lands and waste places.

Field Collections: Taki, *K.N. - 1016;* Basanti, *K.N. - 1274.*

Distributions: Almost throughout India. Sri Lanka, South China Malaya and Polynesia.

Economic Importances: Roots are useful for external application of mammary abscesses.

17. *PUTRANJIVA* Wall.

17.1. ***Putranjiva roxburghii*** Wall. Tent. Fl. Nepal 61. 1826; Hook.f. Fl. Brit. India 3: 336. 1887; Watt Dict. Econ. Prod. 6(1): 372. 1892; Prain Beng. Plants 2: 701. 1903 (Rep. ed. 1963). *Drypetes roxburghii* (Wall.) Airy Shaw, Kew. Bull. 1966 & 1968. *(Vern. Name: Putranjiva).*

Moderate to large, evergreen tree. Leaves 5-10 cm long, alternate, obliquely elliptic-oblong or ovate, obtuse, acute or shortly acuminate. Flowers unisexual, yellowish; male flowers in numerous, minute, axillary clusters. Drupes ellipsoid, tomentose. *Fls. & Frts.:* November-March.

Commonly planted as ornamental avenue tree.

Field Collection: Naihati, *K.N. - 1564.*

Distributions: Almost throughout in India, both wild and planted; Sri Lanka.

Economic Importances: Stones of the fruits are made into neckless and put round the necks of children to save or preserve them from the evil.

18. *RICINUS* L.

18.1. ***Ricinus communis*** L. Sp. Pl. 1007. 1753; Hook.f. Fl. Brit. India 5: 457. 1887; Watt Dict. Econ. Prod. 6(1): 506. 1892; Prain Beng. Plants 2: 701. 1903 (Rep. ed. 1963). *(Vern. Name: Redi* or *Rerri).*

Evergreen small tree. Leaves alternate, palmately lobed, large. Flowers monoecious, cream-coloured, in erect terminal branching. Capsule

3-angled, oblong. Seeds 3, oblong, spotted. *Fls. & Frts.:* Almost throughout the year.

Rarely planted, but very often as escape in the garden-edges and waste lands, near the gardens.

Field Collections: Atghara/Madarhat, *K.N. - 59;* Baduria, *K.N. - 318.*

Distributions: Almost throughout India; introduced into India at an early time. A native of Abyssinia.

Economic Importances: The oil extracts from its seeds are purgative, also as a lubricate and an illuminant. The leaves are used for cure of boils.

19. *SAPIUM* R. Brown

Key to the Species

1a	Leaves short-petioled; serrate, green beneath	1. *S. indicum* Willd.
1b	Leaves long-petioled; entire, glaucous beneath	2. *S. sebiferum* Roxb.

19.1. ***Sapium indicum*** Willd. Sp. Pl. 4: 572.1805; Naskar & Guha Bakshi Mangr. Swamps Sundarbans 113.1987. *Excoecaria indica* (Willd.) Muell.-Arg. Linnaea 32: 123.1863. *(Vern. Name: Hurnabatul.).* (Fig.46).

Small glabrous thorny trunk; evergreen tree upto 15 m high. No aerial roots. Leaves coriaceous, elliptic or elliptic lanceolate or less commonly acute at apex, acute or obtuse at base, regularly crenulate margin. Spikes terminal solitary; male flowers crowded, sub-sessile. Female flowers large, pedicelled. Fruit almost globose, smooth, capsule fruit upto 2.5 cm in diam. *Fls. & Frts.:* April-July.

A back-mangrove species; occasionally grow on the ridge forests of Sundarbans; rarely near the road-sides of the village shrubberies.

Field Collection: Rangabelia, *K.N. - 1512.*

Distributions: Coastal and deltaic regions of India.

Economic Importances: Leaves produce greenish-yellow dye. Fruits are poisonous to fishes. Stem bark is used for insanity and hydrophobia.

19.2. ***Sapium sebiferum*** Roxb. Fl. Ind. 3: 693. 1832; Hook.f. Fl. Brit. India 5: 470. 1887; Watt Dict. Econ. Prod. 6(2): 472.1893; Prain Beng. Plants 2: 714. 1903 (Rep. ed. 1963). *Excoecaria sebifera* Muell.-Arg. DC. Prodr. 15 (2) 1210. 1866. *(Vern. Name: Manchina).*

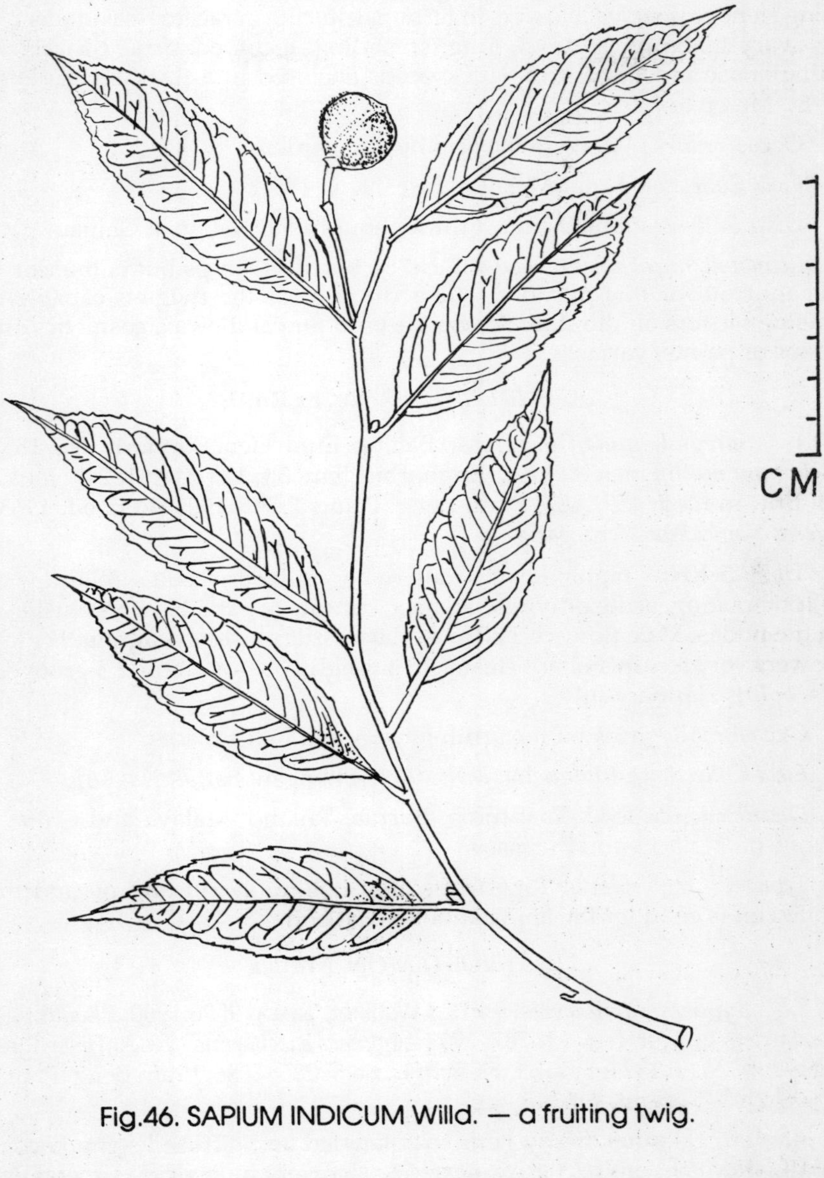

Fig.46. SAPIUM INDICUM Willd. — a fruiting twig.

Small to moderate sized evergreen tree. Leaves rhomboid, entire on long slender petioles. Flowers in terminal spiciform racemes; stamens 2 to 3; ovary glabrous; styles 3, long, spreading, recurved. Fruit coriaceous, subglobose indehiscent capsules; seeds enclosed in a thick layer. *Fls. & Frts.:* March to July.

Occasionally planted in the ornamental gardens.

Field Collection: Sagar Mela Premises, *K.N. - 451.*

Distributions: Planted almost throughout India. Native of China.

Economic Importances: Seed oil can be burnt in lamps but is inferior to coconut oil for this purpose. The acrid juice of the twigs is capable of raising blisters on the skin. Seed oil is used for candlewax, cosmetics and soaps or paints/varnishes.

20. *SUREGADA* Roxb. ex Rottl.

20.1 ***Suregada multiflora*** (Juss.) Baillon, Etud. Gen. Euphorb. 396. 1858. *Gelonium multiflorum* A. Juss., Euphorbia Tent. 3.t. 10.f.31A. 1824; Hook.f. Fl. Brit. India 5: 459.1887; Prain Beng. Plants 2: 702.1903 (Rep. ed. 1963). *(Vern. Name: Ban-Naranga).*

Tree 5-10 m high. Leaves alternate, sub-coriaceous, elliptic or elliptic-oblong, acute or obtuse at apex, cuneate at base. Flowers clustered in the nodes. Male flowers 5-20 in axillay sessile clusters, fragrant. Female flowers very few in axillary cluster. Fruit slightly 3-gonous and 3-grooved. *Fls. & Frts.:* January-July.

Occasionally grow on the scrub-jungles and moist places.

Field Collections: Malancha, *K.N. - 774;* Bhangankhali, *K.N. - 161.*

Distributions: North-East India. Burma, Tailand, Malaya and China a native of Malacca and Burma.

Economic Importances: Ripe fruits after boiling in mustard oil and the boiled oil is good for the application to heat boil.

21. *SYNOSTEMON* Muller

21.1. ***Synostemon bacciforme*** (L.) Webster, Taxon 9: 26.1960. *Phyllanthus bacciformis* L. Syst. (ed. 13) 707. 1774. *Agyneia bacciformis* (L.) A. Juss. Tent. Euphrob. 24.t.6.1824; Hook.f. Fl. Brit. India 5: 285. 1886; Prain Beng. Plants 2: 698.1903 (Rep. ed. 1963).

Annual, glabrous, fleshy herb with tangled and diffused stem. Leaves small, alternate, entire, fleshy, nerveless. Flowers monoecious apetalous, males in axillary clusters; female solitary larger. Capsule splitting into 3, 2-valved cocci. *Fls. & Frts.:* Throughout the year, mostly in November-March.

Common on the dampy places.

Field Collection: Sagar Island, *A.K. Mukherjee - 5930.* (CAL).

Distributions: Almost throughout India.

22. *TRAGIA* L.

Key to the Species

1a	Climbing shrub	1. *T. hispida* Willd.
1b	Stout herb	2. *T. involucrata* L.

22.1. ***Tragia hispida*** Willd. Sp. Pl. 4: 323. 1805. *T. involucrata* sensu Hook.f. Fl. Brit. India 5: 465. 1887 (pro-parte); Prain Beng. Plants 2: 714.1903 (Rep. ed. 1963). *T. involucrata* var. *augustifolia* Hook.f. Fl. Brit. India 5: 465.1887.

Mostly climber, young shoot villous, purberulous on maturity. Leaves lanceolate or linear lanceolate, acuminate at apex, in entire or toothed only at the base, cordate at base, 3-5 nerved from the base. Flowers terminal or leaf-opposed bracteate racemes. Capsules 3-lobed. Seeds globose, smooth. *Fls. & Frts.:* August-March.

Common in the shrubberies along the sides of the road or railway tracks.

Field Collection: Goranbose, *K.N. - 1874.*

Distributions: Throughout in India. Burma, Sri Lanka, China and S.E. Asia.

22.2. ***Tragia involucrata*** L. Sp. Pl. 980. 1753; Hook.f. Fl. Brit. India 5: 465. 1887 (Pro-parte); Watt Dict. Econ. Prod. 6(4): 71. 1893; Prain Beng. Plants 2: 714. 1903 (Rep. ed. 1963). *(Vern. Name: Bichhuti).*

Stout herb. Leaves variable, ovate-oblong, elliptic or rhomboid, acuminate, sharply serrate, hairy. Flowers yellowish, usually in racemes; male flowers-sepals 3, stamens 3. Capsule 3-lobed. *Fls. & Frts.:* November-April.

Commonly grow on waste lands; its touch have burning sensation to skin.

Field Collections: Baruipur, *K.N. - 21;* Canning, *K.N. - 196.*

Distributions: Throughout India. Sri Lanka, South-East Asia.

23. *TREWIA* L.

23.1. ***Trewia nudiflora*** L. Sp. Pl. 1193. 1753; Naskar & Guha Bakshi Mangr. Swamps Sundarbans 114. 1987. *(Vern. Name: Petali).*

Medium tree, deciduous. Common on salty soils of village surrounding of the Sundarbans; Naskar & Guha Bakshi (1.c.) described this taxon after collecting from Gosaba *(K.N. - 658)* and Bhebe/near Basirhat *(K.N. - 720).*

CXIX. ULMACEAE

(Splitted out from Urticaceae)

1. *TREMA* Lour.

1.1. ***Trema orientalis*** (L.) Blume, Mus. Bot. Lugd.-Bat. 2: 62. 1856; Hook.f. Fl. Brit. India 5: 484. 1888; Watt Dict. Econ. Prod. 6(4): 75. 1893; Prain Beng. Plants 2: 720. 1903 (Rep. ed. 1963). *Celtis orientalis* L. Sp. Pl. 1044. 1753; *(Vern. Name: Jilan* or *Chikan).*

Small to medium evergreen tree. Leaves obliquely ovate, ovate-lanceolate, acuminate, crenate serrulate, rough on the upper surface. Flowers greenish white, in lax spreading cymes. Drupes ovoid glabrous, black when ripe. *Fls. & Frts.:* Almost throughout the year.

Common on the road-side, railway tracks, waste lands, village thickets and village shrubberies.

Field Collections: Dakshin Chtra, *K.N. - 519;* Haroa, *K.N. - 599.*

Distributions: Almost throughout India. A native of most of the damper parts of India, including Bengal. Nepal, Sikkim Himalayas, Sri Lanka, Malaya Islands, South China, South to North eastern Australia.

Economic Importances: Said to be used in epilepsy; fruits are sweet and edible, but very rarely.

CXX. URTICACEAE

Key to the Genera

1a	With stinging hairs	2. *FLEURYA* Gaud. *nom. cons.*
1b	Without stinging hairs	(2)
2a (1)	Leaves opposite, less than 1 cm in length	3. *PILEA* Lindl. *nom. cons.*
2b (1)	Leaves alternate, more than 1 cm in length	(3)

3a (2)	Leaves oblique; female perianth 3	1. *Elatostema* Forsst.
3b (2)	Leaves not oblique; female perianth	4. *Pouzolzia* Gaud.

1. *ELATOSTEMA* Forsst.

1.1. ***Elatostema cuneatum*** Wight Ic. Pl. Ind. Or. 6: 35. t. 2991, f. 3.1853; Hook.f. Fl. Brit. India 5: 568.1888.

Annual herb, pubescent 10-20 cm high; stem much branched, often rooting near the base. Leaves sub-sessile, falcately cuneate, tip round or sub-acute. Female heads sessile, solitary, at each node; perianth segments of female 3, very short. Achenes ellipsoid, ribbed, yellow. *Fls. & Frts.:* July-October.

Rarely grow on the shady areas.

Field Collection: Beliachandi, *K.N. - 1937.*

Distributions: Sikkim Himalaya and W. Peninsula. Throughout the Tropics.

Notes: D. Prain (1903a) did not mention this taxon from Bengal.

2. *FLEURYA* Gaud. *nom. cons.*

2.1. ***Fleurya interrupta*** (L.) Gaud., Freyc. Voy. Bot. 497.t. 83.1826; Hook.f. Fl. Brit. India 5: 548. 1888; Prain Beng. Plants 2: 721. 1903 (Rep. ed. 1963). *Urtica interrupta* L. Sp. Pl. 985. 1753. *Laportea interrupta* (L.) Chew., Gard. Bull. 21: 200. 1965. *(Vern. Name: Lal-Bichuti).*

Annual erect herb with stinging hairs. Leaves alternate, ovate, acuminate and toothed. Flowers greenish white, dioecious, aggregated on the elongated axis. Achines oblique and compressed. *Fls. & Frts.:* July-November.

Common on the dampy waste places; road-sides and even on the old wall. Cause painful irritation on contact of this plant.

Field Collections: Madhyampur/near Taki, *K.N. - 949.*

Distributions: Throughout India; Khasi Hills and Deccan Peninsula. Sri Lanka, Malaya, Pacific Islands, China and Abyssinia.

Notes: Chew (1.c.) has treated *Fleurya* Gaud. under the synonyme *Laportea* Gaud. as there is no marked differences between the two.

3. *PILEA* Lindl. *nom. cons.*

3.1. ***Pilea microphylla*** (L.) Liebm., Vidensk. Selsk. Skr. 5(2): 296.1851; Hook.f. Fl. Brit. India 5: 551. 1888; Prain Beng. Plants 2: 722.1903 (Rep. ed. 1963). *Perietaria microphylla* L. Syst. (ed. 10) 1309.1759. *Pilea muscosa* Lindl. Coll. Bot. t. 4. 1821; Hook.f. Fl. Brit. India 5: 551. 1888. *Urtica minima* Roxb. Fl. Ind. 3: 85. 1832. *(Gun-powder Plant).*

Very minute, erect or ascending weed. Stem often creeping at base, simple or branched. Leaves elliptic or oblanceolate, acuminate at apex, entire, petiole filiform shorter than the acuminate at apex, entire, petiole filiform, shorter than the blade. Flowers minute in biparius cymes on short stalked clusters. Achenes between ellipsoid smooth, glabrous. *Fls. & Frts.:* July-August.

Everywhere common on the dampy areas and on old brick wall.

Field Collection: Dakshin Barasat, *K.N. - 1649.*

Distributions: Almost throughout India, naturalized throughout. A native of south America.

Economic Importances: Useful for the stomach and intestine trouble and leaves are also useful for sores and bruises.

4. *POUZOLZIA* Gaud.

4.1. ***Pouzolzia zeylanica*** (L.) Burm., Pl. Jav. Rar. 67.1828; *Perietarin zeylanica* L. Sp. Pl. 1052.1753. *P. indica* L. Mant. 128.1767. *Pauzolzia indica* Gaud. Freye. Voy. Bot. 503. 1826; Hook.f. Fl. Brit. India 5: 581. 1888; Prain Beng. Plants 2: 724.1903 (Rep. ed. 1963). *Urtica suffruticosa* Roxb. Fl. Ind. 3: 584.1832.

Small, erect or prostrate perennial herb. Leaves ovate, petiolate with obscure veins; stipules free, persistent, alternate, base acute. Flowers dioecious, axillary, minute in biperaous cymes. Achenes broadly ovoid, shining. *Fls. & Frts.:* July-August.

Very common weed on the waste places and garden edges.

Field Collections: Dhopagachi, *K.N. - 38;* Benipur, *K.N. - 223.*

Distributions: Throughout India. A native of Tropical America; Sri Lanka, Malaya Islands and China.

Economic Importances: Roots are edible. Leaves poultice is used to sores, boils and stomach troubles.

CXXI. MORACEAE

(Splitted out from URTICACEAE)

Key to the Genera

1a	Anthers in bud reversed; stamens inflexed; ovules pendulous, anatropous	(2)
1b	Anthers in bud erect; style undivided or 2-fid; flowers unisexual	(4)
2a (1)	Style undivided, elongated; male flowers spicate, female in globose heads; achenes stipitate	2. *BROUSSONETIA* Vent
2b (1)	Style 2-fid or 2-partite	(3)
3a (2)	Female sepals fleshy in fruit and enclosing the achenes	4. *MORUS* L.
3b (2)	Female sepals not fleshy; female sepals coriaceous, clasping the ovary	5. *STREBLUS* L.
4a (1)	Flowers all exposed	1. *ARTOCARPUS* Forsst.
4b (2)	Flowers monoecious or rarely dioecious on the inner walls of closed receptacle	3. *FICUS* L.

1. *ARTOCARPUS* Forsst.

Key to the Species

1a	Large evergreen tree, glabrous, stipules large, glabrous; fruits large hanging on short stalk from the trunk or from the branches, anthocarpus pyramidal	1. *A. heterophyllus* Lamk.
1b	Large deciduous tree, above glabrous or puberulous; stipules not like above, fruit smaller, not hanging from the trunk like above, with smooth or winged surface	2. *A. lakoccha* Roxb.

1.1. ***Artocarpus heterophyllus*** Lamk. Encycl. 3: 210. 1789. *Artocarpus integrifolia* Linn.f. Suppl. 412.1781; Hook.f. Fl. Brit. India 5: 541.1888; Watt Dict. Econ. Prod. 1: 330. 1889; Prain Beng. Plants 2: 729. 1903 (Rep. ed. 1963). *(Vern. Name: Kanthal).*

Large evergreen tree, glabrous, except the youngest shoots. Leaves elliptic or obovate, acuminate, entire; stipules; large glabrous. Fruits large hanging on short stalked from the trunk and large branches; anthrocarpus pyramidal. Seeds oblong or reniform only. *Fls. & Frts.:* Throughout the year.

Very common throughout the lower-Ganga delta. Sometimes self grown in the orchards or the village areas.

Distributions: Almost throughout India; a native of the India and Malaya. Throughout the Tropical Countries.

Field Collection: Amtala, *K.N. - 370.*

Economic Importances: Green fruits are cooked as vegetable, while the yellow pulp of the fruits are eaten in ripening. The tribal people of Tripura and Manipur fully depend upon the ripe fruits as their staple food during the season. The ripe pulp of the fruits are full of alcoholic properties. The wood is extensively used to prepare furnitures and building materials. Green leaves are much liked by goat and ships.

1.2. ***Artocarpus lakoccha*** Roxb. Fl. Ind. 3: 524.1832; Hook.f. Fl. Brit. India 5: 543.1888; Watt Dict. Econ. Prod. 1: 333. 1889; Prain Beng. Plants 2: 729.1903 (Rep. ed. 1963). *(Vern. Name: Dephal.).*

Deciduous tree. Leaves thinly coriaceous, above glabrous or puberulous, reticulate beneath. Flower-heads shortly peduncled, pubescent. Fruits globose, sub-lobed smooth or wrinkled glabrous. *Fls. & Frts.:* March-November.

Frequently planted or naturalized in the village shrubberies.

Field Collection: Chemagari/Sagar Islands, *K.N. - 458.*

Distributions: Throughout India. A native of South-East Asia.

Economic Importances: Ripe fruits are edible, but very low qualities; wood used as fire-wood; the bark is a remedy for skin-diseases and fruits are regarded as liver tonic.

2. *BROUSSONETIA* Vent.

2.1. ***Broussonetia papyrifera*** Vent, Table Regne Veg. 3: 547; Hook.f. Fl. Brit. India 5: 490. 1888; Prain Beng. Plants 2: 726.1903 (Rep. ed. 1963).

Small tree. Leaves dentate, often lobed, membranous, petiolate. Male spike cylindric, peduncled, pubescent. Female heads longer peduncled, globose,tomentose. Fruit fleshy, red, shining. *Fls. & Frts.:* January-March.

Occasionally planted.

Field Collection: Specimens not preserved.

Distributions: Throughout India. A native of Burma, Malaya and Pacific Islands.

3. *FICUS* L.

Key to the Species

1a	Climbing shrub; leaves variously broad and lobed	4. *F. heterophylla* Linn.f.
1b	Erect shrubs or trees	(2)
2a (1)	Leaves with very many fine parallel side nerves	3. *P. elastica* Roxb.
2b (1)	Leaves not as above	(3)
3a (2)	Receptacle smaller, less than 2 cm	(4)
3b (2)	Receptacle larger, more than 2.5 cm across	2. *F. benjamina* var. *comosa* Kurz
4a (3)	Male, female and gall flowers in the same receptacle	(5)
4b (3)	Male and gall flowers in separate receptacle to the female	(8)
5a (4)	Leaves tomentose or pubescent beneath	1. *F. benghalensis* L.
5b (4)	Leaves glabrous	(6)
6a (5)	Leaves elliptic or oblong	11. *F. virens Ait.*
6b (5)	Leaves sub-rotund, 10-12.5. cm across	6. *F. microcarpa* Linn.f.
6c (5)	Leaves ovate	(7)
7a (6)	Leaves gradually acuminate, rarely caudate	9.*F. rumphii* Blume
7b (6)	Leaves caudate with cuspidate	8. *F. religiosa* L.
8a (4)	Stamens 1	(9)
8b (4)	Stamens 2-3; leaves not lobed	7. *F. racemosa* L.
9a (8)	Leaves opposite	5. *F. hispida* Linn.f.
9b (8)	Leaves alternate	10. *F. semicordata* Buch. Ham. ex Smith.

3.1. ***Ficus benghalensis*** L. Sp. Pl. 1059. 1753; Hook.f. Fl. Brit. India 5: 499. 1888; Prain Beng. Plants 2: 735. 1903 (Rep. ed. 1963). *F. indica* L. Sp. Pl. 1060. 1753. *F. contonaefolia* Vahl, Enum. Pl. 2: 189.1806. *(Vern. Name: Bat* or *Bar).*

Large, evergreen tree. Leaves coriaceous. Both male and female and gall flowers enclosed in hypanthodium. Receptacle fleshy, orange-red when ripe. *Fls. & Frts.:* June-March.

Occasionally planted along road-sides and in the vicinity of temples. Seedling of this plant commonly sown by bird droppings in the crevices of old buildings.

Field Collection: Canning, *K.N. - 170.*

Distributions: Native of India, wild in the sub-Himalayan forests and on the lower slopes of the Deccan Hills. Tropical World.

Economic Importances: Hindus worship this plant and have a common religious belief; used as shade plant along the road sides.

3.2. ***Ficus benjamina*** L. var. ***comosa*** Kurz. For. Fl. 2: 446; Hook.f. Fl. Brit. India 5: 508. 1888; Prain Beng. Plants 2: 734.1903 (Rep. ed. 1963). *F. comosa* Roxb. Cor. Pl. II.t.125. *(Vern. Name: Pakur).*

Trees, branches drooping. Leaves ovate, thinly coriaceous. Receptacles sessile in axillary pairs, globose or ovoid, smooth and blood-red when ripe, *Fls. & Frts.:* April-August.

Occasionally planted as avenue tree or in the lawns; also naturalised.

Field Collection: Bamanpukur/Minakhan, *K.N. - 1077.*

Distributions: Eastern Himalaya, Assam, Deccan Peninsula and Andaman Islands. Bangladesh, Burma, Sumatra, Celebes and Timor.

Economic Importances: Leaves decoction after mixing with oil is said to be useful as dressing for ulcers and milky juice is used for treating the eye of young babies.

3.3. ***Ficus elastica*** Roxb. Fl. Ind. 3: 541. 1832; Hook.f. Fl. Brit. India 5: 506. 1884; Watt Dict. Econ. Prod. 3: 350. 1890. *(Vern. Name: Attah-bar).*

Large tree, usually epiphytic. Leaves shortly petioled, coriaceous shining oblong to elliptic rather abruptly, obtusely caudate, entire, base rounded or narrowed. Male flowers scattered over interior of receptacle, pedicelled; sepals 4, ovate, anther ovate, sessile. *Fls. & Frts.:* February-April.

Rarely planted in gardens as ornamental plants.

Field Collection: Bahuru, *K.N. - 1630.*

Distributions: Sikkim Himalaya eastwards. Khashi Hills and Assam. Burma and Malayan Archipelago.

Economic Importances: The milky white sap provide the material from which erasers are made.

3.4. ***Ficus heterophylla*** Linn.f. Suppl. 442. 1781; Hook.f. Fl. Brit. India 5: 532.1888; Watt Dict. Econ. Prod. 3: 353. 1890; Prain Beng. Plants 2: 736.1903 (Rep. ed 1963). *F. acutiloba* Miq., Hook. Lond. Journ. Bot 7: 227. 1848. *(Vern. Name: Gaori-Sheora).*

Scabrid or pubescent, creeping shrub; branchlets shortly pubescent. Leaves polymorphous, petiolate, ovate or orbicular-ovate to lanceolate, more or less acuminate. Receptacle pedunculate, solitary, axillary, spherical or pyriform. Achenes sub-globose. *Fls. & Frts.:* October-May.

Occasionally on the dampy area or near the water courses or in the hedges.

Field Collection: Budge Budge, *M.K. Ghosh-2626* (CAL).

Distributions: Throughout India. Sri Lanka and Malaya Islands.

3.5. ***Ficus hispida*** Linn.f. suppl. 442. 1781; Hook.f. Fl. Brit. India 5: 522. 1888; Watt Dict. Econ. Prod. 354.1890; Prain Beng. Plants 2: 736. 1903 (Rep. ed. 1963). *F. oppositifolia* Roxb. Cor. 124 & Fl. Ind. 3: 561. 1832. *(Vern. Name: Dumur).*

Hispid shrub or small tree. Leaves opposite, elliptic or obovate. Receptacle hairy, yellowish when ripe. Achene ovoid. *Fls. & Frts.:* March-July.

Common on the village hedges and thickets, everywhere, as weeds.

Field Collection: Piyalinagar/Baruipur. *K.N. - 101.*

Distributions: Common throughout India. Malay Islands, China.

Economic Importances: Fruits are eaten as vegetable and has some medicinal values, poor people cook the green fruits during crisis; the plant commonly planted as hedge plant in the village gardens.

3.6. ***Ficus microcarpa*** Linn.f. Suppl. 442. 1781. *F. retusa* L. Mant. 129. 1767; Hook.f. Fl. Brit. India 5: 511.1888; Watt Dict. Econ. Prod. 3: 360. 1890; Prain Beng. Plants 2: 735. 1903 (Rep. ed. 1963). *(Vern. Name: Jir-Kamrup).*

Large spreading evergreen tree, with a few aerial roots. Leaves blunt small. Receptacle in pairs at the base of leaf stalk, usually white or purple when ripe. *Fls. & Frts.:* February-May.

Rare on the village shrubberies and in the Sundarbans area.

Field Collection: Tapsia, *K.N. - 1755.*

Distributions: Himalaya, Khasia Hills, Assam and Bengal, Burma, Malaya Islands, China and New Caledonia.

Economic Importances: Leaves and barks are applied medicinally to wounds and bruises and for headache.

3.7. ***Ficus racemosa*** L. Sp. Pl. 1060. 1753. *F. glomerata* Roxb. Fl. Ind. 3: 558. 1832; Hook.f. Fl. Brit. India 5: 535. 1888; Watt Dict. Econ. Prod. 3: 351. 1890; Prain Beng. Plants 2: 737. 1903 (Rep. ed. 1963). *Covellia glomerata* Miqs., Hook. Lond. Journ. Bot. 7: 465. 1848. *(Vern. Name: Jagya Dumur).*

Medium to large evergreen tree. Leaves ovate-oblong or elliptic-lanceolate, entire, glabrous on both surface when mature, base acute. Receptacle shortly pedunculate on short leafless warted branches. *Fls. & Frts.:* March-August.

Sometimes in the village shrubberies and on the embankment of village ponds or in the thickets.

Field Collection: Satzelia, *K.N. - 2163.*

Distributions: Throughout India; indigenous in most parts of India, Sri Lanka, Burma and Malaya.

Economic Importances: Leaves, barks and the fruits are used medicinally; the sap is also used especially as an application in case of mumps.

3.8. ***Ficus religiosa*** L. Sp. Pl. 1059.1753: Hook. f Fl. Brit. India 5: 513. 1888; Watt Dict. Econ. Prod. 3: 357. 1890; Prain Beng. Plants 2: 735. 1903 (Rep. ed. 1963). *(Vern. Name: Aswatha).*

Large, glabrous tree. Leaves long-petioled, abruptly cordate, coriaceous shining above, acuminate, forming a tail about 1/3 rd of the blade. Receptacle in axillary pairs, sessile, smooth, depressed spherical. *Fls. & Frts.:* March-June.

Common in the village shrubberies, jungles; sometime epiphytic in nature; grow on the old walls or on the tree trunks.

Field Collection: Jugdia-Benipr, *K.N. - 221.*

Distributions: Wild in the sub-Himalayan Forests, in Bengal and in the Central India. Planted in Sri Lanka.

Economic Importances: A good shade plant and the travellers relax themselves under its shade during summer. Hindus worship the tree and nurshing both *F. religiosa* L. and *F. benghalensis* L. together and observe a marriage cermony taking in mind the former as Bride and the latter as the Bridge- groom.

3.9. ***Ficus rumphii*** Blume, Bijd. 437.1825; Hook.f. Fl. Brit. India 5: 512. 1888; Prain Beng. Plants 2: 735.1903 (Rep. ed. 1963). *F. cordifolia* Roxb. Fl. Ind. 3: 548. 1832. *(Vern. Name: Gaiaswattha).*

Large tree, often epiphytic. Leaves long-petioled, sub-coriaceous, minutely tuberculate above, broadly ovate, gradually acuminate, margin entire sub-undulate, base broad, slightly narrowed towards the petiole. Receptacles sessile in axillary pairs, black when ripe. *Fls. & Frts.:* Throughout the year.

Common on the Sundarbans areas and village shrubberies of L. Ganga.

Field Collection: Nurpur, *K.N. - 1935.*

Distributions: Punjab, Northern, Western and Central India and Assam. Burma, Malay Peninsula and Malay Islands.

Economic Importances: The juice and fruits are reputed to have medicinal quantities as emetics and anthelmintics.

3.10. ***Ficus semicordata*** Buch. Ham ex Smith, Rees Cyclop. 14: 71. 1810. *F. cunia* Buch Ham. Roxb. Fl. Ind. 3: 561. 1832; Hook.f. Fl. Brit. India 5: 523.1888; Prain Beng. Plants 2: 736.1903 (Rep. ed. 1963). *(Vern. Name: Jagya Dumur).*

Trees. Leaves alternate, oblong or elliptic lanceolate, serrate to crenate, scabrid; petiole short. Receptacle in pairs or in clusters on drooping, mostly leafless branches, sometimes near the base of the trees or form longer branches 1-2 cm across. *Fls. & Frts.:* July-November.

Occasionally on the village shrubberies.

Field Collection: Beliachandi, *K.N. - 265.*

Distributions: Central India. Burma, Yunnan, Tonkin, Malay and Thailand.

3.11. ***Ficus virens*** Ait. Hort. Kew 3; 451. 1789. *F. lutescens* Blume. *Ficus infectoria* Roxb. Fl. Ind. 3: 551.1832; Hook.f. Fl. Brit. India 5: 515.1888; Prain

Beng. Plants 2: 736.1903 (Rep. ed. 1963). *F. lucescens* Blume, Bijdr. 444. 1825. *(Vern. Name: Pakur).*

Large spreading deciduous tree; all parts glabrous; bark grey, smooth, scaly. Leaves membranous, ovate or ovate-oblong, base usually rounded, slightly cordate. Receptacle axillary, in pairs, sessile, globose whitish, flushed with red and dotted when ripe. *Fls. & Frts.:* Almost throughout the year.

Occasionally on the village shrubberies and as an avenue tree.

Field Collection: Durgapur, *K.N. - 1720.*

Distributions: Plains and low-hills of India from the salt range of Sikkim, Bengal and Assam. Burma, Ceylon and Malaya.

4. *MORUS* L.

4.1. ***Morus indica*** L. Sp. Pl. 986.1753; Hook.f. Fl. Brit. India 5: 492. 1888; Watt Dict. Econ. Prod. 5: 281. 1891; Prain Beng. Plants 2: 726. 1903 (Rep. ed. 1963). *(Vern. Name: Tut).*

Deciduous, monoecious tree, with milky juice and grey white bark. Leaves lobed, rarely entire. Flowers in catkins. Fruits white or red. *Fls. & Frts.:* February-March.

Rarely planted.

Field Collection: Dandirhat/Taki, *K.N. - 989.*

Distributions: Planted in Punjab, North-West Himalaya. Western Tibet, Afghanistan, North and West Asia.

Economic Importances: The leaves are very much liked by the silkworm. It is very essential plant for sericulture.

5. *STREBLUS* L.

5.1. ***Streblus asper*** Lour., Fl. Cochin. 2: 615. 1790; Hook.f. Fl. Brit. India 5: 489. 1888; Prain Beng. Plants 2: 721. 1903 (Rep. ed. 1963). *Trophis asper* Retz., Obs. 5: 30. 1788. *(Vern. Name: Sheora).*

Rigid shrub with milky juice; branches tomentose or pubescent. Leaves rigid, elliptic, rhomboid, ovate or obovate, irregularly toothed. Flowers dioecious, minute. Fruit pisiform, 1-seeded, sub-globose, berry. *Fls. & Frts.:* February-July.

Common on the hedges and village shrubberies, everywhere.

Field Collection: Budge Budge, *K.N. - 754.*

Distributions: Drier parts of India and Andaman & Nicobar Island. Penang, Malay Islands, Cochin, China and Siam.

Economic Importances: Decoction of the leaves applied in leucoderma, it is largely used in the village to rippening fruits. Wood is strong and used for agricultural tools handles and as fire woods.

CXXII. CANNABINACEAE

1. *CANNABINA* L.

1.1. ***Cannabina sativa*** L. Sp. Pl. 1027.1753; Hook.f. Fl. Brit. India 5: 487.1888; Watt Dict. Econ. Prod. 2: 103.1889; Prain Beng. Plants 2: 720.1903 (Rep. ed. 1963). *C. indica* Lamk. Encyc. Method. 1: 695.1783. *(Vern. Name: Ganja, Bhang* or *Siddhi).*

Annual scarcely branched, smelling herb. Leaves alternate or the lower opposite, upper 1-3, lower 5-11 partite, lanceolate, serrate flesy. Flowers small, greensih white. Furits compressed crustaceous nut; seeds flattened. *Fls. & Frts.:* Almost throughout the year.

Cultivated in the gardens, but very few. Its cultivation is strictly prohibited.

Field Collection: Specimens note collected and preserved.

Distributions: Throughout India. Widely distributed in the Tropics; a native of Siberia.

Economic Importances: Contain strong nicotine properties; for that its cultivation is restricted in India. It yields Ganja, bhang and shiddhi.

CXXIII. CASUARINACEAE

1. *CASUARINA* L.

1.1. ***Casuarina equisetifolia*** L. ex J.R. & G. Foster, Char. Gen. Pl. 104, t. 52. 1776; Hook.f. Fl. Brit. India 5: 598.1888; Watt Dict. Econ. Prod. 2: 230.1889; Prain Beng. Plants 2: 739.1903 (Rep. ed. 1963). *(Vern. Name: Belati Jhau).*

Tall, 25-40 m high tree with drooping branches, conical-shaped, seems to be conifer. Branches dark green, hollow, cylindric, sulcate, whorls of 5-10 teeth-like scales, leaves present.Male spike cylindric at the end of the branches, brown coloured, comparatively small. Female spikes globose, greenish, hard and axils of the branches larger than the male spike. Fruiting heads cone-like, hard woody with furrowed surface. *Fls. & Frts.:* January-August.

Commonly planted as avenue tree, on the lawn and ornamental gardens also. On the coastal areas of the Sundarbans—a huge plantation was made especially on Bakkhali, Frasergunj, Kakdwip and even on the reserve forests of the Sundarbans.

Field Collection: Sagar Mela Premises/Sagar Islands. *K.N.* - 427.

Distributions: Planted almost throughout India. It was introduced in Calcutta Botanic Garden in the year 1798. A native of Australia, Malaya and Pacific Islands.

Economic Importances: Ornamental plants, used for decorative purposes; suited very well in the sandy soil of the coastal areas, and help to prevent soil erosion form the strong wind. Yield a good resin and the bark is used for tannin as well as for dyeing fishermen's net. Bark is astringent and is said to be employed in the treatment of diarrhoea and dysentery and also used as tonic.

CXXIV. CERATOPHYLLACEAE

1. *CERATOPHYLLUM* L.

1.1. ***Ceratophyllum demersum*** L. Sp. Pl 992.1753; Hook.f. Fl. Brit. India 5: 639.1888; Prain Beng. Plants 2: 743.1903 (Rep. ed. 1963); Subramaniam Aq. Angios. 52.1962; Naskar Aq. & Semi-Aq. Plants 226.1990. *(Vern. Name: Jhanji).*

Delicate, submerged, aquatic herb. Roots not present. Leaves much dissected into filiform segment.Flowers monoecious solitary, axillary. Involucre 8-12 cleft, resembling perianth. Fruit and achene with basal spines and apical persistant base of the style. *Fls. & Frts.:* August-November.

Common in lakes, ponds and marshes, associated with *Hydrilla verticillata*

Field Collection: Magrahat,*K.N.* - 251.

Distributions: Almost throghout India.Cosmopolitan species.

Economic Importances: It very often chocked the water bodies and use fish culture and also aquaculture. Said to be used in biliousness as cooling agents.

Plate-XVII

A

B

C

Plate-XVIII

A

B

Plate-XIX

Plate-XX

A

B

C

Plate-XXI

A

B

C

Plate-XXII

B

A

Plate-XXIIII

A

B

Plate-XXIV

A

B

Plate-XXV

A

B

C

Plate-XXVI

Plate-XXVII

Plate-XXVIII

A

B

Plate-XXIX

A

B

Plate-XXX

A

C

B

Plate-XXXI

A

B

Plate-XXXII

A

B

Plate-XXXIIII

A

B

Plate-XXXIV

A

B

Plate-XXXV

A

B

MONOCOTYLEDONS

CXXV. HYDROCHARITACEAE

Key to the Genera

1a	Perianth 3 in single whorl	5. *VALLISNERIA* L.
1b	Perianth 3+3 in double whorls	(2)
2a(1)	Stem branching, leafy	(3)
2b (1)	Stem none, with stolons only or a creeping rootstock	(4)
3a (2)	Leaves usually opposite below but whorled above; ovules anatropous	2. *HYDRILLA* Rich.
3b(2)	Leaves scattered, lower opposite; ovules orthotropous	3. *NECHAMANDRA* Planchon
4a (2)	Leaves sessile; male flowers several within the spathe	1. *BLYXA* Nor. ex Thrs.
4b (2)	Leaves petioled; flowers 1 or 2-3 within the spathe	4. *OTTELIA* Person

1. *BLYXA* Nor. ex Thrs.

1.1. ***Blyxa octandra*** (Roxb.) Planch. ex Thw., Enum. P. Zeyl. 332.1864 (excl. *specimina*); Naskar Aq. & Semi-Aq. Plants 227.1990. *(Vern. Name: Pata Syola).*

Tufted, annual submerged, delicate dioecious herb. Naskar (l.c.) described this taxon after collecting from Salipur/Mallickpur (*K.N. - 2149.*) In the fish aquarium, this plant is used for decorative purposes. But in the fish culture ponds this plant is very often found as obnoxious weed.

2. *HYDRILLA* Rich.

2.1. ***Hydrilla verticillata*** (Linn.f.) Royle, Ill. Bot. Himal. t. 376.1839; Naskar Aq. & Semi-Aq. Plants 228.1990. *(Vern. Name: Jhangi).*

Submerged, rooted, leafy dioecious herb. Naskar (l.c.) described this taxon after collecting from Salipur/Mallickpore (*K.N. - 2115)* and Subuddhipur/Baruipur (*K.N. - 2292).* Commonly used as aquarium plant; this submersed weed release free oxygen in the water and help the aquatic animals. This plant contain 17 different amino acids and also considered as good food for herbivorous fish species. Sometime it chockes the water bodies.

3. *NECHAMANDRA* Planch.

3.1. ***Nechamandra alternifolia*** (Roxb.) Thw.,Enum.Fl.Zeyl. 332.1862; Naskar Aq. & Semi-Aq. Plants 229.1990. *(Vern. Name: Rasna-Jhangi).*

Submerged, rooted, dioecious fresh-water herb. Naskar (l.c.) described this taxon after collecting from Palta (*K.N. - 1997*). Grass carp relish this aquatic herb and it also causes weed problem in ponds.

4. *OTTELIA* Pers.

4.1. ***Ottelia alismoides*** (L.) Pers. Syn. Pl. 1: 400.1905; Naskar Aq. & Semi-Aq. Plants 230.1990.*(Vern. Name: Pani-kala).*

Rooted, submersed to partially floating flaccid annual herb. Naskar (l.c.) described this taxon after collecting from Garia (*K.N. - 313)* and Herobhanga *(K.N. - 1859)*.Fruits edible; very often causes adverse effect on chocking the fish culture pond and agriculture fields.

5. *VALLISNERIA* L.

5.1. ***Vallisneria spiralis*** L. var. ***denseserrulata*** Mikino, Bot. Mag. Tokyo 28: 27.1914; Naskar Aq. & Semi-Aq. Plants 231.199. *(Vern. Name: Syola).*

Submerged, stoloniferous, tufted herb. Naskar (l.c.) described this taxon from the lower Ganga delta after collecting form Lakshmikantapur *(K.N. - 1780)*. This plant has several medicinal value, *e.g.*, stomachic, leucorrhoea, haemotherima, giddiness, morbid thrust and rheumatism. Used in domestic aquaria also. It causes weed effect in fish culture and agriculture also.

Notes: The other plants of this family are–*Elodea canadensis* Michx. and *Halophila ovalia* (R.Br.) Hook.f.

CXXVI. ORCHIDACEAE

Key to the Genera

1a	Epiphytic herb; leaves linear, narrow, grooved on the mid-rib region, coriaceous; flowers axillary in simple, lax or dense racemes	3. *VANDA* R. Brown
1b	Terrestrial herbs	(2)
2a (1)	Leafless, saprophytic, delicate herb	1. *DIDYMOPLEXIS* Griff.
2b (1)	Plants with leaves	(3)

3a (2)	Stem sub-terranean; leaves elliptic or oblong elliptic, plicate; flowers corymbose or racemes, white with pink lines	2. *GEODORUM* Jackson
3b (2)	Stem aerial; leaves sessile linear, base clasping; flowers in dense racemes, white	4. *ZEUXINE* Lindl. *nom.cons.*

1. *DIDYMOPLEXIS* Griff.

1.1. ***Didymoplexis pallens*** Griff., Calcutta Journ. nat. Hist. 4: 383.t. 17.1844; Hook.f. Fl. Brit. India 6: 122.1890; Prain Beng. Plants 2: 770.1903 (Rep. ed. 1963).

Terrestrial, leafless, saprophytic, delicate herbs, 8-12 cm long. Root tuberous, ellipsoid. Scape simple; flowers shortly pedicelled, dull yellow or white; sepals connate. Column long, straight; stigma broad; anther 2-celled, depressed; pollinia 4. Capsule elliptic, ridged, with long, accrescent pedicels. *Fls. & Frts.:* April-July.

Rarely grow as saprophytic herb on the decaying leaves of the orchards and bamboo gardens.

Field Collection: Subuddhipur, *K.N. - 1466.*

Distributions: From the foot-hills of Sikkim Himalaya to lower Bengal. Malay Peninsula and Java.

Notes: Prain (1903a) did not mention this taxon from the lower Ganga Plain.

2. *GEODORUM* Jackson

2.1. ***Geodorum densiflorum*** (Lamk.) Schlt., Fedde Report. Beith. 4: 259.1929; Wlh. India 4: 125.1956. *Limodorum densiflorum* Lamk. Encyl. 3: 516.1791 - 1792. *L. recurvum* Roxb. Cor. Pl. 1.t.39.1795. *Geodorum dilatatum* R. Br. Ait. Hort. Kew (ed.2) 5: 207.1813; Hook.f. Fl. Brit. India 6: 17.1890; Prain Beng. Plants 2: 764.1903 (Rep. ed. 1963). *G. purpureum* R. Br., Ait. Hort. Kew (ed.2) 5: 207.1813; Hook.f. Fl. Brit. India 6: 16.1890 (in part); Prain Beng. Plants 2: 764.1903 (Rep. ed. 1963).

Terrestrial herbs. Tubers globular. Leaves elliptic, plicate, acute, entire sheath forming a pseudostem. Scapes terete or sub-terete, glabrous. Sheath few, scarious.Flowers white pink- lines, corymbose/racemes; bracts linear lanceolate, acute, pedicels short. Lip sessile at the base of column, not succate. Pollinia 2. Fruits capsule. *Fls. & Frts.:* July-October.

Grow on the shady moist places.

Field Collection: Sundarbans, *V.S.Rao-456472* (CAL).

Distributions: Native of India. Sri Lanka, Burma and Australia.

Economic Importances: Crushed root stock is said to be useful for killing cattle flies and also given to goat for diarrhoea.

Notes: D. Prain (1903a) has treated this species under the head *Geodorum dilatum* R. Brown in Bengal Plants and *G. purpureum* R. Brown as the synonym; but again in the Rec. bot. Surv. India(1905) Prain treated both as different species. The latter based on the collection from Serampur by Voigt but he expressed the needs of its confirmation.

3. *VANDA* R. Brown

3.1. ***Vanda tessellata*** (Roxb.) Hook.ex G. Don. Loud. Hort.Brit. 372.1830. *Epidendrum tessallatum* Roxb. Fl. Corn. 1: 34.t.42.1795. *Vanda roxburghii* R. Brown, Bot. Reg. 6: t. 506.1820; Hook.f. Fl. Brit. India 6: 52.1890; Prain Beng. Plants 2: 767.1903 (Rep. ed. 1963). *(Vern. Name: Rashna).*

Climbing, epiphytic; stem 40-60 cm long. Leaves linear, narrow, grooved on the mid-rib region, coriaceous, arranged alternately. Racemes sub-erect or rarely pendulous.Flowers showy, variegated, bluish; bracts 2.5-3.5. mm long scrarious, ovate, acute. Capsule 6-9 cm X 1.5 cm, oblong, sharply winged. *Fls. & Frts.:* March-August.

Very common epiphytic of the old trees in the village shrubberies.

Field Collections: Canning, *K.N. - 131;* Ghatakpukur, *K.N. - 1365.*

Distributions: Bengal, Bihar and almost throughout India; this species is declared as rare and endangered in India by the Botanical Survey of India. Sri Lanka, Tenasserium and Indo-Malayasia.

Economic Importance: Sometime cultures as ornamental plants.

4. *ZEUXINE* Lindl. *nom.cons.*

4.1. ***Zeuxine strateumatica*** (L.) Schttr., Fedde, Report, Beih 1: 77.1911. *Orchis strateumatica* L. Sp. Pl. 945.1753. ***Zeuxine sulcata*** Lind. Gen. Sp. Orch. 485.1830; Hook.f. Fl. Brit. India 6: 106.1890; Prain Beng. Plants 2: 773.1903 (Rep. ed. 1963). *(Vern. Name: Swet-huli).*

Terrestrial or marshy herb. Leaves sessile, linear, acuminate, base clasping. Flowers in dense racemes, white; lip sub-orbicular, not longer than the sepals. Seeds deep brown. *Fls. & Frts.:* December-February.

Rarely grow on the moist and shady localities, hidden in grasses and bamboo groves.

Field Collection: Dakshin Barasat, *K.N. - 1654.*

Distributions: Throughout India. Sri Lanka, China, Philippines, Japan and Afghanistan.

CXXVII. ZINGIBERACEAE

Key to the Genera

1a	Plants not aromatic; leaves spiral, lateral staminodes small or obsolate	(2)
1b	Plants aromatic; leaves distichous; lateral staminodes generally large	(4)
2a (1)	Filaments short, not petaloid	7. *ZINGIBER* Fabric *nom.cons.*
2b (1)	Filaments long petaloid	(3)
3a (2)	Anther on its middle	2. *COSTUS* L.
3b (2)	Anther divergent at the apex	1. *ALPINIA* L.
4a (1)	Ovary unilocular; placentation parietal	4. *GLOBBA* L.
4b (1)	Ovary trilocular; placentation axile	(5)
5a (4)	Stem elongated, leafy	5. *HEDYCHIUM* Koenig
5b (4)	Stemless herb, with few leaves	(6)
6a (5)	Connective not spurred at the base; leaves small, corolla-tube long and slender	6. *KAEMPFERIA* L.
6b (5)	Connective spurred at the base, leaves often very large; corolla tube small	3. *CURCUMA* L.

1. *ALPINIA* L.

1.1. ***Alpinia allughas*** Roscoe, Trans. Linn. Soc. 8.346.1807; Naskar Aq. & Semi-Aq. Plants 233.1990. *(Vern. Name: Taro* or *Taruko).*

Stout perennial herb, 1-2m high; occasionnnly found in wet places. Naskar (l.c.) described this taxon after collecting from Gocharan *(K.N. - 240)* and Subuddhipur/Baruipur (*K.N. - 2293)*.

2. *COSTUS* L.

2.1. ***Costus speciosus*** (Koen.) Smith, Trans. Linn. Soc. 1: 249.1791; Naskar Aq. & Semi-Aq.Plants 233.1990. *Banksia speciosa* Koen., Retz. Obs. 3: 75.1783. (*Vern. Name: Kus*t or *Keu*).

Rhizomatous herb; common on the waste, shady places; waterside thickets and roadside hedges.Naskar (l.c.) described this taxon after collecting from Harinbari/Sagar Islands (*K.N. - 479);* Dandirhat/Basirhat *(K.N. - 961)*and Atghora/Sitakundu *(K.N. - 1244).* Rootstocks are considered as raw materials for commercial production of diosgenin and precursor of steroides like sex hormones and oral contraceptives. Root-bitter, astringent, purgative, depurative, stimulant, tonic and anthelmintic.

3. *CURCUMA* L.

Key to the Species

1a	Flowering spikes lateral, apart from and appearing usually before the leaves	2. *C. aromatica* Salisb.
1b	Flowering spikes rising in the centre of a previously formed tufted leaves	(2)
2a (1)	Tubers pale-yellow within; leaves oblong-lanceolate, tapering at both the ends; flowers white or pale-yellow	1. *C. amada* Roxb.
2b (1)	Tubers bright-yellow within; leaves oblong, caudate acuminate, tapering to the base; flowers pale-yellow	3. *C. longa* L.

3.1. ***Curcuma amada*** Roxb. Res. 11: 341.1810; Watt Dict. Econ. Prod. 2: 652.1889; Baker, Hook.f. Fl. Brit. India 6: 213.1890; Prain Beng. Plants 2: 783.1903 (Rep. ed. 1963). *(Vern. Name: Amada).*

Cultivated stemless herb; rootstock large, tuberous, sessile, cylindric, yellow indide. Leaves oblong-lanceolate, tapering at both the ends, petiole as long as the fertile bracts, pale-green. Peduncle hidden by the sheathing leaf-base.Flowers white or pale-yellow; lip semi-elliptic, 3-lobed, middle lobe emerginate. *Fls. & Frts.:* July-November.

Frequently planted in the gardens, likes shady mosit places.

Field Collection: Majerhat, *K.N. - 2250.*

Distributions: Cultivated almost throughout India; specially in Bengal; a native of India. The warmer regions of the World.

Economic Importances: Used as spices; the fresh cut rootstock possesses the smell of green mango and it is said to be used as condiment and vegetable. It is regarded as cooling and carminative.

3.2. ***Curcuma aromatica*** Salisb. Parad. Lond. t. 96.1805; Watt. Dict. Econ. Prod. 2: 655.1889; Baker, Hook.f. Fl. Brit. India 6: 210.1890; Prain Beng. Plants 2: 783.1903 (Rep. ed. 1963). *(Vern. Name: Ban Haldi).*

Stemless herbs; rootstock tuberous, root fibrous not ending in small tubers.Leaves elliptic or lanceolate-oblong, caudate-acuminate.Bracts ovate, recurved, cymbiform, obtuse, fertile ones pale green, the coma more or less tinged with red or pink; flowers pink, lip yellow, ovate, sub-entire or obscurely 3- lobed. *Fls. & Frts.:* June-October.

Frequently cultivated elsewhere, a shade loving rhizomatous herbs.

Field Collection: Titagarh, *K.N. - 2223.*

Distributions: Throughout India. Warmer regions of the world.

Economic Importances: Rhizomes is used as medicine and perfume. It is also used as dye, cosmetic, tonic and carminative.

3.3. ***Curcuma longa*** L. Sp. Pl. 2. 1753; Watt Dict. Econ. Prod. 2: 659. 1889; Baker, Hook.f. Fl. Brit. India 6: 214.1890; Prain Beng. Plants 2: 783.1903 (Rep. ed. 1963). *(Vern. Name: Haldi).*

Stemless herb; rootstock large, tubers sessile, cylindric, bright yellow within. Leaves oblong, caudate-acminate, tapering to the base. Fertile bracts pale green, coma tinged with pink. Flowers pale-yellow, lip ovate sub-entire. *Fls. & Frts.:* July-November.

Commonly cultivated in the village gardens and occasionally self-grown on the shady and waste lands in the garden edges.

Field Collection: Baduria, *K.N. - 1300.*

Distributions: Widely cultivated in Bengal and other parts of India

Economic Importances: Used as important and essential spice.Rhizomes are aromatic, stimulant, tonic, carminative, blood purifier, anti-periodic, alternative, externally applied to wounds, as antiseptic. Yellow colour is also used in pharmacy, confectionary, textile, paints and food-industries. Hindus use this in thier religious ceremony.

Notes: In some recent publications the valid name for this taxon is consdiered as *Curcuma domestica* Val.

4. *GLOBBA* L.

4.1. ***Globba marantina*** L. Mant. 170.1791; Naskar Aq.& Semi-Aq.Plants 234. 1990.*Globba bulbifera* Roxb. Asiat. Res 11: 358.1810 *G. marantina* Wall. Cat 6532. *G. marantinoides* Wight Icon.

Slender, erect or drooping herb with leafy stem. Common on the shady localities.Naskar (l.c.) described this taxon after collecting from Purandarpur/Baruipur (*K.N. - 1226),* Narendrapur (*K.N. - 1988)* and Kholapota/Basirhat *(K.N. - 1261).* Bulbs are edible, spice.

Notes: This genus is also sometime treated as separate family–MARANTACEAE

5. *HEDYCHIUM* Koenig.

5.1. ***Hedychium coronarium*** Koening, Retz. Observ. bot. fasc.3: 73.1783; Naskar Aq. & Semi-Aq. Plants 235.1990.*(Vern. Name: Dulal Champa).*

Tall perennial herb; commonly grows in the garden and succeds best in rich deep soil. Naskar (l.c.) described this taxon after collecting from Goranbose *(K.N. - 1882).* Aerial part is used as paper pulp. The rhizomes contain starch, similar to arrowroot starch. Flowers are used as a soruce of perfume in Hawaii. This plant is also used as a remedy for foetid nostrils. Rhizomes febrifuge, antirheumatic, tonic and excitant.

6. *KAEMPFERIA* L.

6.1. ***Kaempferia rotundus*** L. Sp. Pl. 3. 1753; Baker, Hook.f. Fl. Brit. India 6: 222. 1890; Watt Dict. Econ. Prod. 4: 561.1890; Prain Beng. Plants 2: 780.1903 (Rep. ed. 1963). *(Vern. Name: Bhui-champa).*

Handsome, aromatic herb; rootstock tuberous. Leaves oblong,erect petioled. Flowers fragrant, white; the lip purple or lilac; calyx purplish dotted; corolla white; lobes linear; staminodes oblong, white, lip ovate, deeply bilobed. *Fls. & Frts.:* March-July.

Generally cultivated as ornamental plant, but occasionally as wild.

Field Collection: Mollakhali *K.N. - 1845.*

Distributions: Throughout India. Malay Peninsula and Sri Lanka.

Economic Importances: Tubers are used for tumours, swelling and wounds; considered as stomachic and given in gastric complaints. Help to remove blood clots and other purpulent matter in the body; also used as ointments for wounds.

7. *ZINGIBER* Fabric *nom.cons.*

Key to the Species

1a	Leaves linear, sessile, gradually acuminate, glabrous beneath	2. Z. *officinalis* Rosc.
1b	Leaves lanceolate or oblong-lanceolate	(2)
2a (1)	Leaves glabrous beneath or nearly so; bracts green; stamens pale yellow colour	3. Z. *zerumbet* (L.) Smith.
2b (1)	Leaves pubescent beneath; bracts reddish; stamens yellowish white	1. Z. *casumunar* Roxb.

7.1. ***Zingiber casumunar*** Roxb. Asiat. Res. 11: 347.t. 5.1810; Baker, Hook.f. Fl. Brit. India 6: 248.1892; Watt Dict. Econ. Prod. 6(4): 357. 1893; Prain Beng. Plants 2: 785.1903 (Rep. ed. 1963). *Z.purpureum* Rosc., Trans. Linn. Soc. 8: 348.1807. (*Vern. Name: Ban-Ada*).

Perennial herb; root-stock horizontal, tuberous, yellowish inside, aromatic, testing of camphor. Stem leafy, leaves sub-sessile, oblong-lanceolate, acute, almost hairy beneath, pubescent spikes dense, fusiform or oblong, ellipsoid; bracts broadly ovate, sub-acute; bright red or greensih red, pubescent, margins narrowly membranous; lip 3-lobed. Capsule sub-globe. Seed many and small. *Fls. & Frts.:* May-October.

Sometime grow on the waste and moist lands in the village thickets; mostly cultivated.

Field Collection: Bantala/Salt Lake, *K.N. - 1742.*

Distributions: Almost throughout India.

Economic Importances: Rhizome is used as stimulant, carminative and flavoruing agent; given in dyspepsia and flatulent colic; prescribed as an adjunct to many tonic and stimulating remedies.

7.2. ***Zingiber officinalis*** Rosc., Trans. Linn. Soc. 8: 348.1807; Baker, Hook.f. Fl. Brit. India 6: 246.1892; Watt Dict. Econ. Prod. 6(4) 358. 1893; Prain Beng. Plants 2: 785.1903 (Rep. ed. 1963). *(Vern. Name: Ada).*

Perennial herb; root-stock horizontal, tuberous, aromatic. Stem leafy slender.Leaves linear, sessile, gradually acuminate, glabrous. Spikes terminating, oblong-cylindric, peduncle slender, scale sheathing glabrous, lip 3-lobed. Ovary 3-celled. Fruit oblong. *Fls. & Frts.:* May-October.

Rarely cultivated in the damp places of the garden edges.

Field Collection: Duttapukur, *K.N. - 1706.*

Distributions: Cultivated almost throughout India; native of South-east Asia. Widely cultivated in the Tropical Asia.

Economic Importances: Commonly used as spice, as flavouring agent and also used with pepper as an abortifacient. Rhizome is used as stimulant, carminative and is given in dyspepsia and flatulent colic.

7.3. ***Zingiber zerumbet*** (L.) Smith., Exot. Bot. 2: 105.t.112.1804; Naskar Aq. & Semi-Aq. Plants 236.1990.*Amomum zerumbet* L. Sp. Pl. 1. 1753. *Fls. & Frts.:* July-October.

Gregariously grow on the moist and shady places, on the garden edges, waste lands and village thickets.

Field Collection: Flata, *K.N. - 1998.*

Distributions: Throughout Bengal and other parts of India. A pantropic species.

Economic Importances: Rhizomes is used as stimulant, carimnative; it is given in dyspepsia and flatulent colic.

CXXVIII. CANNACEAE

1. *CANNA* L.

1.1. ***Canna orientalis*** Roscoe. Monandra. Pl. t. 12.1828. *C. indica* L. Sp. Pl. 1.1753; Watt Dict. Econ. Prod. 2: 102.1889; Baker, Hook.f. Fl. Brit. India 6: 260.1892; Prain Beng. Plants 2: 787.1903 (Rep. ed. 1963). *C. indica* L. var. *orientalis* Roscoe Scit. Pl.t.12.1912. *(Vern. Name: Sarbajaya.).*

Perennial, rhizomatous herb.Leaves oblong or elliptic-oblong or ovate-oblong, acute or shortly acuminate at apex; lower leaves larger. Flowers racemes, variegated, yellow, red or sometimes whitish; bracts lanceolate. Fruit ellipsoid, echinate. Seeds many, rounded and black. *Fls. & Frts.:* All round the year, specially June-September.

Field Collection: Dehimedanmalla/Mallickpore, *K.N. - 83.*

Distributions: Throughout India. Sri Lanka, Malaya, Malacca and Tropical America.

Economic Importances: A common ornamental plant for its large, delicate, deep-coloured variegated flowers. Root-diaphoretic, diuretic in fevers and dropsy, demulcent and stimulent.

CXXIX. MUSACEAE

1. *MUSA* L.

1.1. ***Musa paradisiaca*** L. Sp. Pl. 1043.1753; Cheesman, Kew Bull. 1948: 145-153.1948; Wlh. India 6: 452.1962. *Musa sapientum* L. Sp.Pl.1477. 1753; Watt Dict. Econ. Prod. 5: 292.1891; Baker, Hook.f. Fl. Brit. India 6: 262.1892; Prain Beng. Plants 2: 789.1903 (Rep. ed. 1963). *(Vern. Name: Kanthali Champa Kala).*

Stoloniferous, pseudostemed, 2-4 m high, cylindrical. Leaves long petioled, sheathed, large 1.5-2.5 m X 0.5 m oblong. Spikes drooping solitary, bracts ovate, many flowered, deciduous. Calyx 5-toothed at the tip, petal shroter than the calyx. Fruit 7-10 cm long, sub-cylindric, slightly curved, orange-yellow with cream coloured pulp. Seed not formed in the cultivated species. *Fls. & Frts.:* Throughout the year, specially August-March.

Commonly cultivated throughout the lower Ganga delta in the orchard or banks of ponds and ditches.

Field Collection: Amtala/on the side of Diamond Harbour Road, *K.N. - 375.*

Distributions: Cultivated almost throughout India; mainly in Madras, Bombay, Kerala, Mysore, Bengal, Bihar and Assam. Malaya, Burma and Indonesia.

Origin of Banana: This species and majority of the edible bananas are mostly of hybrid origin.The wild progenitors of these edible bananas are *Musa acuminata* Colla & *M. balbisiana* Colla; *M. balbisiana* is a diploid seeded type, while *M. acuminata* Colla includes both diploid and triploid forms. Hybridization among the diploid and triploid tyes of *M. acuminata* Colla and with the type of *M. balbisiana* has resulted the present diploid, triploid and tetraploid cultivated form of *Musa*.

The centre of origin of this banana complex is mainly East Asia, with perhaps Malaya as the primary centre and India as the secondary one.

Economic Importances: Banana is one of the most important fruit in India. Ripe fruits are delicious and green fruits are also cooked as vegetable.The ripe fruits are rich with carbohydrates, minerals and vitamins. It is also enriched with protein of Albumin and globulin. Banana is also a fair source of calcium and iron and a rich source of potassium, magnesium, sodium and phosphorous. The enzymes present in Banana include amylase, invertase, protease, catalase, peroxidase, lipase, oxygenase, phosphatase and ascorbic acid oxidase. The leaves of banana are also used as food plates; and the stock of the inflorescence is also cooked as vegetable.

CXXX. BROMELIACEAE

1. *ANANAS* Mill.

1.1. ***Ananas comosus*** (L.) Merr. Interpr. Herb. Amb. 133. 1917. *Bromelia comosa* L., Stickman Herb. Amb. 21.1754. *B. ananas* L. Sp. Pl. 285.1753. *Ananassa sativa* Lindl. Bot. Reg. 13. sub.t.1068. 1827; Prain Beng. Plants 2: 791.1903 (Rep. ed. 1963). *Ananas sativus* Schult.f. Syst. 7: 1283. 1830; Watt Dict. Econ. Prod. 1: 236. 1889. *(Vern. Name: Anaras).*

Tufted, stemless herb. Leaves long, linear with spinous-serrate edges, acute. Inflorescences dense, capitate, spiral; flowers blue, cone-like, globose, embedded in the axil of each bract. Sepals-3, wide acuminate, closely imbricate, free-above the ovary. Petals-3, erect, slight imbricate. Stamens 6, anther linear. Carpels-3, 3-celled, fleshy ovary, embedded in the thick rachis, many ovules in each cell. Fruits fleshy embedded in the fleshy receptacle. Seeds few, ovoid to oblong. *Fls. & Frts.:* March-August.

Commonly cultivated in the shadow waste land-almost throughout the districts; grow gregariously on the orchard or the side of ponds, ditches or nalas; it likes somewhat loamy soil.

Field Collection: Haroa, *K.N. - 586.*

Distributions: Common on Assam, Shillong, Tripura, Agartala and North Bengal. Introduced in India from West Indies by about the middle of the 16th century. Native of Tropical America.

Economic Importances: The ripe fleshy receptacle is a delicious food and preserved in the form of jelly; leaves are said to be useful in intenstine worms and also used in making 'fine cloth'.

CXXXI. IRIDACEAE

Key to the Genera

1a	Creeping rhizomatous; stem leafy; flowers in branched panicles	1. *BELAMCANDA* Adans.
1b	Tunicates bulb; stem leafless; flowers in axils of narrow spathes	2. *CIPURA* Audl.

1. *BELAMCANDA* Adans.

1.1. ***Belamcanda chinensis*** (L.) DC. Redoute, Lil. 3: t. 121. 1805; Hook.f. Fl. Brit. India 6: 277. 1892; Prain Beng. Plants 2: 794.1903 (Rep. ed. 1963); Wlh. India 1: 173.1948. *Ixia chinensis* L. Sp. Pl.36. 1753. *Pardanthus chinensis* Ker., Koenig. & Sims. Ann. Bot. 1: 246. *Morea chinensis* Murr. Syst. Veg. 93; Roxb. Fl. Ind. 1: 170. 1824. *(Vern. Name: Dasbichandi).*

Erect herb, with creeping rhizomatous rootstock. Stem leafy, leaves ensiforme, equitant. Inflorescence dichotomously branched with green, spathe-like floral leaves. Flowers pedicelled, many in each spathe; perianth petaloid, tube very short, rather wide; stamens 6, oblong, spreading. Fruit obovoid, membranous, loculicidal capsule. Seeds sub-globose; tasta lax. *Fls. & Frts.:* Almost throughout the year.

Common in the orchards.

Field Collection: Dhopagachhi/nera Baruipur, *K.N. - 2140.*

Distributions: Almost throughout India. A native of China and Japan.

Economic Importances: Said to be useful for tonsillitis, given in chest and liver complaints; and also reported to have some purgative properties.

2. *CIPURA* Audl.

2.1. ***Cipura paludosa*** Audl. Pl. Gui. 1: 38. t.13; Prain Beng. Plants 2: 793.1903 (Rep. ed. 1963).

Herbs, with a tunicate bulb; stem slender up to the inflorescences; leafless. Radical leaves few, narrow. Flowers in narrow, fascicled spathes, sessile or shortly peduncled. Perianth petaloid; tube very short; segments rather broad, flat, stamens 3, adnate to base of outer perianth segments; filaments free. Carpels 3, connate in a 3-celled ovary. Fruit an ovoid or oblong. Seeds globose. *Fls. & Frts.:* September-February.

Generally cultivated as ornamental plants, sometime naturalised in the thickets.

Field Collection: Herobhanga. *K.N. - 1857.*

Distributions: Cultivated almost throughout India. A native of America.

Economic Importance: Ornamental garden plants.

CXXXII. AGAVACEAE

Previously the Genus-*Agave* L. was treated under Amaryllidaceae, but recently it is treated under a distinct family- Agavaceae.

1. *AGAVE* L.

1.1. ***Agave cantala*** Roxb. Fl. Ind. 2: 162. 1832; Hook.f. Fl. Brit. India 6: 277. 1892; Prain Beng. Plants 2: 795.1903 (Rep. ed. 1963). *(Vern. Name: Kantala).*

Shrub, usually with a short, woody stock. Leaves large, thick, fleshy, tip and edges with sharp spines, shining. Flowers solitary, on a large, simple pole like scape, bracts rather small. Perianth-tube funnel-shaped, lobes 6, linear, erect or spreading, stamens 6, adnate to base, longer than perianth lobes; carpels 3, connate in a globose, ovoid, or oblong, often fleshy ovary. Fruit an erect, globose, ovoid or oblong-cylindric, coriaceous capsule. Seeds many, flattened obliquely, testa black. *Fls. & Frts.:* February-May.

Gregariously grow on the dry, somewhat sandy soil, on the waste land or roadside thickets.

Field Collection: On the Railway tracks/Icchapur, *K.N. - 1821.*

Distributions: Widely distributed in the drier regions of India. A native of Mexico.

Economic Importance: Leaves produce useful fine fibres.

CXXXIII. HYPOXIDACEAE

This family is sometime treated as a tribe under the family-Amaryllidaceae.

1. *CURCULIGO* Gaertn.

1.1. ***Curculigo orchioides*** Gaertn. Fruct. 1.63.t.13.1788; Hook.f. Fl. Brit. India 6: 279. 1892; Prain Beng. Plants 2: 796. 1903 (Rep. ed. 1963). *C. malabarica* Wight. Icon. 6: 22, t. 2043.f.1.1853. *C. brevifolia* Dry. Ait. Hort. Kew (ed. 2) 2: 253.1811. *(Vern. Name: Tala muli* or *Kua-kenda).*

Rootstock tuberous, elongate, robust. Leaves few, glabrous, plicate linear, elliptic, acuminate at apex. Scape axillary short. Flowers subtended by lanceolate bracts which sheath the hypanthium. Capsule hypogynous with a minute beak. Seeds 1-4, oblong, shining, with deep wavy groves. *Fls. & Frts.:* June-October.

Occasionally on the shady waste lands.

Field Collection: Nimpith/Joynagar, *K.N. - 1616.*

Distributions: Throughout India; Java.

Economic Importances: The roots are bitter and beneficial in blood diseases, piles, jaundice, asthma, diuretic, tonic, aphrodisiac, used as a poultice for itch and skin diseases.

CXXXIV. AMARYLLIDACEAE

Key to the Genera

1a	Leaves all radical from a bulb; scape leafless	(2)
1b	Leaves clustered at the top of a rhizomatous rootstock scale leaf at the base	3. *POLYANTHES* L.
2a (1)	Stamens not united by a membranous corona	1. *CRINUM* L.
2b (1)	Stamens united by a membranous corona	2. *PANCRATIUM* L.

1. *CRINUM* L.

Key to the Species

1a	Bulb narrowed into a neck; leaves linear lanceolate, shortly acuminate many flowered	1. C. *asiaticum* L.
1b	Bulb ovoid; leaves deeply channeled, obtuse comparatively few flowered	2. C. *defixum* Ker-Gawl.

1.1. *Crinum asiaticum* L. Sp. Pl. 292. 1753; Hook.f. Fl. Brit. India 6: 280. 1892; Prain Beng. Plants 2: 797.1903 (Rep. ed. 1963). *C. asiaticum* L. var. *toxicaria* Watt. Dict. Econ. Prod. 2: 589.1889. *C. toxicarium* Roxb. Fl. Ind. 2: 134.1832. *(Vern. Name: Bara Kansur).*

Stem bulbous, caudex. Leaves linear, shortly acuminate, sheathing base. Scape compressed, usually 15-50 flowered. Perianth tube cylindric, slender, greenish-white. Stamens with pinkish filaments and brownish red linear straight anthers. Style exserted. Fruit sub-globose or ellipsoid beaked; 1 or 2 seeded. *Fls. & Frts.:* April-November.

Mostly cultivated as ornamental plants, but sometime as an escape along the banks of ponds, jheels or roadside ditches.

Field Collection: Titagarh, *K.N. - 2211.*

Distributions: Almost throughout India. Malaya and W.Polynesia.

Economic Importances: Tubers are useful in case of vomitting, urinary discharges and tumours.

1.2. ***Crinum defixum*** Kar-Gawl. Quart. Journ. Sci. & Arts. 3: 103.1817; Naskar & Guha Bakshi Mangr. Swamps Sundarbans 83.1987. *(Vern. Name: Sukh-darsan).*

Stout, bulbous herb, bulb ovoid, caudex; often on the margins of the rivers and streams. Naskar & Guha Bakshi (1.c.) described this taxon after collecting from Shyam Nagar *(K.N. - 2083)* and Kulti *(K.N. - 1927).* Leaf juice is used as a cure for ear-sore.

2. *PANCRATIUM* L.

2.1. ***Pancratium biflorum*** Roxb. Fl. Ind. 2: 125.1832; Hook.f. Fl. Brit. India 6: 285. 1892; Prain Beng. Plants 2: 798. 1903 (Rep. ed. 1963).

Herb with tunicate bulbs. Leaves linear lanceolate. Scape compressed, shorter than the leaves, spathes 3-4; perianth throat not dilated, tube trigonous; staminal cup funnel-shaped, erose. Fruit sub-globose. Seeds many, angular; testa thick, black. *Fls. & Frts.:* April-September.

Gregariously cultivated for its ornamental foliages and beautiful flowers.

Field Collection: Ghoshpur/Kulti, *K.N. - 2128.*

Distributions: Planted almost throughout India. Tropical regions of the world.

Economic Importances: Ornamental garden plants.

3. POLYANTHES L.

3.1. ***Polyanthes tuberosa*** L. Sp. Pl. 316. 1753; Watt Dict. Econ. Prod. 6(1): 312. 1892. *(Vern. Name: Rajani Gandha).*

Rhizomatous herb, rhizome with bulbous swellings. Leaves linear, radical or on the lower part of the stem. Flowers in long terminal simple racemes; white, very fragrant; perianth funnel-shaped, trigonous. *Fls. & Frts.:* April–October.

Commonly planted for its showy and fragrant flowers, sometime as an escape in the waste lands and garden edges.

Field Collection: Subuddhipur/Baruipur, *K.N. - 2294.*

Distributions: More or less throughout India, cultivated. A native of Mexico.

Economic Importances: Largely liked for its highly fragrant white flowered scapes.

Remarks: Zephyranthes tubispatha Herb. is commonly planted as ornamentals; a native of Peru, but now naturalised in India.

CXXXV. DIOSCOREACEAE

1. *DIOSCOREA* L.

Key to the Species

1a	Stem twisting to the left; capsules reflexed upwards	(2)
1b	Stem twisting to the right; capsule facing forwards	(5)
2a (1)	Leaves compound	(3)
2b (1)	Leaves always simple	(4)
3a (2)	Leaflets digitately 3-nerved	6. *D. hispida* Dennst.
3b (2)	Leaflets pinnately veined	8. *D. pentaphylla* L.
4a (2)	Capsules longer than broad; seeds winged at base	3. *D. bulbifera* L.
4b (2)	Capsules boarder than long; seeds winged, but all round	4. *D. esculenta* (Lour.) Burkill
5a (1)	Leaves softly tomentose or coriaceous, with cartilaginous margins	(6)
5b (1)	Leaves quite glabrous, margin not cartilaginous	(7)
6a (5)	Leaves opposite or alternate, ovate or cordate, cuspidate or acuminate, 7-nerved; capsule sub-ovate; seeds orbicular, winged	2. *D. anguina* Roxb.

6b (5)	Leaves opposite or sub-opposite, lanceolate to elliptic-oblong, ovate or sub-orbicular; capsule sub-orbicular; seeds orbicular, with large membranous winged	7. *D. oppositifolia* L.
7a (5)	Stem terete or compressed, smooth or faintly angled	5. *D. glabra* Roxb.
7b (5)	Stem 4-gonous, winged or strongly angled	1. *D. alata* L.

1.1. ***Dioscorea alata*** L. Sp. Pl. 1053. 1753; Watt Dict. Econ. Prod. 3: 126. 1890; Hook.f. Fl. Brit. India 6: 296.1892; Prain Beng. Plants 2: 802.1903 (Rep. ed. 1963). Wlh. India 3: 69. 1962. *D. atropurpurea* Roxb. Hort. Beng. 72.1814, *nomen.D. globosa* Roxb. Fl. Ind. 3: 797.1832. *D. purpurea* Roxb. Fl. India 3: 797.1832. *D. rubella* Roxb. Fl. Ind. 3: 797. 1832. *(Vern. Name: Chupri Alu).*

Large climber, stem quadrangular winged, twining to the right. Leaves opposite or rarely alternate, usually 5-nerved. Bulbils globose, ovoid, sometime elongated or flattened. Male flowers in slender fascicled spikes. *Fls. & Frts.:* October-January.

Sometimes cultivated, but mostly grow gregariously on the village thickets or as twiner to village jungle.

Field Collections: Joynagar, *K.N. - 870;* Barasat, *K.N. - 929.*

Distributions: Cultivated in all the States of India. A native of S.E. Asia and is considered to be closely allied to the wild species *D. persimilis* Prain & Burkill and *D. hamiltonii* Hook.f.

Economic Importances: The tuber is used as vegetable; it is starchy and can be dried and ground into meal. It is considered anthelmintic and useful in leprosy, piles and gonorrhoea.

1.2. ***Discorea anguina*** Roxb. Fl. Ind. 3: 803.1832; Hook.f. Fl. Brit. India 6: 293.1832; Prain Beng. Plants 2: 802.1903 (Rep. ed. 1963). *(Vern. Name: Kukur-Alu).*

Unbranched climber; stem twisted, purbescent; rootstock woody; tuber cylindrical, straight. Leaves opposite or alternate, ovate or cordate, cuspidate. Male flowers numerous in small panicle; rachis slender, pubescent; bracts ovate-lanceolate. Female flowers small in axillary peduncle. Capsule sub-ovate. Seeds orbicullar, winged. *Fls. & Frts.:* October-January.

Rarely grow on the village shrubberies and road-side jungles.

Field Collection: Media/near Gobardanga, *K.N. - 1694.*

Distributions: Common throughout India.

Economic Importances: Tubers and bulbs are edible, during famine.

1.3. ***Dioscorea bulbifera*** L. Sp. Pl. 1033.1753; Watt Dict. Econ. Prod. 3: 128.1890; Hook.f. Fl. Brit. India 6: 296.1892; Prain Beng. Plants 2: 801.1903 (Rep. ed. 1963). *D. sativa* Thunb. Fl. Jap. 151. 1784 (*non* L. 1753); Hook.f. Fl. Brit. India 6: 295.1892 (excl. *D. cliffortiana* auct., non L.). *D. pulchella* Roxb. Fl. Ind. 3: 801.1832. *(Vern. Name: Ban-alu).*

Stem unbranched, quadrangulate. Leaves alternate, cordate, entire, emarginate petiole 5-10 cm long. Flowers pale-green in slender spikes; bracts ovate, acuminate; female flowers subtended by minute bracts. Capsule winged, broad, rounded at both ends. Seeds winged at the lower end. *Fls. & Frts.:* June- October.

Commonly grow in the village thickets as a huge twiners; occasionally planted for its bulbils and tubers.

Field Collection: Subuddhipur/Baruipur, *K.N. - 2295.*

Distributions: Common throughout India, ascending upto 2000 m in the Himalayas; but it does not thrive in the drier parts of India. A native of the tropical world, China and Malaya.

Economic Importances: Bulbs and tubers are used as vegetable, especially as famine food. The tubers are useful for digestion and urinary discharges, ulcers, dysentery and syphilis. Bulbils are also used in sores. In Kashmir, the tuber is used for washing wool and as a fish bait.

1.4. ***Dioscorea esculenta*** (Lour.) Burkill, Gard. Bull. 1. 396.1917; Burkill, Fl. Males. I, 4: 307.1951; Wlh. Ind. 3: 72.1952. *Oncus esculenta* Lour. Fl. Coch. 194.1790. *D. aculeata* Roxb. Fl. Ind. 3: 800. 1832. *D. fasciculata* Roxb. Fl. Ind. 3: 801.1832; Watt Dict. Econ. Prod. 3: 125.1890; Hook.f. Fl. Brit. India 6: 296.1892; Prain Beng. Plants 2: 802.1903 (Rep. ed. 1963). *D. spinosa* Roxb. ex Wall. Cat. 5103.1828; Hook.f. Fl. Brit. India 6: 291.1892. *(Vern. Name: Sushni-alu).*

Prickly climber, twining to the left. Leaves alternate, simple, broadly ovate, acuminate, cordate at base, entire. Seeds winged. *Fls. & Frts.:* October-January.

It is generally cultivated, rarely seen in the village thickets.

Field Collection: Sarberia/Gocharan, *K.N. - 233.*

Distributions: Mostly on the plains of India, ascending upto 1,000 m in Eastern Himalayas. Probably a native in Siam and Indo-China; Malaya and Polynesia.

Economic Importances: The tubers are starchy and free from dioscorine; its taste is sweet and closely resemble to potatoes. Backward classes depend upon it in the period of crisis. Tubers are also said to be used as an application for swellings.

1.5. ***Dioscorea glabra*** Roxb. Fl. Ind. 3: 804.1832; Watt Dict. Econ. Prod. 3: 131.1890; Hook.f. Fl. Brit. India 6: 294.1892; Prain Beng. Plants 2: 802.1903 (Rep. ed. 1963). *D. nummularia* (non Lamk.) Roxb. Fl. Ind. 3: 803.1832; Prain Beng. Plants 2: 802.1903(Rep. ed. 1963). *(Vern. Name: Sora-alu).*

Glabrous, stem often clothed with prickles, especially at the base, twining to the right. Bulbils absent. Leaves generally opposite, rarely alternate, simple, usually elliptic-ovate or ovate-oblong, acuminate, base cordate, petiole 4-10 cm long. *Fls. & Frts.:* October-January.

Gregariously grow on the village thickets or on the moist waste lands, associated with *Lantana* sp., *Smilax* sp. and *Coccinia* sp.

Field Collection: Baruipur, *K.N. - 2253.*

Distributions: Almost throughout India; Andaman & Nicobar Islands. Burma, Malaya Peninsula.

Economic Importances: Tubers are edible in Andaman Islands and Khasi Hills.

1.6. ***Dioscorea hispida*** Dennst., Schluss. Hort. Malab. 15.1818; Burkill Fl. Males. 1,4: 318.1951; Wlh. India 3: 75.1952. *D. daemona* Roxb. Fl. Ind. 3: 805.1832; Watt Dict. Econ. Prod. 3: 129.1890; Hook.f. Fl. Brit. India 6: 289.1892; Prain Beng. Plants 2: 801.1903 (Rep. ed. 1963). *(Vern. Name: Kurukandu* (Hindi)).

Stem clothed with prickles; twining to the left. Root tuberous, lobed, biennial. Leaves alternate, trifoliolate, petiole as long as the leaflets. Bulbils absent. Flowers bi-sexual. Capsule 2, quardrate oblong; pedicles short. Seeds membranous, oblong, obtuse. *Fls. & Frts.:* September-December.

Rarely grow on the village thickets and hedges or waste lands.

Field Collection: Falta, *K.N. - 1999.*

Distributions: Throughout India. A native of the Tropical Asia and occurs in the rain forests from India to Formosa, South West China, Malaya to New Guinea.

Economic Importances: The tubers may produce the industrial starch and edible flour. A clean white starch is readily prepared from the tubers.

1.7. ***Dioscorea oppositifolia*** L. Sp. Pl. 1033.1753; Watt Dict. Econ. Prod. 3: 132.1890; Hook.f. Fl. Brit. India 6: 292.1892; Wlh. India 3: 75. 1952.

Climber with glabrous or finely pubescent; stem twining to the right. Leaves alternate, sometimes opposite. Bulbils absent. Capsule winged. *Fls. & Frts.:* October-January. (Fig.47).

Rarely grow on the village thickets.

Field Collection: Brace Bridge, *K.N. - 2226.*

Distributions: A native of South India and occurs throughout the hills of Deccan at an elevations of 7,00 m - 4,400 m., Coastal areas of Sri Lanka.

Economic Importances: The tubers are useful to reduce swellings.

Notes: D. Prain (1903a &1905) did not mention this species from Bengal; but it is not found to grow in the natural conditions on the village shrubberies.

1.8. ***Dioscorea pentaphylla*** L. Sp. Pl. 1032.1753; Watt Dict. Econ. Prod. 3: 132.1890; Hook.f. Fl. Brit. India 6: 289.1892; Prain Beng. Plants 2: 801.1903 (Rep. ed. 1963). *D. triphylla* L. Sp. Pl. 1032.1753. *D. jaccuemontii* Hook.f. Fl. Brit. India 6: 290.1892. *(Vern. Name: Kanta-alu).*

Large twining herb, twining to the left. Stem clothed with prickles bulbils present on the nodes. Leaves alternate, 3-7 foliolate, glabrous, median leaflets largest, acuminate, obtuse. Bracts and bracteoles enclosed the buds. Stamens inserted at the base of the sepals. Capsule dark. Seeds winged. *Fls. & Frts.:* August-December.

Commonly grow on the village thickets or hedges.

Field Collection: Mallickpore, *K.N. - 1410.*

Distributions: Throughout the plains of India, ascends to 2000 m in the Himalaya and in Andaman Islands. A native of Tropical Asia, extending eastwards to the remotest islands of the Pacific; Malaya and Guinea.

Economic Importances: Underground root-stocks and the leaves are cooked and used as vegetables; medicinally also used to disperse swelling and as tonic.

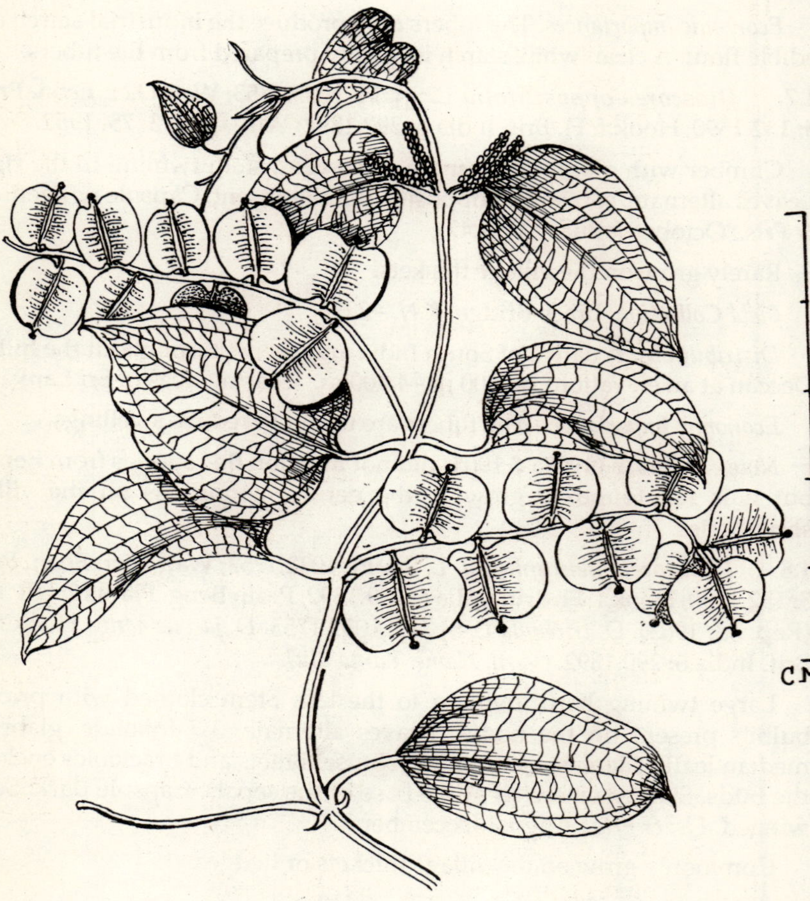

Fig.47. DIOSCOREA OPPOSITIFOLIA L.
Twig with flowering spikes and winged fruits

CXXXVI. LILIACEAE

Key to the Genera

1a	Leaves minute; often spinescent scales	2. *ASPARAGUS* L.
1b	Leaves well developed; without spinescent scales	(2)
2a (1)	Climbing herbs; leaves end in a tendril; flowers larger	5. *GLORIOSA* L.
2b(1)	Erect undershrubs or herbs; leaves not ending with tendril	(3)
3a (2)	Leaves thick, cartilaginous or fleshy	(4)
3b (2)	Leaves not very thick, not cartilaginous, not fleshy	(5)
4a (3)	Margin of the leaves spinous, perianth tubular, shortly toothed, reddish yellow	2. *ALOE* L
4b (3)	Margin of the leaves not spinous, perianth narrowly tubular, divided into linear segments, white or cream coloured	6. *SANSEVIERIA* Tunb. *nom. cons.*
5a (3)	Flowers in heads or umbels sheathed	1. *ALLIUM* L.
5b (3)	Flowers racemose or fascicled in racemes	4. *ASPHODELUS* L.

1. *ALLIUM* L.

Key to the Species

1a	Bulb simple, more fleshy, larger, brown to reddish in colour; leaves fistular, cylindric	1. *A. cepa* L.
1b	Bulb compound, less fleshy, smaller, compressed, pure white; leaves flat	2. *A. sativum* L.

1.1. ***Allium cepa*** L. Sp. Pl. 300. 1753; Watt Dict. Econ. Prod. 1: 169.1889; Hook.f. Fl. Brit. India 6: 337.1892; Prain Beng. Plants 3: 809.1903 (Rep. ed. 1963). *(Vern. Name: Piyaj).*

Bulbous herb with pungent odour. Leaves fistular, sub-distichous. Head dense, with cymose umbellate flowers; flowers trimerous. *Fls. & Frts.:* January-March.

Commonly cultivated as winter vegetable, throughout the lower Ganga Plain.

Field Collection: Garia/on a public garden, *K.N. - 321.*

Distributions: Cultivated throughout India. A native of Iran.

Economic Importances: Bulbs are largely used as spices and delicious vegetable to the Bengalees and Punjabi. The flowering green stocks are also cooked as vegetable. Medicinally, it acts as heart stimulant, so that it increases pulse volumes and also helps the coronary flow. It also stimulates smooth masculature; promotes bile production and reduce blood sugar. Fresh juice is bactericidal. Regular intake of onion help coronary flow.

1.2 ***Allium sativum*** L. Sp. Pl. 296.1753; Watt Dict. Econ. Prod. 1: 172.1889; Hook.f. Fl. Brit. India 6: 337.1892; Prain Beng. Plants 2: 809.1903 (Rep. ed. 1963). *(Vern. Name: Rasun).*

Annual herb, with strong pungent odour. Bulb smaller, compressed, pure white. Leaves flat. Heads lax, bearing both flowers and bulbils. *Fls. & Frts.:* January-March.

Mostly cultivated as winter annuals, throughout the lower Ganga delta.

Field Collection: Diamond Harbour, *K.N. - 397.*

Distributions: Cultivated throughout India. Indigenous to Central Asia.

Economic Importances: Used as spices or in condiment. Juice is rubeficient and helpful as ear drop against deafness. Allicin is effective against gram-negetive organism. It is excellent in atonic dyspepsia and in pulmonary phthisis. It is useful in cough and is a good expectorant; it is also useful in gout and rheumatism.

2. *ALOE* L.

2.1. ***Aloe vera*** L. Sp. Pl. 320.1753; Dict. Econ. Prod. India 1: 186. 1889; Hook.f. Fl. Brit. India 6: 264.1892; Wlh. India 1: 61.1948. *A. barbadensis* Mill. *(Vern. Name: Ghrita-Kumari).*

Acaulescent herb. Leaves thick, fleshy with spiny margins and mucilagenous contents. Flowers racemose, borne on long scapes, violent. *Fls. & Frts.:* March-July.

Commonly cultivated in the ornamental gardens.

Field Collection: Subuddhipur/Baruiur, *K.N. - 2296.*

Distributions: Cultivated throughout India. A native of South America.

Economic Importances: Ornamental potted plant; its leaves are useful in the treatment of blood-pressure.

3. *ASPARAGUS* L.

3.1. ***Asparagus racemosus*** Willd. Sp. Pl. 2: 152.1799; Watt Dict. Econ. Prod. 1: 345.1889; Hook.f. Fl. Brit. India 6: 316.1892; Prain Beng. Plants 2: 805.1903 (Rep. ed. 1963). *Asparagopsis javanica* Kunth. Enum. 5: 100. 1850. *(Vern. Name: Satamuli).*

Scandent undershrub, root tuberous, fasciculated. Stem terete, woody, branched, erect or decurved spines. Cladodes glabrous, sulcate. Leaves scale-like, spinescent. Flowers white, small, fragrant, numerous in simple or branched racemes. Fruit berry, globose. Seeds black, ellipsoid, globose. *Fls. & Frts.:* December-March.

Commonly cultivated/planted, but occasionally as a garden escape.

Field Collection: Harinbari/Sagar Island, *K.N. - 492.*

Distributions: Throughout the Tropical and Sub-Tropical regions of India, ascending to Himalaya and Kashmir. Sri Lanka and Tropical regions of the world.

Economic Importances: Roots are used as food, spices as well as condiment and are also antidysenteric.

4. *ASPHODELUS* L.

4.1. ***Asphodelus tenuifolius*** Caven, Anal Cienc. nat. 3: 46.t.27.f. 2.1810; Naskar Aq. & Semi-Aq. Plants 237.1990.

Erect annual herb; gregariously grow on the wet and waste places. Naskar (l.c.)described this taxon from the lower Ganga delta after collecting from Shyamnagar *(K.N. - 2090).* Its seeds are diuretic, applied externally to inflamed parts.

5. *GLORIOSA* L.

5.1. ***Gloriosa superba*** L. Sp. P. 305. 1753; Watt Dict. Econ. Prod. 3: 506. 1890; Hook.f. Fl. Brit. India 6: 358.1892; Prain Beng. Plants 2: 807.1903 (Rep. ed. 1963). *Methonica superba* Lam. Encyc.Method 4: 133. 1797. *(Vern. Name: Bishalanguli* or *Ulatchandal).*

Glabrous climber; stem slender, terete. Leaves opposite to alternate, sub-sessile, ovate-lanceolate, entire, parallel nerved, mid-nerve strong. Capsule linear oblong. *Fls. & Frts.:* July-November.

Frequently found, scrambling among bushes and associated with *Zingiber* spp., *Dioscorea alata, Cleome* sp.

Field Collection: Sagar Mela/Sagar Island, *K.N. - 438.*

Distributions: Throughout India. Sri Lanka, Malaya Peninsula, China and Tropical Africa.

Economic Importances: The tuber is reported to be useful in ulcers, piles, leprosy, eye-disease, rheumatism, itches, asthma. It is also purgative, anthelmintic, and used in scorpion sting.

6. *SANSEVIERIA* Thunb. *nom.cons.*

6.1. ***Sansevieria roxburghiana*** Schultz.f. Syst. Veg.7: 357. t. 12.1829; Hook.f. Fl. Brit. India 6: 271.1892; Prain Beng. Plants 2: 792.1903 (Rep. ed. 1963). *(Vern. Name: Murga).*

Creeping, rhizomatous shrub. Leaves straight, linear, flat. Racemes, densely flowered; flowers sub-erect, bracteate, arranged in 2-4 on small tubucles on the racemose rachis. Perianth tubular, split half way down into revolute lobes. *Fls. & Frts.:* June-October.

Generally cultivated, but rarely noticed as garden escape.

Field Collection: Kestopur/Baguihati, *K.N. - 1892.*

Distributions: Throughout India. Sri Lanka, and the warmer parts of the World.

Economic Importances: Rhizome is said to be used in consuptive complaints and coughs. Juice of the tender shoot is used to clear throats of viscid phlegm of the children.

Notes: Previously the genus-*Sansevieria* Thunb. was treated under the family-Haemodoraceae.

Remarks: Other cultivated species noted are *Cordyline terminalis* Kunth., *Dracaena spicata* Roxb., *Hemerocallis fulva* L.

CXXXVII. SMILACACEAE

Previously, the genus-*Smilax* L. was treated under Liliaceae, but now it is recognised as a separate family.

1. *SMILAX* L.

Key to the Species

1a	Umbels 1-3, rarely 5	1. *S. macrophylla* Roxb.
1b	Umbels many, usually verticillate on rachis	2. *S. perfoliata* Lour.

1.1. ***Smilax macrophylla*** Roxb. Fl. Ind. 3: 793.1832; Hook.f. Fl. Brit. India 6: 310.1892; Watt Dict. Econ. Prod. 6 (3): 255.1893; Prain Beng. Plants 2: 806. 1903 (Rep. ed. 1963). S. *ovalifolia* Roxb. Fl. Ind. 3: 794.1832. *(Vern. Name: Kumarika).*

Climbing shrubs with prickles. Leaves alternate, broadly ovate or sub-orbicular, acuminate or cuspidate, glarbous, shining, base rounded; 5-7 nerved, reticulate; presence of long, slender tendrils. Flowers in pedunculate, 1-3 rarely 5 flowered umbels. Berry globose, 1-3 seeded, seeds biconvex. *Fls. & Frts.:* July-November.

Very common on the village shrubberies, thickets and hedges.

Field Collections: Beliachandi/Sarberia, *K.N. - 267;* Taki, *K.N. - 1033.*

Distributions: Throughout India. Both Tropical and Sub-Tropical regions of the world.

Economic Importances: Root is applied for rheumatism and pains in the lower extremities.

1.2. ***Smilax perfoliata*** Lour., Fl.Cochinch. 622.1790; Koyama Adv. Front. Pl. Sci. 4: 59.1863. *S. prolifera* Wall. ex Roxb. Fl.Ind. 795.1832; Hook.f. Fl. Brit. India 6: 312.1892; Prain Beng. Plants 2: 806.1903 (Rep. ed. 1963). *S.laurifolia* Roxb. Fl. Ind. 3: 793.1832. *(Vern. Name: Kumarika).*

Stem prickled, recurved. Leaves broadly elliptic-oblong or ovate-oblong, rounded or obtuse or slightly emerginate with a mucronate tip at apex, entire, with prominant parallel veins. Inflorescence in axillary racemose, developing umbels, many flowered. Flowers fragrant. Fruits globose or sub-globose, red when ripe. *Fls. & Frts.:* July-December.

Rarely found on the dry waste lands of village shrubberies and scrub-jungles.

Distributions: Major parts of India, especially on the dry regions. Common in Sri Lanka and Burma.

CXXXVIII. PONTEDERIACEAE

Key to the Genera

1a	Tepals forming a distinct tube below; petioles swelling, spongy, anther equal	1. *EICHHORNIA* Kunth. *nom. cons*
1b	Tepals nearly free; petioles not swelling; anther unequal	2. *MONOCHORIA* Presl.

1. *EICHHORNIA* Kunth. *nom. cons.*

1.1. ***Eichhornia crassipes*** (Mart.) Solms. Laub., DC. Mon. Phan. 4: 527.1883; Naskar Aq.& Semi-Aq. Plants 238.1990. *Pontederia crassipes* Mart. Nov. Gen. Sp. Pl. 9.t. 4.1856. *P. azurea* Hook. Bot. Mag. t. 2932.1829. *Eichhornia speciosa* Kunth., Enum. Pl. 4: 131.1843. *E. crassicaulis* Sch., Abh. Natur. f. Ges. Halle 4: 177. 1862. *(Vern. Name: Kachuripana/Bilatipana).*

Perennial, rooted floating herb, with offset. Very common everywhere. Naskar (l.c.) described this taxon from the lower Ganga delta.

2. *MONOCHORIA* Presl.

Key to the Species

1a	Root-stock creeping, branced; leaves almost triangular-ovate, usually with a sagittate or hastate base	1. *M. hastata* (L.) Solms.
1b	Root-stock sub-erect or oblique usually very short; leaves varying from broadly ovate to ovate-oblong from an obtuse round truncate or a cordate base	2. *M. vaginalis* (Burm.) Presl.

2.1. ***Monochoria hastata*** (L.) Solms, DC. Monogr. Phan. 4: 523.1883; Naskar Aq. & Semi-Aq. Plants 240.1990. *Pontederia hastata* L. Sp. Pl. 1753. *Monochoria hastaefolia* Presl., Reliq. Haenk. 1: 128.1827. *(Vern. Name: Nukha).*

Robust swampy herb. Naskar (l.c.) described this taxon after collecting from Madhyampur/Taki *(K.N. - 954);* Baruipur *(K.N. - 2252);* Titagarh *(K.N. - 2199)* and Taki, *(K.N. - 1004).*

2.2. ***Monochoria vaginalis*** (Burm.) Presl. Reliq. Haeule. 1: 128.1827; Naskar Aq. & Semi-Aq. Plants 241.1990. *Pontederia vaginalis* Burm.f. Fl. Ind. 80.1768. *(Vern. Name: Nukha).*

Swampy herb. Root stock short, sub-erect. Naskar (l.c.) described this taxon after collecting from Ghushighata/ near Kulti *(K.N. - 1113)*.

2.2./1. ***Monochoria vaginalis*** (Burm.f.) Presl. *var.* ***plantaginea*** (Roxb.) Solms. A. & C. DC. Monogr. Phan. 4: 524.1883; Naskar Aq.& Semi-Aq. Plants 242.1990.

Naskar(l.c.) described this taxon after collecting from L. Bengal.

CXXXIX. XYRIDACEAE

1. *XYRIS* L.

Key to the Species

1a	Leaves flat, obtuse	1. *X. indica* L.
1b	Leaves linear, acute	2. X. pauciflora Willd.

1.1. ***Xyris indica*** L. Sp. Pl. 42.1753; Naskar Aq. & Semi-Aq. Plants 242.1990. *(Vern. Name: Chine-Ghas).*

Erect, annual 15-50 cm high; tufted aquatic herb. Naskar (l.c.) described this taxon after collecting from Ghoshpur/near Kulti *(K.N. - 2129).* This plant cure ring-worm, itch and leprosy.

1.2. ***Xyris pauciflora*** Willd., Phytogr. 1: 2, t.f.1 & Sp. Pl. 1: 255.1797; Naskar Aq. & Semi-Aq. Plants 243.1990.

Tufted erect annual herb on the marshy and wet lands and even on the cultivated fields. Naskar (l.c.) described this taxon after collecting from Bantala/Salt Lake *(K.N. - 1741).*

CXL. COMMELINACEAE

Key to the genera

1a	Fertile stamens 3; staminodes present	(2)
1b	Fertile stamens 6; staminodes absent	(3)
2a (1)	Cymes 1 or 2; arising from a spathaceous bracts	2. *COMMELINA* L.
2b (1)	Cymes panicled; not arising from a spathaceous bract	4. *MURDANNIA* Royle *nom. cons.*
3a (1)	Bracteoles conspicous	3. *Cyanotis* D. Don. *nom. cons.*
3b (1)	Bracteoles inconspicous	1. *AMISCHOPHA-CELUS* Rolla Rao et Kammathy

1. *AMISCHOPHACELUS* Rolla Rao et Kammathy

1.1. ***Amischophacelus axillaris*** (L.) Rolla Rao & Kammathy, Journ. Linn. Soc. (Bot.) 59 (379) : 306.1966; Naskar Aq. & Semi-Aq. Plants 244. 1990. *Commelina axillaris* L. Sp. Pl. 42. 1753. *Tradescantia axillaris* (L.) L. Mant. 321.1771. *Cyanotis axillaris* (L) J.A. & J. H. Sch. f. Syst. Veg. 7(2): 1154.1830. *Zygomenses axillaris* (L.) Salisbury, Trans. Gort. Soc. 1: 271. 1872. *(Vern. Name: Soltraj* or *Bhaghanula).*

Erect annual, diffused herb with fibrous roots. Leaves thick, elongate, linear or linear-lanceolate, acute, pubescent. Commonly grow on the marshy waste lands. Naskar (l.c.) described this taxon after collecting from Beliachandi *(K.N. - 264)* It is used for tympanitis and as external application in ascites.

2. *COMMELINA* L.

Key to the Species

1a	Leaves ovate to ovate-lanceolate, ovules -5	(2)
1b	Leaves linear or linear-lanceolate; ovules 2-5, rarely 4 or 5	(3)
2a (1)	Leaves ovate, sub-acute to round apex	1. *C. benghalensis* L.
2b (1)	Leaves ovate-lanceolate, acute or acuminate at apex	2. *C. diffusa* Burm.f.
3a (1)	Spathes complicated with dorsal margin, free to the base, lanceolate, usually acuminate	4. *C. salicifolia* Roxb.
3b (1)	Spathes funnel-shaped, connate nearly to the top, never acuminate	(4)
4a (3)	Capsule 3-celled, rarely 2	(5)
4b (3)	Capsule 2-celled; seeds with large pit	5. *C. suffruticosa* Blume
5a (4)	Spathe 2 cm; seeds white margined	6. *C. undulata* R. Brown
5b (4)	Spathe 2-5 cm; seed grey smooth, compressed	3. *C. obliqua* Buch. Ham.

2.1. ***Commelina benghalensis*** L. Sp. Pl. 41.1753; Naskar Aq. & Semi-Aq. Plants 246.1990. *(Vern. Name: Anchra* or *Kanchira).*

Creeping, glabrous or pubescent herb; common on the moist places. Naskar (l.c.) described this taxon after collecting it from Santoshpur, *(K.N. - 747)*; Goshpur*(K.N. - 1121)* . It is bitter and useful in dermotology as refrigerant, laxative, beneficial in leprosy. Its root is remedial for headache.

2.2. ***Commelina diffusa*** Burm.f.Fl. Ind. 18.t.7.f.2.1768; Naskar Aq. & Semi-Aq. Plants 247.1990. *C. nudiflora* auct. plur. non. Linn. Sp. Pl. 41.1753.

Creeping herb on the moist places. Naskar (l.c.) described this taxon after collecting from Fultala/Baruipur *(K.N. - 46)*.

This plant is useful in burns, itches and boils; leaves are also used for poulticing sores.

2.3. ***Commelina obliqua*** Buch-Ham ex Don, Prodr. 45.1825 (not of Vahl); Naskar Aq. & Semi -Aq. Plants 247.1990. *C. paludosa* Blume, Pl. Java 1: 2. 1827. *(Vern. Name: Jata-Kanchira).*

Stout, branched herb. Commonly found on the village roadsides, swampy areas, thickets and garden edges. Naskar (l.c.) described this taxon after collecting from Madarhat/Baruipur *(K.N. - 12)* and Joygram/ Minakhan *(K.N. - 1133)* Root is antidote to snake poison; and also useful in fever and bilious affections, laxative and refrigerant.

2.4. ***Commelina salicifoia*** Roxb. Fl. Ind. 1: 172. 1832. *C. longifolia* Lamk. Ill. Gen. 1: 129. 1791; Rola, Taxon 10: 254.1961; Naskar Aq. & Semi-Aq. Plants 248. 1990. *(Vern. Name: Pani-Kanchira).*

Slender stemed, diffused or spreading herbs with long internodes. Very common found in the roadside hedges and wall of the old houses. Naskar (1.c.) described this taxon after collecting from Fultala/Baruipur *(K.N. - 46)*. and Metia/Dhanyakuria *(K.N. - 570)*. This plant is useful in dysentery and insanity.

2.5. ***Commelina suffruticosa*** Blume, Hook.f. Fl. Brit. India 6: 374.1892; Naskar Aq. & Semi-Aq. Plants 249.1990.

Stout, branched and scrambling herb with creeping rhizomes; commonly grow on the moist, humus soil. Naskar (1.c.) described this taxon after collecting it form Brace Bridge *(K.N. - 1217)*.

2.6. ***Commelina undulata*** R. Brown 270. 1810; Hook.f. Fl. Brit. India 6: 373.1892; Rolla, Notes Roy. bot. Gard. Edin. 26(4): 352.1965. *C. kurzi* Clarke, J. Linn. Soc. 6: 444.1871; Hook.f. Fl. Brit. India 6: 373. 1892. *C. erecta* L. Sp. Pl. 41. 1753; Morton, J. Linn. Soc. London 60(382): 183.1967; Naskar Aq. & Semi-Aq. Plants 247.1990.

Erect, stout, branched herbs; rarely found on the village thickets and hedges. Naskar (1.c.) described this taxon after collecting from Bantala/Salt Lake *(K.N. - 1745)*.

Notes: Prain (1903a) did not mention this taxon from the lower Ganga delta.

3. *CYANOTIS* D. Don *nom. cons.*

3.1. *Cyanotis cristata* (L.) D. Don, Prodr. 45.1825; Naskar Aq. & Semi-Aq. Plants 250. 1990.

Stem branched, erect or creeping; glabrous with hairs; rarely grow on the marshy places, on the roadside ditches or thickets. Naskar (1.c.) described this taxon from the lower Ganga delta after collecting from Shyamnagar *(K.N. - 2101)*.

4. *MURDANNIA* Royle *nom. cons.*

(Now it is treated as distinct genus, though Baker & Bakhunizen f. (Fl. Java 3: 17.1968) has treated it under *Aneilema* R. Brown).

Key to the Species

1a	Ovary 1-ovuled	3. *M. vaginatum* (L.) Brueckner.
1b	Ovary with 2-ovules or more	(2)
2a (1)	Ovary 3 or more ovules; capsule 3-7 seeded; panicles terminating leafy stem, leaves large, narrowly oblong or ensifrom, panicle large	2. *M. spirata* (L.) Bruechner
2b (1)	Ovary-2 ovules; capsules 2-seeded; panicle sub-terminal, rarely axillary, leaves linear or lanceolate	1. *M. nudiflora* (L.) Brenan

4.1. *Murdannia nudiflora* (L.) Brenan, Kew Bull. 189. 1952; Naskar Aq. & Semi-Aq. Plants 251. 1990. *Commelina nudiflora* L. Sp. Pl. 41.1753. *Tradescantia malabarica* L. Sp. Pl. (ed.2) 412.1762. *Aneilema nudiflora* R. Brown, Prodr. 271. 1810. *A. nudiflorum* (L.) Wall. List 182. No. 5224.1890. *Murdannia malabarica* (L.) Brueckner, Pfam. (ed.2) 15A: 173. 1930.

Annual herb, branches decumbent, common on the moist, shady places and also as weed on the agricultural fields. Naskar (1.c.) described this taxon after collecting from Ghoramara/Sagar Islands *(K.N. -408);* Sankchura/near Taki *(K.N. - 987)* and Minakhan *(K.N. - 1999).*

It is cooked by the villagers and very useful for the treatment of leprosy. It is also useful for headache, asthma, piles, stomach and kidney troubles and some other diseases.

4.2. ***Murdannia spirata*** (L.) Brueckner, Pfamilien (ed.2) 15A: 173.1930; Naskar Aq. & Semi-Aq. Plants 252. 1990. *Commelina spirata* L. Mant. 176. 1767. *Aneilema spiratum* R. Brown Prodr. 271. 1810.

Branched, procumbent herb. Occasionally grow on the shady, waste places and village thickets or roadsides. Naskar (1.c.) described this taxon after collecting from Madhyampur/near Taki *(K.N.- 952)* and Sankchura/near Taki *(K.N. - 990).*

4.3. ***Murdannia vaginatum*** (L.) Brueckner, Pfamilien (ed.2) 15A: 173. 1930; Naskar Aq. & Semi-Aq. Plants 252. 1990. *Aneilema vaginatum* R. Brown Prod. 271. 1810.

Aquatic herb. With fibrous root and decumbent stem. Naskar (1.c.) described this taxon after collecting from Dhopagachi *(K.N. - 2142).*

Notes: Rheo discolar Hance, an American species is also cultivated largely as ornamental potted plant throughout the lower Ganga delta.

CXLI. FLAGELLARIACEAE

1. *FLAGELLARIA* L.

1.1. ***Flagellaria indica*** L. Sp. Pl. 33. 1753; Naskar & Guha Bakshi Mangr. Swamps Sundarbans 127. 1987; Naskar Aq. & Semi-Aq. Plants 253. *(Vern. Name: Ban-Chanda).*

Tall, reed-like climber; climbing by the leaf-tendrils. Common on the hedges and thickets. Naskar & Guha Bakshi (1.c.) and Naskar (1.c.) described this taxon after collecting from Sundarbans/Frazergunj *(K.N. - 1912).*

CXLII. ARECACEAE

(*nom. alt.* Palmaceae)

Key to the Genera

1a	Leaves palmatisect or fan like	2. *BORASSUS* L.
1b	Leaves pinnatisect or pinnate	(2)
2a (1)	Stem erect, tall and stout	(3)

2b (1)	Stem climbing, rarely erect with much armed	3. *CALAMUS* L.
3a (2)	Dwarf palm of tidal swamps, stem much-branched and buried	6. *NYPA* Steck
3b (2)	Tall palms, with erect and unarmed stems	(4)
4a (3)	Fruit large ovoid, nut with fibrous husk	5. *COCOS* L.
4b (3)	Fruit small, round or ovoid, 1·or 2 seeded berry	(5)
5a (4)	Spadix 3-4 m long; branches simple, forming a dense tassel drooping from the stout short peduncle	4. *CARYOTA* L.
5b (4)	Spadix less than 1 m; branches compound	(6)
6a (5)	Fruit with fibrous husk	1. *ARECA* L.
6b (5)	Fruit with no fibrous husk	7. *PHOENIX* L.

1. *ARECA* L.

1.1. ***Areca catechu*** L. Sp. Pl. 1189. 1753; Watt Dict. Econ. Prod. 1: 291. 1889; Beccari, Hook.f. Fl. Brit. India 6: 405. 1892; Prain Beng. Plants 2: 826. 1903 (Rep. ed. 1963); Wlh. Ind. 1: 110. 1948. *(Vern. Name: Supari).*

Trunk quite straight, solitary upto 25 m height and about 15 cm thick. Leaves pinnate, 1.5- 2.5 cm long, soft, tufted on the head with sheathing leaf base; leaflets numerous. Spathe double, compressed; spadix much branched; male flowers very numerous, sessile, stamens 6, sagittate; female flowers solitary. Fruit small, ovoid 2.5-3.0 cm diam, numerous. *Fls. & Frts.:* Almost throughout the year.

Mostly planted, throughout the lower Ganga delta, common on the garden boundary.

Field Collection: Dhopagachhi/Baruipur, *K.N. - 2141.*

Distributions: Planted almost throughout India, mostly on the moist Tropical belt and coastal regions. A native of S.E. Asia.

Economic Importances: Dry leaves are used as fire-wood; ripe and dried fruits are commonly used with *'pan'* or 'mukhsuddhi'. Nut is anthelmintic and is useful against both tape-worms and round-worms. It has also digestive, astringent and tonic properties.

2. *BORASSUS* L.

2.1. ***Borassus flabellifer*** L. Sp. Pl. 1187. 1753; Beccari, Hook.f. Fl. Brit. India 6: 482. 1893 (excl. Syn. *B. aethiopum);* Prain Beng. Plants 2: 822.1903 (Rep. ed. 1963). *B. flabelliformis* Murr. Syst. (ed. 13) 827.1774. *(Vern. Name: Tal).*

Unbranched, 20-30 m tall, 20-30 cm radius; leaf-scar present on the stem. Leaves palmately compound, stip, leaf-base hard, woody with bifurcate sheathing base and with dented sharp margin. Male spadix branched clustered 20-30 cm x 2-3 cm. Female flowers ± 2.5 cm. Fruit 12-18 cm diam. *Fls. & Frts.:* January-September.

Rarely planted, generally self-sown on the embankments of village ponds, canal-sides, road-sides and on the waste lands.

Field Collection: Kachuberia/Sagar Islands, *Guha Bakshi & Sen-4.*

Distributions: Very common in all the hot and damp parts of India, including lower Bengal. Burma, Sri Lanka, Malaya and Tropical Africa. Considered a native of Africa.

Economic Importances: Leaves are largely used in fans, carpets, mats and umbrella manufacturing. Both green and ripe fruits are edible. Juice of the young inflorescences is collected and consumed as delicious 'ras' or 'toddy'. This inflorescences juice is used to prepare mollasses after boiling. Fruits are much used as an aperient and improve digestion. The juice is a stimulant and an aperient.

3. *CALAMUS* L.

3.1. ***Calamus viminalis*** Willd. var. ***fasciculata*** Becc. ex Hook.f. Fl. Brit. India 6: 444. 1892; Prain Beng. Plants 2: 827. 1903 (Rep. ed. 1963). *(Vern. Name: Bent)..*

Stout, tufted ratten shrub. Leaflets pinnatifid, many inequidistant or fascicled, linear lanceolate, petioles short, traingular to sub-quadrangular with sharp spines. Rachis terete. Spadix long, branched. Spikes distichous. Flowers greenish-yellow. Fruit sub-globose, covered by broad and shining scales. *Fls. & Frts.:* September-May.

Gregariously grow in the forest thickets or in the scrub-jungles and waste lands.

Field Collections: Magrahat (*K.N. - 249*); Salipur/Baruipur (*K.N. - 2152*).

Distributions: Throughout the plains and low hills of India. A native of Malaya and Burma.

Economic Importances: Used for basket and furniture makings.

4. *CARYOTA* L.

4.1. ***Caryota urens*** L. Sp. Pl. 1189. 1753; Watt Dict. Econ. Prod. 2: 206. 1890; Beccari & Hook.f. Fl. Brit. India 6: 422. 1893; Prain Beng. Plants 2: 822. 1893 (Rep. ed. 1963). *(Vern. Name: Gol-sago).*

Trunk straight, unarmed, 15-20 m high, 30-45 cm diam., cylindric annualate, smooth. Leaves bipinnate, 5-6 m X 3-4 m, leaflets 10- 15 cm long, fasciculate or alternate; petiole very stout, with fibrous netted margins. Spadix 3-4 m long; branches simple, forming a dense tassel drooping from the stout short peduncle. Male flowers-buds narrowly cylindric, sepals cordate ciliate; petals coriaceous, concave, reddish. Stamens about 40; filaments short, white, anther acuminate. Female flowers: sepals rather broader than the male. Staminodes usually 3. Ovule solitary, erect. Fruit about 2 cm in diam., reddish; pericarp acrid. *Fls. & Frts.:* Almost throughout the year.

Planted only as an ornamental plant, on the avenue, lawn and gateways.

Field Collection: Specimen not preserved.

Distributions: Throughout the hotter parts of India. Sri Lanka, Tropical Asia and Malaya.

Economic Importances: Besides the ornamental uses, its leaf-sheath fibre is useable. The pith yields a good sago. The inflorescences is also tapped for toddy.

5. *COCOS* L.

5.1. ***Cocos nucifera*** L. Sp. Pl. 1188. 1753; Watt Dict. Econ. Prod. 2: 415. 1890; Beccari & Hook.f. Fl. Brit. India 6: 482. 1893; Prain Beng. Plants 2: 824. 1903 (Rep. ed. 1963). *(Vern. Name: Dab* or *Narikel).*

Unarmed, 20-30 m tall, with 20-30 cm radius trunk. Leaves pinnately compound long tufted at the top, leaf-base sheathing, woody. Inflorescences compound spadix with a fleshy spathe. Fruit large 20-30 cm x 25-35 cm, ovoid, trigonous. *Fls. & Frts.:* Throughout the year.

Commonly cultivated throughout the delta in the orchards, roadsides and embankments of the ponds.

Field Collection: Specimen not preserved.

Distributions: Throughout India, mainly in the coastal regions. Believed to be indigenous of Cocos Islands and North Andamans.

Economic Importances: It exhibited more than 80 articles of commerce. Fibrous pericarp yields coir, used for ropes and mats. The coconut-oil is edible and also used for hair oil. Coconut juice and the pericarp is delicious drink and food, respectively. Mid rib of the leaves is also used as house duster.

Notes: The coconut plantation is also common practice here.

6. *NYPA* Steck

6.1. ***Nypa fruticans*** (Thunb.) Wurmb., Verh. Batav. Genootsch 156. 1781; Naskar & Guha Bakshi Mangr. Swamps Sundarbans 86. 1987. *(Vern. Name: Golpata).*

Looks like sunken coconut tree; no well developed aerial roots; leaf looks like coconut leaf; leaf-base serve the function like aerial roots, pneumatophores for the gas exchange. Leaves 5-6 m long, pinnatisect, large, midrib woody, round. Flowers monoecious on terminal erect branches, spadix. Fruit large 10-15 cm, globose 1-seeded, seed length upto 10 cm, modified hypogeal germination, cryptoviviparous. *Fls. & Frts.:* Throughout the year.

Grow on the inter-tidal river flats or ridge tidal inundated swampy forest as mangrove associated species.

Field Collection: Kalistan/Bakkhali, *K.N. - 2390.*

Distributions: Swampy mangrove delta of the Ganga-Brahmaputra estuary, Sundarbans, Andaman & Nicobar Islands. In Sri Lanka, Burma, Malay Peninsula through Indonesia to New Guinea and Solomon Islands, Philippines, Ryukyu Islands and Queensland. Its isothermic leaves are very much durable for house thatching and also used in domestic purposes. For these rapid exploitation, it is now endangered in many parts of the natural mangrove forests.

Economic Importances: Its inflorescences sap or juice is also important for alcohol and mollasses preparation and exploited in Indonesia, Malaysia and Philippines.

Notes: At present in the Sundarbans Mangrove forests this species, *Nypa fruticans* is only restricted in the S.E. portion on the core-area of the reserve forests. Some workers considered it as the endangered species in Sundarbans.

7. *PHOENIX* L.

Key to the Species

1a	Stem unarmed, slender, 10-20 m tall, 20-25 cm radious; clothed with persistent leaf-base, grow almost everywhere	2. *P. sylvestris* (L.) Roxb.
1b	Stem soboliferous; 3-10 m tall, 8-10 cm radious; confined to muddy maritime swamps	1. *P. paludosa* Roxb.

7.1. ***Phoenix paludosa*** Roxb. Fl. Ind. 3: 789. 1832: Naskar & Guha Bakshi Mangr. Swamps Sundarbans 87.1987. *(Vern. Name: Hental).*

Stem stoloniferous, 3-4 m tall and 2.5- 3.5 cm diam., covered with a dark fibrous network of the sheath and long pointed base; occasionally developed slender pneumatophores like structures. Leaves 1-2 m long, leaflets bifarious. Spadix with numerous spreading branches, upto 1 m long; stamens 6, short; female flowers with 6 scale like staminodes. Fruits drupe, shining black when ripe. Seeds cartilaginous, germination hypogeal, cryptoviviparous. *Fls. & Frts.:* March-July.

Commonly grow on the tidal and inter-tidal mangrove forests of the Sundarbans; grow on the porous and mature soil of the riversides or forming ridge dense forests.

Field Collections: Sagar Mela Premises, *K.N. - 445;* Mouli/near Minakhan, *K.N. - 1357.*

Distributions: Almost throughout the coastal mangrove forests on inter-tidal zone of the Indian coasts, Andaman & Nicobar Islands. Burma, Sri Lanka, Malaya and Tropical Asia.

Economic Importances: Villagers in the Sundarbans area are largelly use the stem to build their cottage and also used as fire-wood. For these rural people uses, this plant is now exploited very rapidly.

7.2. ***Phoenix sylvestris*** (L.) Roxb. Fl. Ind. 3: 787. 1832; Watt Dict. Econ. Prod. 6(1): 208. 1892; Beccari & Hook.f. Fl. Brit. India 6: 425. 1892; Prain Beng. Plants 2: 825. 1903 (Rep. ed. 1963). *Elate sylvestris* L. Sp. Pl. 1189. 1753, proparte. *(Vern. Name: Khenjur).*

Stem unarmed, solitary and stout 10-20 m tall. Leaves-scars tufted throughout the stem; petiole short, compressed with modified leaflets as spines. Male spadix scented, white, branches slender, peduncle compressed; spathe separating into 2- boat shaped valves; flowers many. Female spadix short, flowers scattered. Fruit-drupe, orange to reddish brown, with sweet edible flesh. *Fls. & Frts.:* March-July.

Rarely planted but common as self-sown in the village thickets, roadsides, scrub-jungles and village gardens as well as on the embankments of the ponds, lakes, canals. Prefer dry sandy soil, somewhat zerophytic in habitat.

Field Collection: Sagar Island, *K.N. - 2420.*

Distributions: Throughout India, mainly on the dry and plain districts; a native of India. Wild in Indus basin, Burma and Sri Lanka.

Economic Importances: Its leaves are used for making mat, baskets, bags, broom & fan making. Fruits are inferior food but commonly edible by the villagers. Its fruits is considerd as remedy for heart diseases, fevers, abdominal complaints and insanity. Root is useful for cure toothache. Its trunk is used for domestic cottage and other domestic uses and as fuel.

Remarks: Other cultivated/ornamental species of Arecaceae are– *Arenga pinnata, Livistona chinensis, L. rotundifolia, Roystonea regia, R. oleracea,* and *Elaeia guinensis.*

CXLIII. PANDANACEAE

1. PANDANUS Linn.f.

Key to the Species

1a	Small tree or shrubby; sweet smelling inflorescence; ovary cell numerous	2. *P. tectorius* Soland ex Park.
1b	Shrubby; foetid inflorescence; ovary cell one	1. *P. foetidus* Roxb.

1.1. ***Pandanus foetidas*** Roxb. Fl. Ind. 3: 742.1832; Naskar & Guha Bakshi Mangr. Swamps Sundarbans 140.1987. *(Vern. Name: Keya).*

Densely branched shrub, without any stem. Gregariously grow in small patches on the thickets. Naskar & Guha Bakshi (1.c.) described this taxon after collecting from Nilgunj *(K.N. - 2102)*. Its leaves are used for mat making, paper pulp and thatching; a valuable fibre is also produced from its leaves.

1.2. ***Pandanus tectorius*** Soland. ex Parkinson, Journal Voy. H.M.S. Endeavour 46.1775; Naskar & Guha Bakshi Mangr. Swamps Sundarbans 141. 1987. *(Vern. Name: Keya Kanta).*

Tall, gregarious dioecious small tree, 4-6 m high. Naturally grow on the waste land. Naskar & Guha Bakshi (1.c.) described this taxon after collecting from the lower Ganga delta.

Its leaves yield strong white fibre; use for making cordage, mat and sacks also, Tender floral leaves are edible as vegetable. Leaves is said to be used for leprosy, small-pox, scabies and the diseases of heart and brain.

CXLIV. TYPHACEAE

1. *TYPHA* L.

Key to the Species

1a	Leaves 1-3 cm broad, semi-cylindric above the sheath	1. *T. domingensis* Pers.
1b	Leaves 2-3 cm broad, trigonous above the sheath	2. *T. elephantina* Roxb.

1.1. ***Typha domingensis*** Pers., Syn. Pl. 2: 532. 1807; Naskar Aq. & semi-Aq. Plants 254. 1990. *T. angustata* Bory & Chaub. Exp. Sci. Morce. Bot. 2(1): 338. 1832. *T. angustifolia* Sibth. & Sm. F. Grace. Prodr. 2: 226. 1815 (not. L.) *T. elephantina* Grah. Cat. 227. (not of Roxb. 1832). *(Vern. Name: Hogla).*

Tall, 2-3 m high aquatic or marshy herb. Gregariously grow on the shallow ponds, jheels, lakes, canals and along the sides of the marshy lands. Naskar (1.c.) described this taxon after collecting from Dighirpar/Canning *(K.N. - 199)* and Gosaba *(K.N. - 677).*

1.2. ***Typha elephantina*** Roxb. Fl. Ind. 3: 566. 1832; Naskar Aq. & Semi-Aq. Plants 255. 1990. *(Vern. Name: Pati* or *Hogla).*

Tall, 2.5 - 5.0 m high, aquatic or marshy herb, robust than *T. domingensis.* Gregariously grow on the shallow ponds, jheels, lakes, canals and roadside ditches. Field Collections were made from Panihati *(K.N. - 1608)* and Rudranagar/Sagar Island *(Guha Bakshi & Sen-4).* Naskar (1.c.) described this taxon after collecting from lower Ganga delta.

Its leaves used in making mats, screens, nursery corners and thatching, also used in paper industry as raw materials. Young shoots and root stocks are edible. Root-stocks astringent, diuretic and sedative, used in measles, dysentery and gonorrhoea. Shoot decoction applied in leprosy and enlargement of spleen. Dry fruits used as an alternative to cotton swab. The plant possesses water purifying properties.

CXLV. ARACEAE

Key to the Genera

1a	Flowers bisexual	(2)
1b	Flowers unisexual, monoecious or rarely dioecious	(4)
2a (1)	Epiphytic, climbing herbs	(3)
2b (1)	Stout, spinous, aquatic or marshy herbs	5. *LASIA* Lour.
3a (2)	Perianth 4-6 segmented; ovule solitary, plant unarmed	7. *POTHOS* L.
3b (2)	Perianth absent, berries free, ovule solitary	8. *SCINDAPSUS* Schott
4a (1)	Aquatic herbs; ovaries few	(5)
4b (1)	Terrestrial herbs	(6)
5a (4)	Rooted on the earth; leaves linear or lanceolate; ovary in a single basal whorl	4. *CRYPTOCORYNE* Fisch ex Wydler
5b (4)	Floating small herb; leaves obovate cuneate, ovary single basal	6. *PISTIA* L.
6a (4)	Tuberous	(7)
6b (4)	Herbs, usually tall, coarse; leaves and scapes arising from the rhizome, leaves peltate; spadix with a barren appendage, not adnate to the spathe, spathe yellow or orange	(8)
7a (6)	Leaves entire, 3-5 lobed, cordate or hastate or pedate; spathe deep red; flowers and leaves present together	9. *TYPHONIUM* Schott.
7b (6)	Leaves solitary, 3-partite, segments pinnatisect, flowers appearing without leaves	2. *AMORPHOPHAL-LUS* Bl. ex Decne *nom. cons.*

8a (6)	Ovules few, basal	1. *ALOCASIA* D. Don *nom.cons.*
8b (6)	Ovules many on a triparietal placentae	3. *Colocasia* Clus. ex Fabric

1. *ALOCASIA* G. Don *nom. cons.*

Key to the Species

1a	Self grown, leaves comparatively small, petiole shorter and thin	1. *A. fornicata* (Roxb.) Schott.
1b	Generally cultivated; leaves larger, petiole large and thick	2. *A. indica* Schott.

1.1. ***Alocasia fornicata*** (Roxb.) Schott., Osterr. Bot. Wochenbl. 4: 410. 1854; Naskar Aq. & Semi-Aq. Plants 257. 1990. *(Vern. Name: Salu-Kachu).*

Caudex erect or sub-erect 20-25 cm high. Gregariously grow on the roadside ditches, nalas and/or shady, dampy, waste places of the village thickets. Naskar (1.c.) described this taxon after collecting from Nilgunj *(K.N. - 2103).* Rustic people cook the petiole and the slender rhizomes during crisis or femine.

1.2. ***Alocasia indica*** Schott., Osterr. Bot. Wochenbl 4: 410. 1854; Watt Dict. Econ. Prod. 1: 178. 1889; Hook.f. Fl. Brit. India 6: 525. 1893; Prain Beng. Plants 2: 836. 1903 (Rep. ed. 1963). *Arum indicum* Roxb. Fl. Ind. 3: 498. 1832. *(Vern. Name: Man Kachu).*

Stem stout with bulbiferous suckers. Leaves large, ovate, deeply sagittate, cordate repand, lobes rounded; tube of the spathe many times shorter than the very long, linear oblong, subtruncate, cuspidate limb, stigma sessile. *Fls. & Frts.:* May- September.

Mostly cultivated in the dampy, shady places throughout the lower Ganga delta.

Field Collection: Specimen not preserved.

Distributions: Bengal, Assam and throughout India. Tropical Asia.

Economic Importances: Root-stock and leaves are commonly used as vegetable. In Bengal, it forms an important vegetable.

2. *AMORPHOPHALLUS* Blume ex Decne. *nom. cons.*

Key to the Species

1a	Forks of leaves bulbiferous, petiole smooth, appendage rose or light pink	1. *A. bulbifer* (Roxb.) Blume

1b	Forks of the leaves not bulbiferous, petiole minutely tuberculate; appendage dark-purple	2. *A. companulata* (Roxb.) Blume ex Decne

2.1. ***Amorphophallus bulbifer*** (Roxb.) Blume, Rumphia 1: 148.1835; Watt Dict. Econ. Prod. 1: 225. 1889; Hook.f. Fl. Brit. India 6: 515.1893; Prain Beng. Plants 2: 835.1903 (Rep. ed. 1963). *Arum bulbiferum* Roxb. Fl. Ind. 3: 510. 1832. *(Vern. Name: Amla Bele).*

Tuber globose, 5-15 cm diam. Leaves bulbiferous at forks; lobes vary in size and shape. Spathe erect, ovate when unrolled acute, pinkish yellow without, rosy or pinkish within. Appendage flesh coloured, oblong cylindric, apex obtuse. *Fls. & Frts.:* April-August.

Common on the roadside and railway tracks, prefer dampy situations.

Field Collection: Dhopagachi/near Baruipur, *K.N. - 2143.*

Distributions: Throughout India, Burma.

Economic Importances: Rhizomatous roots and petioles are edible.

2.2. ***Amorphophallus companulatus*** (Roxb.) Blume ex Decne Nouv. Ann. Mus. Hist. Nat. Paris 3: 366. 1834; Watt Dict. Econ. Prod. 1: 225. 1889; Hook.f. Fl. Brit. India 6: 513. 1893; Prain Beng. Plants 2: 835. 1903 (Rep. ed. 1963). *Arum companulatum* Roxb. Corom. Pl. 3: 68,t.272.1819. *(Vern. Name: Ol).*

Tuber depressed, bulbiferous sphere. Leaves broad, segments simple or forked, leaflets oblong, acute; petiole dark green with blotches. Peduncle elongating in fruit; sheaths linear oblong. Spathe campanulate, limb sub-erect or recurved, waved and crenulate, spadix hardly longer than the spathe, appendage globose amorphous, sinously lobed. *Fls. & Frts.:* March-November.

Mostly cultivated, rarely as a garden escape in the village thickets or waste places, garbages.

Field Collection: Specimen not preserved.

Distributions: On the plains of India. Malaya to Polynesia.

Economic Importance: Corm (ol) is edible as vegetable, it also functions like purgative.

3. *COLOCASIA* Clus. ex Fabric

Key to the Species

1a	Leaves with bronze margin; flowers yellow	1. *C. esculenta* (L.) Schott.

1b	Leaves without bronze margin; flowers pale yellow	2. *C. nymphaeifolia* Kunth.

3.1. ***Colocasia esculenta*** (L.) Schott., Melet. 1: 18. 1832; Naskar Aq. & Semi-Aq. Plants 258. 1990. *Arum esculentum* L. Sp. Pl. 965.1753. *A. colocasia* L. Sp. Pl. 965. 1753. *Colocasia antiquorum* Schott., Melet. 1: 18. 1832. *(Vern. Name: Kachu).*

Rhizomes variable. Gregariously grow on the roadsides, waste lands and pond or ditch sides. Naskar (1.c.) described this taxon after collecting from Atghora/Madarhat *(K.N. - 62).* Poor villagers often cook the petiole and the rhizome during scarcity.

3.2. ***Colocasia nymphaeifolia*** Kunth, Enum. 3: 37. 1850; Naskar Aq. & Semi-Aq. Plants 258. 1990. *(Vern. Name: Sar Kachu).*

Stout herb. Common on the dampy waste lands, canal sides or road-side thickets. Naskar (1.c.) described this taxon after collecting from Malatipur/near Deganga *(K.N. - 1591).* Petioles and rhizomes are edible.

4. *CRYPTOCORYNE* Fisch. ex Wydler

Key to the Species

1a	Spathe densely fimbriate, not twisted	1. *C. ciliata* (Roxb.) Fisch. ex Wydler.
1b	Spathe glabrous, tube both closely twisted	2. *C. retrospiralis* Fisch.

4.1. ***Cryptocoryne ciliata*** (Roxb.) Fisch. ex Wydler, Linn. 5: 428. 1830; Naskar Aq. & Semi-Aq. Plants 259. 1990. *Ambrosinia ciliata* Roxb. Corm. Pl. 7: 90, t.294. 1819. *(Vern. Name: Kerali).*

Erect, unarmed aquatic herb with terranean stolons. Occasionally grow on the water side, on the bank of the ponds, ditches or roadsides. Naskar (1.c.) described this taxon after collecting from Sarisha *(K.N. - 1944).*

4.2. ***Cryptocoryne retrospiralis*** Fisch. ex Wydler, Linn. 5: 428. 1830; Naskar Aq. & Semi-Aq. Plants 260.1990. *C. roxburghii* Dalz. & Gibs. Bomb. Fl. 259. 1861 (not of Schott.)

Aquatic, submerged herb with fleshy fibrous roots. Occasionally grow on the marshy places as submerged herb. Naskar (1.c.) described this taxon after collecting from Ghoshpur *(K.N. - 2130).*

5. *LASIA* Lour.

5.1. ***Lasia spinosa*** (L.) Thw., Enum. Pl. Zeyl. 336. 1864; Naskar Aq. & Semi-Aq. Plants. 260.1990. *Dracontium spinosum* L. Sp. Pl. 967.1753. *Lasia heterophylla* Schott., Melet. 1: 21. 1832. *(Vern. Name: Kanta-Kachu).*

Stout, prickly, aquatic herb; on the bank of roadsides ditches. Naskar (1.c.) described this taxon after collecting from Kakdweep *(K.N. - 1923).*

6. *PISTIA* L.

6.1. ***Pistia stratiotes*** L. Sp. Pl. 963. 1753; Naskar Aq. & Semi-Aq. Plants 261. 1990. *(Vern. Name: Khudipana* or *Takapana).*

Small, floating, stemless, stoloniferous herb. Gregariously grow and cover the surface of the stagnent water ponds, ditches and canals. Naskar (1.c.) described this taxon after collecting from the lower Ganga delta. A very obnoxious weed in Bengal.

7. *POTHOS* L.

7.1. ***Pothos scandens*** L. Sp. Pl. 968. 1753; Roxb. Fl. Ind. 2: 457. 1832; Hook.f. Fl. Brit. India 6: 551.1892; Prain Beng. Plants 2: 839.1903 (Rep. ed. 1963).

Unarmed, epiphytic climber. Leaves obovate or lanceolate, bracts very small. Spadix globose ovoid or shortly oblong, floating upwards; spathe cymbiform, cuspidate. *Fls. & Frts.:* June-November.

Frequently grow on the waste lands, generally grow on the village-scrub jungles.

Field Collection: Mazilpur, *K.N. - 1618.*

Distributions: Almost throughout India. Malay Islands and China.

Economic Importance: Occasionally planted as ornamental plant.

8. *SCINDAPSUS* Schott.

8.1. ***Scindapsus officinalis*** (Roxb.) Schott, Melet 1: 21.1832; Hook.f. Fl. Brit. India 6: 541.1893; Prain Beng. Plants 2: 839. 1903 (Rep. ed. 1963). *Pothos officinalis* Roxb. Fl. Ind. 1: 431.1820. *(Vern. Name: Gaj-pipul).*

Robust, climbing shrub, wrinkled when dry. Leaves ovate, elliptic, ovate or nearly orbicular, caudate acuminate, base rounded or slightly cordate, primary nerves distinct, petiole broadly winged up to the knee.

Spathe coriaceous, oblong, green outside, yellow within. Spadix dense flowered, cylindrical, peduncle stout, short, terete. Seeds ovate cordate. *Fls. & Frts.:* Almost throughout the year.

Rarely grow on the tree trunk.

Field Collection: Lakshmikantapur, *K.N. - 1781.*

Distributions: Almost throughout India. Burma and Sri Lanka.

Economic Importances: Ornamental; dried fruit is said to be carminative and anthelmintic; the Santals applied this plant for rheumatism.

9. *TYPHONIUM* Schott.

Key to the Species

1a	Spathe passes into a long slender apical tail; neuters dissimilar in shape	1. *T. flagelliforms* Blume
2b	Spathe without an apical tail, neuters all similar	2. *T. trilobatum* (L.) Schott.

9.1. ***Typhonium flagelliforme*** (Roxb.) Blume, Rumphia 1: 134. 1835; Naskar Aq. & Semi-Aq. Plants 262. 1990. *Arum flagelliforms* Roxb. Hort. Beng. 65.1814. *A. cuspidatum* Blume, Cat. Hort. Buit. 101. 1823.

Tuber sub-globose. Leaves glabrous, highly variable in shape; common on the marshy waste lands, side of the ponds or ditches. Naskar (1.c.) reported and described this taxon after collecting from Subuddhipur *(K.N. - 2297).*

9.2. ***Typhonium trilobatum*** (L.) Schott., Wien. Zeitschr. 3: 72. 1829; Naskar Aq. & Semi-Aq. Plants 263. 1990. *Arum trilobatum* L. Sp. Pl. 965. 1753. *(Vern. Name: Ghet Kachu* or *Bhetkal).*

Tuberous herb, globose or sub-globose. Common on the waste, moist and grassy lands. Naskar (1.c.) described this taxon after collecting from Joynagar-Mazilpur *(K.N. - 861).*

CXLVI. LEMNACEAE

Key to the Genera

1a	Fronds rootless, each one with a solitary basal reproductive pouch	1. *WOLFFIA Horkel ex Wimmer*
1b	Fronds rooted, each one with 2 lateral reproductive pouch	(2)

2a	Fronds with a root, nerveless or faintly 1-3 nerved	1. *LEMNA* L.
2b	Fronds more than one roots, conspicously 5-18 nerved	2. *SPIRODELA* Schleiden

1. *LEMNA* L.

Key to the Species

1a.	Frond entire symmetric; branches hastate, persistent	1. *L. perpusilla* Torrey
1b	Frond serrate symmetric; branches hastate, persistent	2. *L. trisulca* L.

1.1. ***Lemna perpusilla*** Torrey, Fl. N.Y. 2: 245. 1843; Naskar Aq. & Semi-Aq. Plants 264. 1990. *L. paucicostata* Hegelm., Lemm. 138.1868. ***(Vern.** Name: Khudi Pana).*

Small, free-floating aquatic herb. Commonly found as free floating herb. Naskar (1.c.) described this taxon after collecting from the lower Ganga delta.

Ducks and herbivorous grass-carp relish this aquatic plant as delicious food. Its fronds are reported to possess cooling, astringent and diuretic properties; useful in skin diseases and as opthalmic wash.

1.2. ***Lemna trisulca*** L. Sp. Pl. 970. 1753; Naskar Aq. & Semi-Aq. Plants 265.1990. *L. cruciata* Roxb. Fl. Ind. 3: 566. 1832.

Small submerged herb. Common in the ponds, tanks, jheels and raodside ditches. Naskar (1.c.) described this taxon after collecting from Narendrapur Ramkrishna Mishram pond *(K.N. - 1989)*.

2. *SPIRODELA* Schleiden

2.1. ***Spirodela polyrhiza*** (L.) Schleiden, Linnaea 13: 392.1839; Naskar Aq. & Semi-Aq. Plants 266. 1990. *Lemna orbiculata* Roxb. Fl. Ind. 3: 565. 1832. *(Vern. Name: Matar Dal Pana).*

Free-floating, tufted herb with round or ovate fronds. Abundant in stagnant waters; thrives for some period at the bottom of ponds. Naskar (1.c.) described this taxon after collecting from the lower Ganga delta.

3. *WOLFFIA* Horkel ex Scheiden

3.1. ***Wolffia arrhiza*** (L.) Horkel ex Wimmer, Fl. Schleiden 3: 140.1857; Naskar Aq. & Semi-Aq. Plants 267. 1990. *Lemna arrhiza* L. Mantiss. 2: 294.

1771. *L. globosa* Roxb. Hort. Beng. 66.1814. *Wolffia michelii* Sch. Beitr. Bot. 233. 1844. *(Vern. Name: Suji Pana).*

Smallest among the flowering plants. Free floating on the water surface. Naskar (1.c.) described this taxon after collecting from the lower Ganga delta.

An excellent fish food and ducks relish this weed in fresh conditions.

CXLVII. ALISMATACEAE

Key to the Genera

1a	Flowers unisexual; receptacle globose; stamens 6-many	2. *SAGITTARIA* L.
1b	Flowers bisexual; receptacle flat; stamens 6 in 2 seriate	1. *LIMNOPHYTON* Miquel.

1. *LIMNOPHYTON* Miquel.

1.1 ***Limnophyton obtusifolium*** (L.) Miq. Fl. Ind. Bat. 3: 242. 1855; Naskar Aq. & Semi-Aq. Plants 269. 1990. *Sagittaria obtusifolia* L. Sp. Pl. 993. 1753.

Robust aquatic herb. Leaves whorls, sagittate. Occurs in freshwater pools, ditches and canal sides. Naskar (1.c.) described this taxon after collecting from Ghoshpur *(K.N. - 2131).*

2. *SAGITTARIA* L.

Key to the Species

1a	Leaves floating, broadly ovate, deep cordate, obtuse with a cordate base. Lower flowers bisexual; stamens 6-10; ripe carpels turgid	1. *S. guayanensis* ssp. *lappula* (D.Don) Bogin
1b	Leaves rising above the surface of water, hastate or sagittate base. Lower flowers female; stamens 24; ripe carpels laterally compressed	2. *S. sagittafolia* L.

2.1. ***Sagittaria guayanensis*** H.B.K. ssp. ***lappula*** (D.Don) Bogin, Mem. N.Y. Bot. Garden 9: 192,f.5.1955; Naskar Aq. & Semi- Aq. Plants 269. 1990. *Sagittaria lappula* D. Don, Prodr. Fl. Nep. 22. 1825. *S. guayanensis* H.B.K.; Hook.f. Fl. Brit. India 6: 561.1893.

Leaves floating, glabrous, ovate or sub-orbicular, with a triangular sinus at the base. Common to the marshy places, ponds, jheels and in the fields. Naskar (1.c.) described this taxon after collecting from Minakhan/near the bank of Bidyadhari *(K.N. - 1105).*

2.2. ***Sagittaria sagittifolia*** L. Sp. Pl.2: 993.1753; Naskar Aq. & Semi-Aq. Plants 270. 1990. *(Vern. Name: Chhota Kut).*

Rhizome thick, stoloniferous, scapigerous herb. Common on the riverbeds, paddy fields and side of the ditches. Naskar (1.c.) described this taxon after collecting from Bamanpukur/near Minakhan *(K.N. - 1143).* This plant is antiseptic; the fleshy corms are edible by the poor villagers.

CXLVIII. BUTOMACEAE

Previously, this family was treated under Alismataceae as a sub-family Butomoidaceae; but now it is treated as separate distinct family.

1. *TENAGOCHARIS* D. Don

1.1. ***Tenagocharis latifolia*** (D.Don) Buchen., Abh. Naturw. Ver.Bremen 2: 2.1868; Naskar Aq. & Semi-Aq. Plants 272.1990. *Butomus latifolius D. Don Prodr. Fl. Nep. 22.1825. Butomopsis lanceolata* Kunth, Enum. Pl. 3: 165.1841. *Butomus lanceolatus* Roxb. Fl. Ind. 2: 315.1832.

Erect 20 cm-60 cm high with milky juice. Common in the wet places and rice fields. Naskar (l.c.) described this taxon after collecting from Kestopur/near Baguihati *(K.N. - 1893).*

CXLIX. APONOGETONACEAE

1. *APONOGETON* Linn.f. *nom. cons.*

Key to the Species

1a	Leaves submerged, linear oblong to lanceolate, 3-7 nerved, translucent; sepals 1-3, white, 1-2 seeded follicles	1. *A. crispum* Thunb.
1b	Leaves floating, partly submerged, oblong, 3-5 nerved; sepals 2, pink or white, follicles smooth, 4-8 seeded	2. *A. natans* (L.) Engl. & Krause

1.1. ***Aponogeton crispum*** Thunb., Nov. Gen. 1: 73.1781; Naskar Aq. & Semi-Aq. Plants 273.1990. *(Vern. Name: Ghenchu).*

Perennial, stoloniferous and aquatic herb. Common on all the ponds, pools, jheels and roadside ditches. Naskar (l.c.) described this taxon after collecting from the lower Ganga delta. The stolons are edible by the poor villagers during famine. Occasionally it is used as aquarium plant.

1.2. *Aponogeton natans* (L.) Engl. & Krause, Engler Pflreich 24: 22. 1906; Naskar Aq. & Semi-Aq. Plants 174.1990. *Saururus natans* L. Mant. 2: 227.1767. *Aponogeton monostachyus* Linn. f. Suppl. 214.1781. *(Vern. Name: Ghenchu).*

Aquatic, stoloniferous herb; gregariously grow in the village ponds pools and roadside ditches. Naskar (l.c.) described this taxon after collecting from Minakhan *(K.N. - 2134).* The rootstock of this plant contains much starch and is edible by the poor villagers.

CL. POTAMOGETONACEAE

1. *POTAMOGETON* L.

Key to the Species

1a	Leaves submerged, lanceolate, membranous, semiampexicaul at base, crisped and serrulate 3-nerved	1. *P. crispus* L.
1b	Upper leaves only floating, floating leaves large, coriaceous; submerged leaves lanceolate, membranous, many nerved	2. *P. nodosus* Poir.

1.1. ***Potamogeton crispus*** L. Sp. Pl. 126.1753; Naskar Aq. & Semi-Aq. Plants 275.1990. *P. tuberosus* Roxb. Hort.Beng. 121. 1814.

Submerged herb; stem branched. Benthos or common pond weed. Naskar (l.c.) described this taxon after collecting from the lower Ganga delta, Dhopagachi *(K.N. - 2144).*

1.2. ***Potamogeton nodosus*** Poir., Encycl. Meth. Bot. Suppl. 4: 535.1810; Naskar Aq. & Semi-Aq. Plants 275. 1990. *P. indicus* Roxb. Fl. Ind. 1: 471.1820.

Aquatic herb, gregariously grow on the stagnant water. Naskar (l.c.) described this taxon after collecting from the lower Ganga delta, Sundarbans *(K.N. - 2299).*

CLI. NAJADACEAE

Key to the Genera

1a	Commonly found in the freshwater ponds, jheels, lakes and vast water bodies	1. *NAJAS* L.

1b	Commonly in the brackish water fisheries of the Sundarbans area	2. *RUPPIA* L.

1. *NAJAS* L.

1.1. ***Najas graminea*** Del., Fl. Egypt. 282, t. 50,f.3.1812; Naskar Aq. & Semi-Aq. Plants 276.1990. *(Vern. Name: Kanta-Janjhi).*

Short, slender grass-like, often grace-fully plumose; commonly on the pools, ponds and ditches. Naskar (l.c.) described this taxon after collecting from Baruipur *(K.N. - 2251).*

1.2. ***Ruppia maritima*** L. Sp. Pl. 182.1753; Naskar, *et al.* J. Ind.Soc. Coastal agric Res.3(1) 69.1985. *R. restellata* Koch. *(Vern. Name: Nona Jhanjhi).*

Submerged, aquatic herb, looks very much similar to the filamentous algae, much branched, dull-olive green; stem 60-70 cm long; leaves long 5-10 cm, filiform. Peduncles short, 5-7 mm long, not spiral. Flowers very small, hermaphrodite 2-6 enclosed within leaf sheath; perianth absent, stamen-2, anther sessile, 2-celled. Achenes 2-3 mm long,beaked, endocarp hard. Seed uncinate, embryo straight. *Fls. & Frts.:* June-October.

Commonly found on the brackish water fisheries in the Sundarbans.

Field Collection: Jharkhali, *K.N. - 2390.*

Distributions: Throughout the brackishwater in the tropical Asia; and in the coastal regions of India.

Economic Importances: Brackish water fish and prawn species relish this weeds.

CLII. ERIOCAULACEAE

1. *ERIOCAULON* L.

Key to the Species

1a	Stemless or stem not conspicous	(2)
1b	Stem erect, root-stock elongated, slender or stout	(4)
2a (1)	Anthers white, leaves tufted	1. *E. cinereum* R. Br.
2b (1)	Anthers grey or blackish	(3)
3a (2)	Leaves tufted; scapes few or solitary	4. *E. reductum* Ruthland
3b (2)	Leaves flat, tapering to the tip; scapes many	5. *E. sollyanum Royle*

4a (1)	Stem stout or slender 10-25 cm high; leaves caespitose	2. *E. luzulaefolius* Mart.
4b (1)	Stem smaller	(5)
5a (4)	Leaves rosette, linear; head-globose or sub-globose	3. *E. quinqueangulare* L.
5b (4)	Leaves ensiform, many-nerved, glabrous; head small depressed, hemispheric	6. *E. truncatum* Ham.ex Mart.

The following 6 species of *Eriocaulon* L. were described by Naskar, in the book entitled; Aquatic & Semi-Aquatic Plants of the Lower Ganga Delta' -pp.277 - 280.As such,in this book only the correct nomenclatures and place of collections of the flora are only mentioned.

1.1. ***Eriocaulon cinereum*** R. Brown 254.1810; Naskar Aq. & Semi-Aq. Plants 277.1990. *E. sieboldianum* Sieb.& Zucc. ex Steud. Syn. Pl. Cyp. 2: 272. 1855. *E. sexangulare* Mart., Wall. Pl. As. Rar. 3: 28.1832 (not of Linn.)

Field Collection: Narendrapur/on the roadside, *K.N. - 1982.*

1.2. ***Eriocaulon luzulaefolium*** Mart., Wall. Pl. As. Rar. 3: 28.1832; Naskar Aq. & Semi-Aq. Plants 278.1990.

Field Collection: Duttapukur, *K.N. - 1708.*

1.3. ***Eriocaulon quinqueangulare*** L. Fl. Zey.48.1747 & Sp. Pl. 87. 1753; Naskar Aq. & Semi-Aq. Plants 278.1990. *(Vern. Name: Guri).*

Field Collection: Bahuru, *K.N. - 1623.*

1.4. ***Eriocaulon reductum*** Ruhland, Pfreich 13: 113. 1903; Naskar Aq. & Semi-Aq. Plants 279.1990.

Field Collection: Salipur/near Malancha, *K.N. - 2150.*

1.5. ***Eriocaulon sollyanum*** Royle, Ill. 409.t.97.f.1.1839; Naskar Aq.& Semi-Aq. Plants 279.1990. *E. trilobatum* Buch. Ham. ex Kovin, Linn. 27: 645 1856. *E. infaustum* Brown, Dyer Fl. Trop. Africa 8: 253.1902.

Field Collection: Bhasila/near Barasat, *K.N. - 1577.*

1.6. ***Eriocaulon truncatum*** Ham. ex Mart, Wall. Pl. As. Rar. 3: 29. 1832; Naskar Aq. & Semi-Aq. Plants 280.1990.

Field Collection: Naihati, *K.N. - 1593.*

All these 6 species of the *Eriocaulon* L. grow in the moist waste lands and marshy lands, as weed flora.

CLIII. CYPERACEAE

Key to the Species

1a	Nut enclosed in a bottle shaped utricle	2. *CAREX* L.
1b	Nut not enclosed in an utricle	(2)
2a (1)	Flowering glumes distichous; spikelets many flowered	3. *CYPERUS* L.
2b (1)	Flowering glumes spirally arranged	(3)
3a (2)	Style base dilated and constricted or articulated above the nut	(4)
3b (2)	Style base not dilated, continuous with the nut	(6)
4a (3)	Stem leafless; bristles usually present	4. *ELEOCHARIS* R. Brown
4b (3)	Stem leafy at the base; bristle absent	(5)
5a (4)	Style persistent, flat and hairy; leaves glabrous	5. *FIMBRISTYLIS* Vahl
5b (4)	Style linear glabrous; leaves with fine hairs	1. *BULBOSTYLIS* Kunth.
6a (3)	Leaves hairy	6. *FUIRENA* Rottb.
6b (3)	Leaves glabrous	7. *SCIRPUS* L.

The following species of the family-Cyperaceae were collected and studied by Naskar (1981) and published in the book entitled, 'Aquatic and Semi-Aquatic Plants of the Lower Ganga Delta' by Naskar (1990). **As such,** the nomenclatures, and field collection notes of all these species **are** mentioned here, only.

1.1. ***Bulbostylis barbata*** (Rottb.) Kunth. Enum. 2: 208.1837; Naskar Aq. & Semi-Aq. Plants 281.1990. *Scirpus barbatus* Rottb. Descr. Pl. Rar. Progr. 27.1772. *Fimbristylis barbata* (Rottb.) Benth., Fl. Austral. 7: 321.1878.

Field Collections: Sonarpur, *K.N. - 995;* Ultadanga, *R. Mallick-633.*

2.1. ***Carex fedia*** Nees, Wt. Contrib. 129.1834; Naskar Aq. & Semi-Aq. Plants 282.1990. *C.wallichiana* Press. ex Nees, Wt. Contrib. 129.1834. *(Vern. Name: Mutha Ghash).*

Field Collection: Malancha, *K.N. - 2156.*

3. *CYPERUS* L.

(Including the genera-*Pycreus* Beauv, *Juncellus* Clarke, *Mariscus* Vahl and Kyllinga Rottb.).

Key to the Species

1a	Floating herb in the still water; nut corky below	3. *C. cephalotes Vahl*
1b	Rooting in the soil or mud; nut not corky at the base	(2)
2a (1)	Spikelets 1-2 flowered	(3)
2b (1)	Spikelets more than 2-flowered	(5)
3a (2)	Creeping herb with stoloniferous rhizome; stem several	2. *C.brevifolius* Hassk.
3b (2)	Erect, terrestrial herb; rhizome short, elongated	(4)
4a (3)	Stem solitary	16. *C. kyllinga* Endl.
4b (3)	Stem tufted	29. *C. triceps Endl.*
5a (2)	Stigma-2	(6)
5b (2)	Stigma-3	(10)
6a (5)	Fruit-dorsally compressed nut	(7)
6b (5)	Fruit-laterally compressed nut	(8)
7a (6)	Dwarf annual herb; spikelets many, densely packed in a terminal head	18. C *michelianus* Link.
7b (6)	Stout perennial herb; spikes umbellate, large	1. *C. alopecuroides* Rottb.
8a(6)	Stem clothed by leaf-sheath, decumbent at the base; rhizome creeping	26. *C. sanguinolentus* Vahl
8b (6)	Stem enveloped in leaf-sheaths only a little way above the base, erect; roots fibrous	(9)
9a (8)	Annual herb; stamens-1, nut 2.5 mm long	24. *C. pumilus* L.
9b (8)	Perennial shrub; stamens usually-2; nut 1.5 mm long	22. *C. polystachyos* Rottb.

10a (5)	Rachilla of spikelets persistent	(11)
10b (5)	Rachilla of spikelets deciduous	(26)
11a (10)	Spikelets digitate or clustered	(12)
11b (10)	Spikelets spicately, rarely sub-racemose, disposed; stamens usually-3	(15)
12a (11)	Annual herb (expt. *C. haspan)*	(13)
12b (11)	Perennial shrub with woody, root-stock; stamens usually -3	19. *C. niveus* Retz.
13a (12)	Nut white at maturity; globose obovoid	11. *C. flavidus* Retz.
13b (12)	Nut brown	(14)
14a (13)	Nut nearly as long as the obovate oblong glume	8. *C. difformis* L.
14b (13)	Nut 1/3 rd the length of the glume	12. *C. haspan* L.
15a (11)	Rachilla of spikelets not or scarcely winged	(16)
15b (11)	Rachilla of spikelets distinctly winged	(21)
16a (15)	Annual herbs with tufted stem	(17)
16b (15)	Perennial shrubs, rhizomes woody and often with stolons	(19)
17a (16)	Glumes with prolonged mid-rib into a distinct mucro or bristle, nut length about 1/3rd of the glume	(18)
17b (16)	Glumes muticous or scarcely apiculate; nut nearly as long as glume	14. *C. iria* L.
18a (17)	Glume boat shaped, keel acute	5. *C. compressus* L.
18b (17)	Glume ovate, lanceolate, acuminate	27. *C. squarrosus* L.
19a (16)	Spikelets compressed, bracts divaricate	23. *C. procerus* Rottb.
19b (16)	Spikelets spicate	(20)
20a (19)	Leaves more than 30 cm long	20. *C. nutans* Vahl

20b (19)	Leaves not more than 15 cm long	17. *C. malaccensis* Lamk.
21a (15)	Leaves short, occasionally 1/2 as long as the stem	(22)
21b (15)	Leaves long, more than half as long as the stem or longer	(23)
22a (21)	Spikelets pale straw coloured; bracts very short, rarely half as long as the umbel rays; stem not articulate when dry	28. *C. tegetiformis* Roxb.
22b (21)	Spikelets rusty brown, bracts more than half as long as the umbel-rays; stem more or less articulate when dry	6. *C. corymbosus* Rottb.
23a (21)	Rhizome stoloniferous, stolons 10-12 cm long	25. *C. rotundus* L.
23b (21)	Rhizome not stoloniferous	(24)
24a (23)	Spikelets much compressed; ascending; wings persistent	(25)
24b (23)	Spikelets terete; spreading horizontally; wings deciduous	9. *C. digitatus* Roxb.
25a (24)	Spikelets digitate, hardly peduncled, spikelets dull yellow or brown	13. *C. imbricatus* Retz.
25b (24)	Spikelets at least some peduncled; spikelets bright yellow or brown	10. *C. exaltatus* Retz.
26a (10)	Umbels simple; lowest glume of spikelets very acute	(27)
26b (10)	Umbels compound; glumes all obtuse	(28)
27a (26)	Rhizome emitting slender stolons	21. *C. paniceum* Boech.
27b (26)	Rhizome short, stolon-0	7. *C. cyperinus* Valck.
28a (26)	Spikelets bearing 4-14 nuts	4. *C. compactus* Retz.
28b (26)	Spikelets bearing 3-6 nuts	15. *C. javanicus* Houtt.

3.1. ***Cyperus alopecuroides*** Rottb. Descr. Pl. Rar. Programme 20. & Ic. 38.t.8.f.2.1773; Naskar Aq. & Semi-Aq. Plants 286.1990. *Juncellus alopecuroides* (Rottb.) Clarke, Hook.f. Fl. Brit. India 6: 595.1893.

Field Collection: Garia, *K.N. - 334.*

3.2. ***Cyperus brevifolius*** (Rottb.) Hassk., Cat. Hort. Bog.24.1844; Naskar Aq. & Semi-Aq. Plants 287.1990. *Kyllinga brevifolia* Rottb. Descr. et. Icon 13,t.4,f.3,1773.

Field Collection: Naihati, *K.N. - 1535.*

3.3. ***Cyperus cephalotes*** Vahl, Enum.2: 311. 1806; Naskar Aq. & Semi-Aq. Plants 287.1990. *(Vern. Name: Gothubi).*

Field Collection: Rangabelia/Sundarbans, *K.N. - 1515.*

3.4. ***Cyperus compactus*** Retz., Obs.Bot. 5: 10.1789; Naskar Aq. & Semi-Aq. Plants 288.1990. *C. dilutus* Vahl, Enum.2: 357.1806. *Mariscus microcephalus* Presl., Rel. Haenk 1: 182.1830. *Cyperus spinulosus* Roxb. Fl. Ind. 1: 203.1832. *Mariscus dilutus* (Vahl) Nees et Wight, Contrib. Bot. Ind. 90.1834. *Mariscus compactus* (Retz.) Druce, Rep. Bot. Exch. Club. Brit. Is. 6: 634.1917.

Field Collection: Salt Lake, *M.K.Ghosh-4031* (CAL).

3.5. ***Cyperus compressus*** L. Sp. Pl. 1: 46.1753; Naskar Aq. & Semi-Aq. Plants 289.1990. *(Vern. Name: Chuncha).*

Field Collection: Taki, *K.N. - 1013;* Bamanpukur, *K.N. - 1101;* Joygram, *K.N. - 1135;* Lake Town, *K.N. - 1197;* Dum Dum, *R. Mallick-8* (CAL).

3.6. ***Cyperus corymbosus*** Rottb., Descr.& Ic. 42.t.7, f.4,1773; Naskar Aq. & Semi-Aq. Plants 289.1990. *C. diphyllus* Retz. Obs. fasc. 5: 11. 1789 (not of Benth.).

Field Collection: Malatipur/near Barasat, *K.N. - 1589.*

3.7. ***Cyperus cyperinus*** (Retz.) Valck.Sur. Gesl. Cyperaceae Mal. Arch. 154.t.6.f.10.1898; Naskar Aq. & Semi-Aq. Plants 290.1990. *Kyllinga cyperina* Retz. Obs. 6: 21.1791. *Mariscus cyperinus* (Retz.) Vahl, Enum. 2: 377. 1806.

Field Collection: Mollakhali, *K.N. - 1847.*

3.8. ***Cyperus difformis*** L. Cent. Pl. 2: 6. 1756; Naskar, Aq. & Semi-Aq. Plants 291.1990. *(Vern. Name: Behua Ghash).*

Field Collection: Minakhan, *K.N. - 1097.*

3.9. ***Cyperus digitatus*** Roxb. Fl.Ind. 1: 209.1820; Naskar Aq. & Semi-Aq. Plants 291.1990. *C. neesii* Boeck ., Linn. 36.315.1870. *C. auricomus* Clarke, J. Linn. Soc. 21: 188.1884.

Field Collection: Bireshwarpur/near Lakshmikantapur, *K.N. - 1758.*

3.10. ***Cyperus exaltatus*** Retz. Obs. fasc. 5: 11. 1789; Watt Dict. Econ. Prod. 2: 684.1889; Naskar Aq. & Semi-Aq. Plants 292.1990.

Field Collections: Sonarpur, *K.N. - 996;* Thuba/Taki, *K.N. - 1040;* Dum Dum, *K. Mallick-74* (CAL).

3.11. ***Cyperus flavidus*** Retz. Obs. fasc. 5: 13.1789; Naskar, Aq. & Semi-Aq. Plants 293. 1990. *C.haspan* var. *indicus* Boeck, Linn. 35: 574. 1868.

Field Collections: Ichhapur, *K.N. - 1798;* Bongaon, *R. Mallick-406.*

3.12. ***Cyperus haspan*** L. Sp. Pl. 45. 1753; Naskar Aq. & Semi-Aq. Plants 293.1990.

Field Collections: Dhopagachi, *K.N. - 33;* Majherhat, *K.N. - 1212;* Baduria, *K.N. - 1305;* Bongaon, *K.N. - 1315;* Barasat, *R. Mallick-460* (CAL).

3.13. ***Cyperus imbricatus*** Retz., Obs. 5: 12.1789. *C. radiatus* Vahl, Enum. 2: 369.1806; Naskar Aq. & Semi-Aq. Plants 294.1990. *(Vern. Name: Berethi).*

Field Collection: Duttapukur, *K.N. - 1710.*

3.14. ***Cyperus iria*** L. Sp. Pl. 1: 45.1753; Naskar Aq. & Semi-Aq. Plants 294.1990. *(Vern. Name: Bara Chuncha).*

Field Collections: Dakshineswar, *K.N. - 682;* Khodarbazar, *K.N. - 1160;* Barasat, *K.N. - 917;* Dandirhat, *K.N. - 965.*

3.15. ***Cyperus javanicus*** Houtt., Nat. Hist. 2: 13.1782; Naskar Aq. & Semi-Aq. Plants 295.1990. *Mariscus albescens* Gaud., Freyce. Voy. Bot. 415.1826. *Cyperus pennatus* Lamk. Iiiustr. 1: 144. 1791. *(Vern. Name: Halaiya).*

Field Collections: Ganga Sagar/Sagar Islands, A.K.M-5906.

3.16. ***Cyperus kyllinga*** Endl., Cat. Hort. Ac. Vindb. 1: 94.1842; Naskar Aq. & Semi-Aq. Plants 296. 1990. *Kyllinga monocephala* Rottb., Desc. & Icon. 13, t. 4, 1773. *(Vern. Name: Swet Gothubi).*

Field Collections: Dhapdhapi, *K.N. - 37;* Joynagar, *K.N. - 863.*

3.17. ***Cyperus malaccensis*** Lamk. Ill. 1: 146.1791; Naskar Aq. & Semi-Aq. Plants 296.1990. *C. pangorei* Roxb. Fl. Ind. 1: 202. 1832. *(Vern. Name: Chumati Pati).*

Field Collections: Haroa *K.N. - 591.*

3.18. ***Cyperus michelianus*** (L.) Link., Hort. Bot. Berol. 1: 303. 1827; Naskar Aq. & Semi-Aq. Plants 297.1990.

Field Collection: Basanti, *K.N. - 2276.*

3.18/1. ***Cyperus michelianus*** (L.) Link. subsp. ***pygmaeus*** (Rottb.) Aschers. & Graebn. Syn. 2 (2): 273.1930. C. *pygmaeus* Rottb. Descr. & Ic. 20, t.14, f.5,1773. *C. squarrosus* Roxb. Hort. Beng. 81.1814 (not of L.) . *Juncellus pygmaeus* Clarke, Hook.f. Fl. Brit. India 6: 596. 1893.

Erect, annual herb, 15-20 cm high. Stem densely tufted. Leaves numerous, shorter than the stem, flaccid, green, accuminate. Spikelets many 4-5 m long, sessile, densely packed in a terminal heads, ovate, compressed. Nuts ellipsoid, surrounded by a 2- armed style. *Fls. & Frts.:* Almost throughout the year.

On the embankments of ponds, ditches and paddy fields.

Field Collection: Dhakuria, *K.N. - 1966.*

Distributions: more or less throughout India . South East Asia.

3.19. ***Cyperus niveus*** Retz. Obs. fasc. 5: 12. 1789; Naskar Aq. & Semi-Aq. Plants 298. 1990. *(Vern. Name: Birmutha).*

Field Collections: Minakhan, *K.N. - 1196;* Salt Lake, *R. Mallick-59* & Agarpara, *R. Mallick-117* (CAL).

3.20. ***Cyperus nutans*** Vahl, Enum. 2: 363.1806; Naskar Aq. & Semi-Aq. Plants 298.1990. *C. distans* Grah., Cat.232 (not of Linn.).

Field Collections: Amtala/Bhodduri, *K.N. - 883;* Basirhat, *K.N. - 939* and Hasnabad, *K.N. - 1048.*

3.21. ***Cyperus paniceum*** (Rottb.) Boech., Linnaea 36: 381.1870; Naskar Aq. & Semi-Aq. Plants 299.1990. *Kyllinga panicea* Rottb. Descr. & Ic. 15,t.4,f.1, 1773. *Mariscus paniceus* (Rottb.) Vahl, Enum. 2: 373.1806.

Field Collection: Goranbose/Sundarbans, *K.N. - 1879.*

3.22. ***Cyperus polystachyos*** Rottb., Progr. 21.1772; Naskar Aq. & Semi-Aq. Plants 299. 1990. *C. odoratus* L. Sp. Pl. 46.1753. *C. polystachyos* (Rottb.) Beauv., Pl. Oware & Benin 2: 48, t. 86, f.2.1807. *Pycreus odoratus* Urbn, Symb. Antill. 2: 164.1900 (excl. basionym).

Field Collections: Gosaba, *K.N. - 679;* Taki/Ghoshpara, *K.N. - 1006.*

3.23. ***Cyperus procerus*** Rottb. Descrip. et. Icon. 29, t. 5, f. 3. 1773; Naskar Aq. & Semi-Aq. Plants 300. 1990. *C. heynei* Boeck., Linnaea 35: 600.1868.

Field Collection: Sagar Islands, *A.K. Mukherjee-8749* (CAL).

3.24. ***Cyperus pumilus*** L. Cent. Pl. 2: 6.1756; Naskar Aq. & Semi-Aq. Plants 301.1990. *C. nitens* Retz. Obs. 6: 13. 1789. *Pycreus nitens* (Retz.) Nees, Nova Acta. Ac. Nat. Cur. 19, Suppl. 1: 53.1843. *P. pumilus* (L.) Domin, Biblioth. Bot. 85: 417.1915.

Field Collection: Gobardanga, *R. Mallick-308* (CAL).

3.25. ***Cyperus rotundus*** L. Sp. Pl. 1: 45. 1753; Naskar Aq. & Semi-Aq. Plants 391. 1990. *C. tuberosus* Rottb. Descr. & Ic. 28.t.7.f.1, 1773.

Field Collections: Sonarpur, *K.N.-349;* Dewanganj, *K.N.-620;* Bhadduri/Amtala, *K.N. - 882;* Madhyampur/near Taki, *K.N.-955.*

3.26. ***Cyperus sanguinolentus*** Vahl, Enum.2: 351.1806; Naskar Aq. & Semi-Aq. Plants 302. 1990. *Pycreus sanguinolentus* (Vahl) Nees., Linnaea 9: 283.1833.

Field Collection: Kestopur/Baguihati, *K.N. - 1896.*

3.27. ***Cyperus squarrosus*** L. Cent. Pl. 2: 6. 1756; Naskar Aq. & Semi-Aq. Plants 303. 1990. *C. aristatus* Rottb., Progr. 22. 1772. *Mariscus squarrosus* (L.) Clarke, Hook.f. Fl. Brit. India 6: 623.1893.

Field Collection: Kestopur/Baguihati, *K.N. - 1897.*

3.28. ***Cyperus tegetiformis*** Roxb. Hort. Beng. 6.1814; Naskar Aq. & Semi-Aq. Plants 303.1990. *(Vern. Name: Gola- Methi).*

Field Collection: Dum Dum Park, *R. Mallick-32* (CAL).

3.29. ***Cyperus triceps*** (Rottb.) Endl., Cat. Hort. Ac. Vindob. 1: 94. 1842; Naskar Aq. & Semi-Aq. Plants 304. 1990. *Kyllinga triceps* Rottb., Descr. et Ic. 14, t. 4,f.6, 1773.

Field Collection: Falta, *K.N. - 1830.*

4. *ELEOCHARIS* R. Brown

Key to the Species

1a	Stems usually slender; glumes membranous, usually 1-3 nerved	(2)
1b	Stems usually stout; glumes sub-rigid, smooth or striate	(4)
2a (1)	Roots fibrous, rhizome short or absent	(3)
2b (1)	Roots rhizomatous, creeping; bristles rusty-brown	4. *E. palustris* R. Brown

3a (2)	Bristles glistening, white	1. *E. atropurpurea* Kunth.
3b (2)	Bristles not glistening, brown or rusty	3. *E. geniculata* Roem. & Schult
4a (1)	Stem terete, transversely septate when dry; not smooth	2. *E. dulcis* Trin. ex Hen.
4b (1)	Stems sub-triquetrous below the spikelets, not septate; nut longitudinally striate	5. *E. spiralis* Brown

4.1. ***Eleocharis atropurpurea*** (Retz.) Kunth., Enum.2: 151.1837; Naskar Aq.& Semi-Aq. Plants 305. 1990. *Scirpus atropurpureus* Retz. Obs. 5: 14 1788. *Eleocharis atropurpureus* Nees, Wight. Contrib. 113. 1834. *Heleocharis atropurpurea* Boeck., Linn. 36: 458. 1870.

Field Collection: Kulpi, *K.N. - 1795.*

4.2. ***Eleocharis dulcis*** (Burm.f.) Trin. ex Henschel, Rum. Ph. 186.1833; Naskar Aq. & Semi-Aq. Plants 306. 1990. *Andropogon dulce* Burm.f., Fl. Ind. 219.1768. *Scirpus plantaginoides* Rottb. Descrip. & Icon. 45, t. 15. f. 2. 1773. *Scirpus plantaginea* Roxb. Fl. Ind. 1: 1832. *Eleocharis plantaginea* (Retz.) R. Br. Prodr. 224. 1810. *E. equisetina* Presl, Rel. Haenk. 1: 195.1828. *Heleocharis plantaginea* Boeck., Linnaea 36: 474. 1870. *(Vern. Name: Chenchka).*

Field Collection: Minakhan, *K.N. - 1108.*

4.3. ***Eleocharis geniculata*** (L.) R. & S., Syst. Veg. 2: 150.1817; Naskar Aq. & Semi-Aq. Plants 307. 1990. *Scirpus geniculatus* L. Sp. Pl. 1: 48. 1753. *Eleocharis capitata* R. Brown Prodr. 225. 1810. *Heleocharis capitata* Boeck., Linn. 35: 461. 1870. *Eleocharis caribaea* (Rottb.) Blake, Rhodora 20: 24. 1918.

Field Collection: Chandpur/Thakurnagar, *K.N. - 1669.*

4.4. ***Eleocharis palustris*** R. Brown Prodr. 224.1810; Naskar Aq. & Semi-Aq. Plants 307. 1990. *Heleocharis palustris* Lindl., Syst. 2: 154.1827.

Field Collection: Thakurnagar, *K.N. - 1676.*

4.5. ***Eleocharis spiralis*** (Rottb.) R. Brown, Prod. 224.1810; Naskar Aq. & Semi-Aq. Plants 308. 1990. *Scirpus spiralis* Rottb. Descr. & Ic. 45, t.15,f. 1,1773.

Field Collection: Titagarh, *K.N. - 2215.*

5. *FIMBRISTYLIS* Vahl

Key to the Species

1a	Lower glumes of the spikelets distichous or sub-distichous, pale, nut large	9. *F. ovata* (Burm.f.) Kern
1b	Lower glumes of the spikelets imbricate	(2)
2a (1)	Style bi-fid; nut bi-convex	(3)
2a (1)	Style tri-fid; nut more or less trigonous	(11)
3a (2)	Stems generally with one spikelets (except *F. polytrichoides*); nut brownish-black	(4)
3b (2)	Stems generally with more than one spikelets (except *F. schoenoides*); nut white	(5)
4a (3)	Nut linear-oblong, curved, reticulate, straw coloured	14. *F. tetragona* R. Brown
4b (3)	Nut obovoid, obtuse, smooth reticulate	10. *F. polytrichoides* Vahl
5a (3)	Stem with single spikelet; glumes obtuse, not incurved at apex; nut smooth white	12. *F. schoenoides* Vahl
5b (3)	Stem with many spikelets; glumes acute, mucronate at aristate, incurved at apex	(6)
6a (5)	Style-base with many long, pendent hairs; glumes shortly aristate, squarrose or sub-squarrose; umbel compound	13. *F. squarrosa* Vahl
6b (5)	Style-base shortly hairy or glabrous	(7)
7a (6)	Nut conspicously longitudinally striate	(8)
7b (6)	Nut smooth	(9)
8a (7)	Spikelets angular from the acute keels of the glumes	4. *F. dichotoma* Vahl

8b (7)	Spikelets terete	5. *F. diphylla* Vahl
9a (7)	Glumes acute, not incurved; spikelets slightly angular from the keeled glumes; umbels lax; nut straw coloured	1. *F. aestivalis* Vahl
9b (7)	Glumes obtuse; incurved; spikelets subterete; umbels usually contracted	(10)
10a (9)	Nut pale; spikelets always solitary	6. *F. ferruginea* Vahl
10b (9)	Nut dusky; spikelets sometimes clustered as well as solitary, sometimes subcapitate	3. *F. cymosa* R. Brown
11a (2)	Spikelets almost all solitary	(12)
11b (2)	Spikelets almost all clustered; glumes ovate obtuse	7. *E. junciformis* Kunth.
12a (11)	Nut slenderly trabeculate	(13)
12b (11)	Nut never trabeculate	2. *F. complanata* Link.
13a (12)	Spikelets ellipsoid-lanceolate, subacute, 6-14 flowered	11. *F. quinqueangularis* Kunth.
13b (12)	Spikelets globose or sub-cylindric, obtuse, many flowered	8. *F. miliacea* Vahl

5.1. ***Fimbristylis aestivalis*** (Retz.) Vahl, Enum. 2: 288. 1806; Naskar Aq. & Semi-Aq. Plants 310. 1990. *Scirpus aestivalis* Retz. 4: 12. 1786.

Field Collection: Bhadduri/Amtala, *K.N. - 879.*

5.2. ***Fimbristylis bisumbellata*** (Forssk.) Bub., Dodec. 30.1850; Naskar Aq. & Semi-Aq. Plants 311.1990. *Scirpus bisumbellatus* Forssk., Fl. Aeg.-Arab. 1: 15. 1775. *Fimbristylis dichotoma* sensu Clarke, Hook. f. Fl. Brit. India 6: 635.1895 (non Vahl).

Field Collection: Titagarh, *K.N. - 2217.*

5.3. ***Fimbristylis complanata*** (Retz.) Limk, Hort. Berol 1: 292.1827; Naskar Aq. & Semi-Aq. Plants 312.1990. *Scirpus complanatus* Retz., Obs. 5: 14. 1789.

Field Collection: Taki/Madhyampur, *K.N. - 956.*

5.4. ***Fimbristylis cymosa*** R. Brown, Prodr. 228. 1810; Naskar Aq. & Semi-Aq. Plants 312. 1990. *F. spathacea* Roth, Nob. Pl. Sp. 24.1821. *F. glomerata* Nees, Linn. 9: 290. 1835. *F. wightiana* Nees, Wight Contrib. 99. 1834.

Field Collection: Ganga Sagar, *A.K. Mukherjee-s.n.* (CAL).

5.5. ***Fimbristylis dichotoma*** (L.) Vahl, Enum. 2: 287.1806; Naskar Aq. & Semi-Aq. Plants 313. 1990. *Scirpus dichotoma* L. Sp. Pl. 1: 50.1753.

Field Collection: Dum Dum, *K.N. - 1187.*

5.6. ***Fimbristylis diphylla*** (Retz.) Vahl, Enum. 2: 289.1806; Naskar Aq. Semi-Aq. Plants 314. 1990. *Scirpus diphylla* Retz, Obs. 5: 15.1789. *Fimbristylis polymorpha* Boeck., Linnaea 37: 14. 1871. *(Vern. Name: Bara Nirbishi).*

Field Collections: Joygram/Minakhan, *K.N. - 1129;* Joygram, *K.N. - 1131.*

5.7. ***Fimbristylis ferruginea*** (L.) Vahl, 2: 291.1806; Naskar Aq. & Semi-Aq. Plants 314.1990. *Scirpus ferrugineus* L. Sp. Pl. 1: 50.1753.

Field Collection: Surjyapur, *K.N. - 292;* Sagar Islands/Naraharipur, *K.N. - 451;* Taki/Thuba Bus Stop, *K.N. - 1038.*

5.8. ***Fimbristylis junciformis*** Kunth., Enum. 2: 239.1837; Naskar Aq. & Semi-Aq. Plants 315. 1990.

Field Collection: Santoshpur, *K.N. - 1954.*

5.9. ***Fimbristylis miliacea*** (L.) Vahl, Enum.2: 287. 1806; Naskar Aq. & Semi-Aq. Plants 315.1990. *Scirpus miliaceus* L., Syst. Veg. (ed.10) 868. 1759. *S. tetragonus* Poir, Encyc. Method 6: 767.1849.

Field Collection: Chandaneswar/Dhosha, *K.N. - 612.*

5.10. ***Fimbristylis ovata*** (Burm.f.) Kern, Blumea 15(1): 26. 1967; Naskar Aq. & Semi-Aq. Plants 316.1990. *Carex ovata* Burm.f., Fl. India 194.1786. *Cyperus monostachya* (L.) Vahl, Enum. Pl. 2: 296. 1806. *Abildgaardia monostachya* Vahl, Enum. 2: 296. 1806. *Fimbristylis monostachyos* (L.) Hassk., Pl. Jav. Rar. 61.1848. *Iriha monostachya* (L.) O. Kuntze, Rev. Gen. Pl. 1: 751.1891. *(Vern. Name: Marmar.).*

Field Collection: Taki/Thuba Bus Stop, *K.N. - 1023.*

5.11. ***Fimbristylis polytrichoides*** (Retz.) Vahl, Enum. 2: 248. 1906; Naskar Aq. & Semi-Aq. Plants 317.1990. *F. juncea* Boeck., Linn. 37: 4.1871. *Scirpus polytrichoides* Retz. Obs. 4: 11.1786. *(Vern. Name: Talnaru).*

Field Collection: Ganga Sagar, *A.K. Mukherjee-8735* (CAL).

5.12. ***Fimbristylis quinquangularis*** Kunth., Enum. 2: 229. 1837; Naskar Aq. & Semi-Aq. Plants 317.1990.

Field Collections: Magrahat, *K.N. - 254;* Haroa, *K.N. - 604.*

5.13. ***Fimbristylis schoenoides*** Vahl, Enum. 2: 286.1806; Naskar Aq. & Semi-Aq. Plants 318.1990.

Field Collection: Rangabelia, *K.N. - 1527.*

5.14. ***Fimbristylis squarrosa*** Vahl, Enum. 2: 289.1806; Naskar Aq. & Semi-Aq. Plants 318.1990.

Field Collection: Sarisha/near Diamond Harbour. *K.N. - 1947.*

5.15. ***Fimbristylis tetragona*** R. Br. Prodr. 226.1810; Naskar Aq. & Semi-Aq. Plants 319. 1990. *F. cylindrocarpa* Boeck., Linnaea 37: 7.1871.

Field Collection: Panihati, *K.N. - 1607.*

6.1. ***Fuirena ciliaria*** (L.) Roxb. Fl. Ind. 1: 184.1820; Naskar Aq. & Semi-Aq. Plants 319.1990. *Scirpus ciliaris* L. Mant. 2: 182.1771; *Fuirena glomerata* Lamk., Tabl. Ency. Meth. Bot.1: 150.1791. *(Vern. Name: Band Kola).*

Field Collection: Naihati, *K.N. - 1594.*

7. SCIRPUS L.

Key to the Species

1a	Large to medium herbs/sedges, 0.2-2.0 m or more high, leafy near the base, rush-like; bristles often present under the nut	(2)
1b	Small, tufted herbs, with very slender stems; bristles-0	(9)
2a (1)	Nut not marked with transverse wavy lines	(3)
2b (1)	Nut transversely wavy	(5)
3a (2)	Hypogynous bristles-0	2. *S. corymbosus* Heyne ex Roth
3b (2)	Hypogynous bristles present	(4)
4a (3)	Glumes bi-fid at the apex	9. *S. tuberosus* Desf.
4b (3)	Glumes not bi-fid at the apex	3. *S. grossus* Linn.f.

5a (2)	Stem triquetrous; bristles 5-6; spikelets terete; nut triquetrous	6. *S. mucronatus* L.
5b (2)	Stem terete	(6)
6a (5)	Glumes distinctly keeled; bristles absent; spikelets angular; nut triquetrous	(7)
6b (5)	Glumes faintly or not keeled and spikelets terete	(8)
7a (6)	Spikelets 7-1 in a single lateral head	8. *S. supinus* L.
7b (6)	Spikelets umbellate, on 3-1 rays	8/1. *S. supinus* L. var. *lateriflorus* Koyama
8a(7)	Nut plano-convex; style sometime bi-fid; glumes ovate, concave, incurved	4. *S. juncoides* Roxb.
8b (7)	Nut tri-quetrous; style always tri-fid; stem very spongy, transversely septate when dry; glumes slightly keeled, not inflated in fruits	1. *S. articulatus* L.
9a (1)	Stem leafy, 2.5-20.0 cm; style as long as nut	5. *S. michelianus* L.
9b (1)	Stem with few short leaves near the base; style hardly one-fourth the nut	7. *S. squarrosus* L.

7.1. ***Scirpus articulatus*** L. Sp. Pl. 1: 47. 1753; Naskar Aq. & Semi-Aq. Plants 321. 1990. *Isolepis articulata* Nees, Wight. Contrib. 108.1834. *Sphoenoplectus articulatus* (L.) Palla, Bot. Jahrb. Syst. 10: 299. 1888. *(Vern. Name: Pappati Chikha* or *Chichora).*

Field Collections: Fultala/Baruipur, *K.N. - 48;* Dakshineswar, *K.N. - 683.*

7.2. ***Scirpus corymbosus*** Heyne ex Roth, Nov. Pl. Sp. 28.1821; Naskar Aq. & Semi-Aq. Plants 322. 1990.

Field Collections: Diamond-Harbour, *K.N. - 2406;* Barasat, *Mallick-530.*

7.3. ***Scirpus grossus*** Linn.f. Suppl. 104. 1781; Naskar Aq. & Semi-Aq. Plants 323.1990. *(Vern. Name: Bara Kashuria).*

Field Collections: Diamond Harbour, *K.N. - 2403;* Barasat, *Mallick-521* (CAL).

7.4. ***Scirpus juncoides*** Roxb. Fl. Ind. (ed. Carey) 1: 1: 228.1820; Naskar Aq. & Semi-Aq. Plants 223.1990. *S. erectus* Poir. sensu C.B. Clarke, Hook.f. Fl. Brit. India 6: 656. 1893.

Field Collection: Dhakuria, *K.N. - 1973.*

7.5. ***Scirpus michelianus*** L. Sp. Pl. 1: 52.1753; Naskar Aq. & Semi-Aq. Plants 324.1990. *Cyperus diffusus* Roxb. Fl. Ind. 1: 189.1832. *(Vern. Name: Chota Gothubi).*

Field Collection: Satzelia/Sundarbans, *K.N. - 1549.*

7.6. ***Scirpus mucronatus*** L. Sp. Pl. 1: 50.1753; Naskar Aq. & Semi-Aq. Plants 325.1990.

Field Collection: Chandpara/Thakurnagar, *K.N. - 1668.*

7.7. ***Scirpus squarrosus*** L. Mant. 2: 181.1771; Naskar Aq. & Semi-Aq. Plants 325. 1990. *Isolepis squarrosa* Roem. & Schult. Syst. 2: 111.1817. *(Vern. Name: Guri).*

Field Collection: Mollakhali, *K.N. - 1848.*

7.8. ***Scirpus supinus*** L. Sp. Pl. 1: 49.1753; Naskar Aq. & Semi-Aq. Plants 326.1990.

Field Collection: Barasat, *R. Mallick-527* (CAL).

7.8/1. ***Scirpus supinus*** L. var. ***lateriflorus*** Koyama, J. Fac. Univ. Tokyo Sect.3, 7: 302.1958; Naskar Aq. & Semi-Aq. Plants 326.1990.

Field Collection: Titagarh, *K.N. - 2216.*

7.9. ***Scirpus tuberosus*** Desf., Pl. Atlant. 1: 50.1798; Naskar Aq. & Semi-Aq. Plants 327.1990. *S. maritimus* L. Sp. Pl. 51.1753. *S. tridentalus* Roxb. Fl. Ind. 1: 225.1832.

Field Collection: Frazergunj, *K.N. - 1911.*

CLIV. POACEAE

(nom.alt. Gramineae)

Key to the Genera

(Modified after Majumdar, R.B., 1956)

1a Spikelets usually 1-many flowered and atleast with a single hermaphrodite floret (2)

1b Spikelets with one perfect florets, when two flowered reduced florets always below the perfect, may be

	unisexual and paired or in different parts of the same plant	(28)
2a (1)	Spikelets many flowered	(3)
2b (1)	Spikelets 1-7 flowered, if 1-flowered, spikelets 3-together	(15)
2c (1)	Spikelets 2-flowered; lowermost floret sterile or male,upper bisexual	(17)
2d (1)	Spikelets 1-flowered	(18)
3a (2)	Inflorescence naked panicles	(4)
3b (2)	Inflorescence not naked; panicles	(14)
4a (3)	Tall, stout woody grasses or shrubs	(5)
4b (3)	Usually herbs, if shrubs mostly halophytic with sessile spikelets	(6)
5a (4)	Tall woody grasses, shrubs, much branched, stem sheath glabrous; pericarp thin	9. *BAMBUSA* Schreb. *nom cons.*
5b (4)	Usually very tall, stout woody, stem unarmed; stem sheath hard, pericarp crustaceous thick	20. *DENDROCALAMUS* Nees
6a (4)	Spikelets awned	(7)
6b (4)	Spikelets not awned	(9)
7a (6)	Aquatic shrubs; lemmas usually 3-nerved; grains broad, concave	44. *MYRIOSTACHYA* Hook.f.
7b (6)	Aquatic, semi-aquatic or terrestrial herbs; grains generally very minute, terete	(8)
8a (7)	Terrestrial herbs; spikelets similar, reddish, imbricate or short and stiff racemes; lemmas 1-nerved	25. DINEBRA Jack.
8b (7)	Aquatic or semi-aquatic herbs; spikelets green in ovoid or oblong, sessile clusters on the rachis; lemmas 3-nerved	29. *ELYTROPHORUS* Beauv.
9a (6)	Spikelets sessile or almost sessile	(10)

9b (6)	Spikelets pedicelled	(13)
10a (9)	Spikes or racemes many, alternately arranged	(11)
10b (9)	Spikes few, more or less digitately arranged	(12)
11a (10)	Spikes short and stiff	21. *DESMOSTACHYA* Stapf.
11b (10)	Racemes long and slender	41. *LEPTOCHLOA* Beauv.
12a (10)	Spikes more than half the length of inflorescence stalk, axis terminating in a spikelets	28. *ELEUSINE Gaertn.*
12b (10)	Spikes never exceeding half the length of the inflorescence stalk, axis terminating into a mucro	19. *DACTYLOCTE-NIUM* Willd.
13a (9)	Terrestrial or semi-aquatic; florets more or less compact	30. *ERAGROSTIS* Beauv.
13b (9)	Aquatic; florets more or less compact	26. *DIPLACHNE* Beauv.
14a (3)	Terrestrial; leaves filiform; rachilla glabrous; flowering glumes dorsally hairy beyond the middle	6. *ARUNDO* L.
14b (3)	Aquatic; leaves lanceolate; rachilla hirsute; flowering glumes glabrous	53. *PHRAGMITES* Trin.
15a (2)	Cultivated herbs; inflorescence symmetrical; spikes not digitate	(16)
15b (2)	Wild herbs; inflorescence open panicles; spikelets awned with bent dorsal awn	7. *AVENA* L.
16a (15)	Spikelets 1-flowered, sessile, 3-together at each node of rachis, glumes nerveless	35. *HORDEUM* L.
16b (15)	Spikelets 2-5 flowered, solitary, placed flatwise at each joint or rachis; glumes 3 to several nerved	67. *TRITICUM* L.
17a (2)	Shrubs or large reed like grass; inflorescence open panicles; spikelets not awned, very minute	65. *THYSANOLAENA* Nees

17b (2)	Tufted creeping herbs; false spikes with adpressed short racemes; spikelets 2-flowered, awnless	5. *ARUNDINELLA* Raddi.
18a (2)	Inflorescence if panicles spikelets with tripartite awn; or spikes with simple awned or awnless spikelets	(19)
18b (2)	Inflorescence open or contracted panicles; spikelets with simple awned or awnless.	(24)
19a (18)	Inflorescence panicles; spikelets with tripartite awn	3. *ARISTIDA* L.
19b (18)	Inflorescence usually digitate spikes; spikelets with undivided awn or if awnless with lower glume	(20)
19c (18)	Inflorescence simple, erect, terminal spikelets with a simple awn or if awnless the lower glume absent	(22)
20a (19)	Spikes solitary, terminal; spikelets not sunk in the rachis, awnless, minute, unilateral	42. *MICROCHLOA* R. Brown
20b (19)	Spikes digitate, spicate or racemed	(21)
21a (20)	Spikelets awnless 1-fid; spikes digitate	17. *CYNODON* Rich. ex Pers.
21b (20)	Spikelets awned	12. *CHLORIS* Sw.
22a (19)	Spikelets fascicled all round a slender tachis, falling in clusters of 2-4; outer glume echinate	66. *TRAGUS* Hall. *nom. cons.*
22b (19)	Spikelets not clustered, falling singly; glumes not echinate	(23)
23a (22)	Spikelets awned and awns always longer than the spikelets; lower glume present	51. *PEROTIS* Ait.
23b (22)	Spikelets if awned, awns minute, lower glume absent	71. *ZOISIA* Willd.

24a (18)	Spikelets flattened; glume-0; lemma 1 and hard or if many than the lower reduced to scaly structures	(25)
24b (18)	Spikelets never awned, terete, glumes present, lemma one, delicate	64. *SPOROBOLUS* R. Brown
25a (24)	Small erect or floating herbs; leafblades ovate-lanceolate, less than 8 mm long; spikelets awned, less than 8 mm long	36. *HYGRORYZA* Nees
25b (24)	Creeping or more or less erect herbs; leaves linear; spikelets awned or not less than 8 mm long	(26)
26a (25)	Culms long creeping on the surface of water; leafblades linear, less than 20 cm long	40. *LEERSIA* Sw.
26b (25)	Culms more or less erect, may be halophytic; leafblades exceeding 20 cm	(27)
27a (26)	Plants hydrophytic, wild or cultivated, ovary single; fertile lemma winged	46. *ORYZA* L.
27b (26)	Plants halophytic, completely wild, fertile lemma dorsally winged	55. *PORTERESIA* Takeoka
28a (1)	Spikelets generally solitary, if 2-flowered the uppermost hermaphrodite, lower male or barren; sometime with a single fertile floret and a lemma white and longer than the glumes; glumes smaller than spikelets	(29)
28b (1)	Spikelets often in pairs or rarely solitary; unisexual female spikelets in accompanied by pedicelled male-spikelets; lemma shorter than glumes; glumes as long as spikelets	(44)
28c (1)	Spikelets unisexual, in different parts of the same inflorescence or in different parts of the same plant	(64)

29a (28)	Spikelets subtended by bristles; spikelets falling singly, or without leaving the bristles on inflorescence axis	(66)
29b (28)	Spikelets not subtended by bristles	(30)
30a (29)	Spikelets awned, pedicel with cilia; lower glume present	(31)
30b (29)	Spikelets never awned, rarely apiculate; lower glume present or absent, pedicel with or without cilia	(33)
31a (30)	Terrestrial creeping herb; leafblades ovate-lanceolate or lanceolate; racemes with rather thinly spaced spikelets	45. *OPLISMENUS* Beauv.
31b (30)	Aquatic or semi-aquatic herbs; leafblades linear or lanceolate	(32)
32a (31)	Racemes with crowded spikelets, arranged in 4 rows	27. *ECHINOCHLOA* P. Beauv.
32b (31)	Panicle simple, produced into an awn like bristle beyond the terminal spikelets, not in 4 rows	57. *PSEUDORAPHIS* Griff.
33a (30)	Lower glume always present	(34)
33b (30)	Lower glume very minute or sometime absent; racemes digitately or rarely alternately arranged	(41)
34a (33)	Inflorescence open panicle, if racemes then with plicate leafblade	(35)
34b (33)	Inflorescence not open panicle	(37)
35a (34)	Lower glume less than half the length of the spikelets	(36)
35b (34)	Lower glume more than half the length of the spikelets	1. *ACROCERAS* Stapf.
36a (35)	Leafblades ovate lanceolate	18. *CYRTOCOCCUM* Stapf.
36b (35)	Leafblades linear, may be plicate and ovate	47. *PANICUM* L.
37a (34)	Inflorescence spiciform panicle	(38)

37b (34)	Inflorescence panicle of adpressed or spreading racemes	(39)
38a (37)	Spikelets ovoid or oblong-conic, acute, saccate	60. *SACCIOLEPIS* Mash.
38b (37)	Spikelets linear-lanceolate, acuminate	37. *HYMENACHNE* Beauv.
39a (37)	Racemes alternate and spreading, pedicels with few stiff cilia	(40)
39b (37)	Racemes adpressed to the rachis, upright, pedicel smooth	48. *PASPALIDIUM* Stapf.
40a (39)	Lower glume facing the rachis	11. *BRACHIARIA Griseb.*
40b (39)	Lower glume away from rachis	68. *UROCHLOA* Beauv.
41a (33)	Lower glume minute, glume 3 with generally a minute pale, nerves straight, prominent	23. *DIGITARIA* Heist. ex Fabr.
41b (33)	Lower glume completely absent	(42)
42a (41)	Glume 2 fimbriate; glumes 3 with a deep-cleft pale and a male flower	8. *AXONOPUS* Beauv.
42b (41)	Glume 2 empty, marginal nerves curved	(43)
43a (42)	Racemes few, atleast uppermost pair digitate; lemma not mucronate; spikelets plano-convex	49. *PASPALUM* L.
43b (42)	Racemes many, alternate; lemma mucronate; spikelets never plano-convex	32. *ERIOCHLOA* H.B. & K.
44a (28)	Spikelets all similar	(45)
44b (28)	Spikelets dissimilar	(47)
45a (44)	Spikelets solitary secund on a filiform or flattened, inarticulated rachis	24. *DIMERIA* R. Brown
45b (44)	Spikelets in open or contracted panicles	(46)
46a (45)	Panicles spiciform, silky, rachis tough	38. *IMPERATA* Cyrill.
46b (45)	Panicles much branched, open, silky; rachis fragile	59. *SACCHARUM* L.

47a (44)	Spikelets sunk in the pits of the rachis	(48)
47b (44)	Spikelets not sunk in the nodes or rachis	(50)
48a (47)	Sessile spikelets accompanied by a dissimilar pedicelled spikelets	(49)
48b (47)	Plants herbaceous; sessile spikelets 2, opposite in each internode, rachis fragile spikes cylindrical	43. *MNESITHEA* Kunth.
49a (48)	Plants shrubby; spikes cylindrical, rachis fragile	58. *ROTTBOELLIA* Linn.f.
49b (48)	Plants herbaceous; spikes compressed, rachis not fragile	33. *HEMARTHRIA* R. Brown
50a (47)	Inflorescence panicles	(51)
50b (47)	Inflorescence made up of 1-several racemes	(57)
51a (50)	Spikelets not enclosed in spathes at maturity, usually in one pairs	(52)
51b (50)	Spikelets enclosed or accompanied by spathes	(55)
52a (51)	Cultivated garden plants; spikelets dorsally compressed, awned	63. *SORGHUM* Moench. *nom. cons.*
52b (51)	Wild plants; spikelets more or less terete, awned	(53)
53a (52)	Erect shrubby plants	(54)
53b (52)	Mostly creeping herb; leafblades falt; awn comparatively long	14. *CRYSOPOGON* Trin. *nom.cons.*
54a (53)	Roots aromatic; leafblades keeled; awn short	69. *VETIVERIA* Lam.
54b (53)	Roots not aromatic; leaf blades not keeled; linear lanceolate; awn long	52. *PHALARIS* L.
55a (51)	Spikelets in pairs, rachis articulate, enclosed in reddish-brown spathe	16. *CYMBOPOGON* Spreng.
55b (51)	Spikelets in three or in two superposed series, rachis inarticulate, spathes green	(56)

56a (55)	Spikelets in condensed racemes	2. *APLUDA* L.
56b (55)	Spikelets in two superposed series	56. *PSEUDANTHISTIRIA* Griff.
57a (50)	Racemes solitary	(58)
57b (50)	Racemes two spiciform or several digitate or umbellate	(60)
58a (57)	Racemes almost in spathes; spikelets without pitted	61. *SCHIZACHYRIUM* Stapf.
58b (57)	Racemes beyond spathes; spikelets without or with pits	(59)
59a (58)	Erect much-branched herb; spikelets without pit	54. *POGONATHERUM* Beauv.
59b (58)	Creeping wild, herbaceous; spikelets pitted	31. *EREMOPOGON* (Hook.f.) Stapf.
60a (57)	Leaf auricled; racemes two	39. *ISCHAEMUM* L.
60b (57)	Leaf not auricled; racemes several	(61)
61a (60)	Racemes umbellate, spikelets with kneed hispid awn	34. *HETEROPOGON* Pers.
61b (60)	Racemes digitate or sub-digitate; spikelets with smooth awn	(62)
62a (61)	Leaves linear; upper glume of sessile spikelets without hairs and more than one nerved	(63)
62b (61)	Leaves lanceolate; upper glume of sessile spikelets with long hairs	4. *ARTHRAXON* Beauv.
63a (62)	Lower glume of sessile spikelets narrow, not clasping the upper glume	10. *BOTHRIOCHLOA* Kuntze
63b (62)	Lower glume of sessile spikelets broad; clusping the upper	22. DICHANTHIUM Will.
64a (28)	Aquatic or semi-aquatic; racemes of male spikelets numerous; caryopsia enclosed in the hard bracts	(65)

64b (28)	Terrestrial, racemes of male spikelets terminal; female flowers axillary	70. *ZEA* L.
65a (64)	Plants glabrous or slightly hairy	15. *COIX* L.
65b (64)	Plants with abundant tubercle based	13. *CHIONACHNE* R. Br.
66a (29)	Spikelets falling with bristles	50. PENNISETUM Pers.
66b (29)	Spikelets falling singly	62. SETARIA Beauv.

1. *ACROCERAS* Stapf.

1.1 ***Acroceras munroanum*** (Balansa) Henr., Blumea 3: 445. 1940; Bor Grass. Burma, Ceyl., Ind. & Pak. 275.1960. *Panicum munroanum* Balansa, Morot J. de Bot. 4: 140.1890. *P. ridleyi* Hack ex Ridley, Trans. Linn. Soc. (ser.2) 3: 400. 1893. *P. latifolium* Hook.f. Fl. Brit. India 7: 39. 1897 (in part). *P. oryzoides* Sw.; Prain Beng. Plants 2: 887. 1903 (Rep. ed. 1963). *P. crassiapiculatum* Merr., Phil. J. Sc.1 (Suppl.) : 356.1906. *Acroceras crassiapiculatum* (Merr.) Alston, Trimen-Handbook Fl. Ceylon 6 (Suppl.): 324. 1931.

Branched, prostrate, rooting at the nodes, supported above the soil by nodal roots or creeping grass. Leaf-blade flat, lanceolate, fine pointed; sheath hairy, mid-vein prominent; ligule obscure. Inflorescence panicles, pedicels longer than spikelets, tip swollen. Caryopsis broadly oblong, plano-convex, dorsally flat, hilum filiform. *Fls. & Frts.:* June- December.

Occasionally grow on the roadsides and high lands.

Field Collection: Bamanghata, *K.N. - 1735.*

Distributions: Eastern part of India. Burma, Malaya, Malaysia and America.

Notes: Prain (1903a) has reported this taxon from the East Bengal (now Bangladesh) only.

2. *APLUDA* L.

2.1. ***Apluda mutica*** L. Sp. Pl. 82.1753; Naskar & Guha Bakshi, Mangr. Swamps Sundarbans 145. 1987. *A. aristata* L. Amoen Acad.4: 303.1756. *Andropogon glaucus* Retz., Obs. Bot. 5, 20,1789. *Calamus gigantea* (Spreng) P. Beauv., Ess. Agrost, 128.t.23.f.1.1812. *(Vern. Name: Chunkarki).*

Tall, annual or perennial grass; common on the grasslands, hedges and thickets and even on the river banks. Naskar & Guha Bakshi (l.c.) described this taxon after collecting from Magrahat *(K.N. - 256)*; Purandarpur/Baruipur *(K.N. - 1146)*. and New Barrackpore (R. Mallick-236).

Buffaloes relish the young grass, but not palatable for cows.

3. *ARISTIDA* L.

Key to the Species

1a	Glumes I and II, not awned	(2)
1b	Glumes I and II long awned; panicle contracted	3. A. *setacea* Retz.
2a (1)	Panicle very slender; spikelets without the awn	2. *A. cumingiana* Trin. & Rupr.
2b (1)	Panicle contracted; spikelets without the awn	1. *A. adscencionis* L.

3.1. ***Aristida adscencionis*** L. Sp. Pl. 82. 1753; Hook.f. Fl. Brit. India 7: 224.1897; Prain Beng. Plants 2: 912.1903 (Rep. ed. 1963); Bor, Grass. Burma, Ceyl. Ind., & Pak. 407.1960. *A. depressa* Retz. Obs. Bot. 4: 22.1786. *A. canariensis* Willd., Enum. Pl. 99. 1809. *Chaetaria adscensionis* (L.) P. Beauv., Ess. Agrost. 30, 151, 158. 1812. *Aristida divaricata* Jacq., Eclog. Gram. Rar. 7, t.6.1813.

Dwarf, annual grass; densely tufted, very slender, erect or ascending. Leaves-blades convolute, filiform; sheaths smooth, with rounded auricles. Panicles 9-30 cm long, compound, contracted, sub-secund, rachis filiform, smooth, branches short; pedicels short, capillary. Spikelets erect, one flowered, acute; glumes whitish with reddish tinge. *Fls. & Frts.:* April-October.

Common grass on waste lands, the dry lands, old brick walls and on poor, shallow, dry infertile soils.

Field Collection: Durganagar, *K.N. - 1716.*

Distributions: Introduced in India during 1696; grow in moist warmer countries and on the sandy river-banks. A native of Africa.

Economic Importance: Young grass is the cattle food.

3.2. ***Aristida cumingiana*** Trin. *&* Rupr., Sp. Gram. Stip. 141.1842; Hook.f. Fl. Brit. India 7: 224.1897; Prain Beng. Plants 2: 912.1903 (Rep. ed.

1963). *A. capillacea* Cavan., Ic. & Descr. Pl. 5, 43, 1799 (*nom.* Lamk. 1797). *Chaetaria trichodes* Nees ex Lindl., Hook. Kew J. Bot. 2, 101,1850.

Dwarf, annual grass; tufted, very slender, leafblades filiform, long. Inflorescence-panicle, sub-cylindric, narrow, slender, loosely branched. Spikelets-green, or purplish, glumes acuminate, light blue with slight reddish tinge except median nerves which are green. *Fls. & Frts.:* April-October.

Common on the dry lands or on the old brick-walls; a pretty little species, which grow sometime gregariously.

Field Collection: Chandpara/near Gobardanga, *K.N. - 1672.*

Distributions: Throughout India. Senegal, Abyssinia, Siam, China, Philippines and Tropical America.

Notes: D. Prain (1903a) did not report this taxon from Bengal proper.

3.3. ***Aristida setacea*** Retz., Obs. Bot. 4: 22.1786; Hook.f. Fl. Brit. India 7: 225.1897; Watt Dict. Econ. Prod. 1: 312.1883; Prain Beng. Plants 2: 913.1903 (Rep. ed. 1963); Bor, Grass. Burma, Ceyl., Ind. & Pak. 412.1960. *Chaetaria setacea* (Retz.) P. Beauv., Res. Agrost. 30, 158, 1812. *Aristida quinqueseta* Steud., Syn. Pl. Glume. 1: 420.1855.

Stout, erect, perennial grass, 9-12 cm high, with woody base; roots fibrous stout wiry, hard but smooth and polished, very slender. Leaves convolute, coriaceous, smooth; sheaths long smooth; ligule with short hairs. Inflorescence panicle 20-36 cm long, inclined, contracted, sub-secund; rachis slender, smooth, branched, long or short; pedicels capillary; spikelets erect. *Fls. & Frts.:* March-July.

Grow on the brick walls, or rather in dry places.

Field Collection: Basanti, *K.N. - 2280.*

Distributions: Throughout India. Sri Lanka, Burma and Mascarene Islands.

Prain: Prain (1903a) did not mention this taxon from Bengal proper.

4. *ARTHRAXON* Beauv.

4.1. ***Arthraxon lancifolius*** (Trin.) Hochst., Flora 39: 188.1856; Hook.f. Fl. Brit. India 7: 143.1897; Prain Beng. Plants 2: 903.1903 (Rep. ed. 1963). Bor Grass. Burma, Ceyl., Ind. & Pak. 100. 1960. *Andropogon lancifolius* Trin., Mem. Acad. Sci. Petersb. (ser.6) 2: 271.1832. *Arthraxon microphyllus* Hochst., Flora 39: 188.1856; Hook.f. Fl. Brit. India 7: 147.1897; Prain Beng. Plants 2: 903.1903 (Rep. ed. 1963).

Annual grass; stem and branches filiform. Leafblades tufted, ovate or elliptic, acuminate membranous, base slightly auricled; sheath short, glabrous, ligule small, hyaline. Panicle branched; rachis capillary; internodes shorter than the slender spikelets. Spikelets solitary. *Fls. & Frts.:* March-September.

Grow on the moderate soil and water, shade loving grass. Profusely grow on the grassy places, roadsides and even on the waste lands and on the old walls.

Field Collection: Dakshineswar, *R. Mallick-171* (CAL).

Distributions: Hilly areas of India. Sri Lanka, Burma, China, Malaysia, Tropical Africa.

Economic Importance: A minor but important fodder grass.

Notes: Prain (1903a) did not report this taxon from lower Bengal.

5. *ARUNDINELLA* Reddi.

Key to the Species

1a	Awn-base slightly twisted	2. *A. pumila* (Hochst.) Steud.
1b	Awn-base not twisted	1. *A. bengalensis* (Spreng.) Druce

5.1. ***Arundinella bengalensis*** (Spreng) Druce, Rep. Bot. Exchng. Club. Brit. Isles. 605. 1916; Bor Grass. Burma, Ceyl., Ind. & Pak. 421. 1960. *Panicum bengalensis* Spreng., Syst. Veg. 1: 311.1825. *Arundinella wallichi* Nees.; Hook.f. Fl. Brit. India 7: 75.1897; Prain Beng. Plants 2: 916.1903 (Rep. ed. 1963). *Panicum strictum* Roxb . Fl. Ind. 1: 306.1820. *Arundinella stricta* (Roxb.) Janowski; Fedde Rep. Sp. Now. 17: 84.1921. *(Vern. Name: Ganga Bena).*

Stout, erect, perennial grass; rhizome woody, creeping hard. Leafblades-radical; sheath lax, often coming out of internodes, hairy. Inflorescence panicles, contracted, strict racemes. Spikelets 3-4 seriate, ovoid, densely imbricate, setuse. *Fls. & Frts.:* March-October.

Common on the dry grasslands, waste places, roadsides and on railway tracks, often creeping and forming a tufted mass.

Field Collection: Bantala, *K.N. - 1740.*

Distributions: Tropical Himalaya and some parts of the plains of eastern India. South East Asia.

Notes: Prain (1903a) did not report this taxon from lower Bengal.

5.2. ***Arundinella pumila*** (Hochst.) Steud. Syn. Pl. Glum. 114.1854; Naskar Aq. & Semi-Aq. Plants 332. 1990. *Acratherum pumilum* Hochst., A Rich. Tent. Fl. Abyss. 2: 414. 1851. *Arundinella tenella* Nees & Wight. ex Steud Syn. Pl. Glum. 115. 1854.

Annual, glabrous grass. Gregariously grow on the moist waste lands. Naskar (l.c.) described this taxon after collecting from Bamanghata (*K.N. - 2309).* Prain (1903a) did not mention this taxon from lower Bengal.

6. *ARUNDO* L.

6.1. ***Arundo donax*** L. Sp. Pl. 81. 1753; Naskar Aq. & Semi-Aq. Plants 333. 1990. *A. sativa* Lamk. Fl. Frane. 3: 616. 1778. *A. bifaria* Retz., Obs. Bot. 4: 21. 1786. *A. latifolia* Salisb., Prodr. Strip. 24. 1796. *Donax arundinaceus* P. Beauv., Ess. Agrost. 78, 152, 161, 1812. *(Vern. Name: Gaba-Nal).*

Tall, stout, perennial grass,. Along water courses, ditches,prefer moist localities. Naskar (l.c.) described this taxon from the lower Ganga delta after collecting from Chaital /Basirhat *K.N. - 724* .It is a fodder grass.

7. *AVENA* L.

Key to the Species

1a	Wild grass	1. *A. fatua* L.
1b	Cultivated grass like cereal plant	2. *A. sativa* L.

7.1. ***Avena fatua*** L. Sp. Pl. 1: 80. 1753; Hook.f. Fl. Brit. India 7: 275.1897; Bor Grass., Burms, Ceyl., Ind. & Pak. 434. 1960. *A. fatua* L. var. *pilossissime* S.F. Gray, Nat. Arr. Pl. 2, 131. 1821. *(Vern. Name: Oat-ghash).*

Medium to tall annual grass. Leaf-blades flat 20-30 cm x 1.5-2.0 cm, many veined, veins with numerous hard, upwardly curved hairs, insertion of blade to sheath oblique. Sheath compressed, smooth. Ligule very obtuse. Inflorescence-panicle. Spikelets 3-fid, rachilla freely dis-articulating below. Caryopsis free, silky all over. *Fls. & Frts.:* February-April.

Rarely grow as weed on the wheat fields, on the railway tracks and on the road-side waste lands.

Field Collection: Barasat, *R. Mallick-673* (CAL).

Distributions: Throughout India. Europe to Central Asia, North Africa.

Economic Importances: Rarely as fodder grass, the grains are low quality cereals and used as fodder.

Notes: Prain (1903a) did not mention this taxon from Bengal.

7.2. ***Avena sativa*** L. Sp. Pl. 79. 1753; Hook.f. Fl. Brit. India 7: 275. 1897; Watt Dict. Econ. Prod. 1: 356.1879; Prain Beng. Plants 2: 917. 1903 (Rep. ed. 1963). *(Vern. Name: Oat).*

Annual or perennial cultivated grass. Leaves flat. Spikelets panicle, terete, 2 or more-flowered, not articulate on their pedicels; glumes 4, dorsally round, awnless. Grain fusiform furrowed. *Fls. & Frts.:* February-April.

Very rarely cultivated, but ocasionally as a wild grass like.

Field Collection: Saktinagar, *K.N. - 553.*

Distributions: Cultivated in South India. In Europe.

Economic Importance: Low quality cereal; grains are fodder for horse.

8. *AXONOPUS* Beauv.

8.1. ***Axonopus compressus*** (Sw.) Beauv. Ess. Agros. 12: 154.167.1812; Bor Gras. Burma, Ceyl., Ind. & Pak. 278.1960. *Milium compressum* Swartz. Prodr. Veg. Ind. Occ. 24.1788. *Paspalum tristachyon* Lamk., Tab. Encyl. Meth. Bot. 2: 176.1791.

More or less erect, terrestrial herb, stoloniferous. Leaf-blades linear-lanceolate. Inflorescence racemes. Spikelets oblong or lanceolate, smooth. Caryopsis-elliptic, white, obtuse, scutellum-rotundate elliptic. *Fls. & Frts.:* January-May.

Common on the roadsides, cultivated lands, railway tracks and moderate waste lands; stolons grow quickly.

Field Collection: Barasat, *R. Mallick-746* (CAL).

Distributions: More or less throughout India. Southern United States, Mexico to Brazil.

Economic Importance: As cattle fodder.

9. *BAMBUSA* Schreb. *nom. cons.*

Key to the Species

1a	Tall tree-like grass	(2)
1b	Shrub; stem not spinous; blade of stem-sheath linear	3. *B. nana* Roxb.
2a (1)	Stem spinous	1. *B. bambos* (L.) Voss.
2b (1)	Stem not spinous	(3)

3a (2)	Ligule distinct	(4)
3b (2)	Ligule narrow, obscure; blade of stem-sheaths triangular or reniform, with cordate base	4. *B. tulda* Roxb.
4a (3)	Ligule of stem-sheaths broad, toothed, leaves broadly triangular, membranous	2. *B. balcoona* Roxb.
4b (3)	Ligule of stem sheaths toothed or fimbriate; leaves short and ciliate	5. *B. vulgaris* Schrader

9.1. ***Bambusa bambos*** (L.) Voss., Vilmorin, Blumengartnerei 1: 1189. 1896. *Arundo bambos* L. Sp. Pl. 81. 1753. *Bambusa arundinacea* Willd. Sp. Pl. 2: 245.1799; Hook.f. Fl. Brit. India 7: 395.1897; Prain Beng. Plants 2: 930.1903 (Rep. ed. 1963). *Bambusa arundo* Klein & Nees, Linnaea 9: 471.1834. *(Vern. Name: Bans).*

Densely caespitose, gregarious, culms graceful and curving, nodes prominent, the lower with long horizontal shoots armed wih recurved spines; stem 20-35 m high and 20-25 cm diam; sheath coriaceous, variable in shape. Leaves linear lanceolate, tip stiff, glabrous or puberulous beneath. Inflorescence an enormous panicle, often occupying the whole stem; branchlets bearing loose clusters of pale, sub-erect, lanceolate acute glabrous spikelets. Lodicules 3, ovate or sub-ovate, hyaline, ciliate. Anthers yellow, obtuse. Style short. *Fls. & Frts.:* Occasionally during the rainy season.

Both wild and cultivated throughout the lower Ganga delta.

Field Collection: Saga Mela Premises/Sagar Islands, *K.N. - 465.*

Distributions: Throughout India, except the Himalaya; indigenous in moist parts of the plains of India; also cultivated in the valley of the Ganges. Sri Lanka, Bangladesh and Burma.

Economic Importances: The culms are used to built up village houses, cottages and also as paper pulp; its grains are occasionally edible. Stem and leaves are used by the Ayurvedic physicians as blood purifier, in leucoderma and inflamatory conditions; liquid accumulated inside the bamboo culm is important in tubercularis, asthma and leprosy. Young shoots contain hydrocyanic acid and are lethal to mosquito larvae.

9.2. ***Bambusa balcooa*** Roxb. Fl. Ind. 2: 196.1832; Gamble, Hook.f. Fl. Brit. India 7: 391. 1897; Prain Beng. Plants 2: 930. 1903 (Rep. ed. 1963). *(Vern. Name: Balku Bans).*

Densely tufted, culms unarmed with thick walls, 18-25 m high and 10-20 cm diam., nodes swollen with a white tinge above, lower branches stiff; sheaths glabrous, triangular. Leafblades pale and soft, pubescent below, lanceolate, or alternate at base, petiole short, hirsute, ligule broadly triangular, membranous. Inflorescence large, compound, panicle, bracteate. Spikelets 6-8 flowered, in dense heads. Lemmas membranous, ovate, many nerved. *Fls. & Frts.:* June-September.

These species are mostly cultivated throughout the lower Ganga delta.

Field Collection: Specimen not preserved.

Distributions: Indigenous in the north east India. Tropical and Sub-tropical Asia, Africa and America.

Economic Importance: Commonly used for house and cottage building purposes and in many other ways.

9.3. ***Bambusa nana*** Roxb. Fl. Ind. 2: 196.1832; Gamble, Hook.f. Fl. Brit. India 7: 390. 1897; Prain Beng. Plants 2: 929. 1903 (Rep. ed. 1963). *(Vern. Name: Chhota Bans).* (Fig.48).

Thick, evergreen dwarf shrub. Stem glabrous, smooth, green when young, then yellow, hard much branched from the base, 2 to 3 m high and 1-2.5 cm diam., branches fasciculate, nodes thickened, internodes 20-40 cm; stem sheath green and turn yellow at maturity, broad, stiff, glabrous. Leafblades distichous mostly small, rounded at the base into a short petioled, ligule short, sheaths ending in a prominent callus. Inflorescence short, diffuse, leafy panicles with few spikelets, solitary or clustered. Spikelets glabrous, straw coloured bearing 5 or more florets, separated by rachilla joints. Ligule ovate, acute, many-nerved, glabrous. Palea shorter, many nerved, 2-keeled, minutely ciliate. Lodicules linear. Stamens exerted, anthers obtusely, shortly apiculate, yellow. Ovary ovoid. Caryopsis elliptic *Fls. & Frts.:* June-May.

Rarely cultivated in the lower Ganga delta, occasionally found wild in the dry waste lands or forest edges.

Field Collections: Specimen not preserved.

Distributions: More or less throughout India. Malay Peninsula and Penang.

Economic Importance: Used as common garden hedges.

9.4. ***Bambusa tulda*** Roxb. Fl. Ind. 2: 193.1832; Gamble, Hook.f. Fl. Brit. India 7: 387. 1897; Prain Beng. Plants 2: 930.1903 (Rep. ed. 1963). *(Vern. Name: Tulda Bans).*

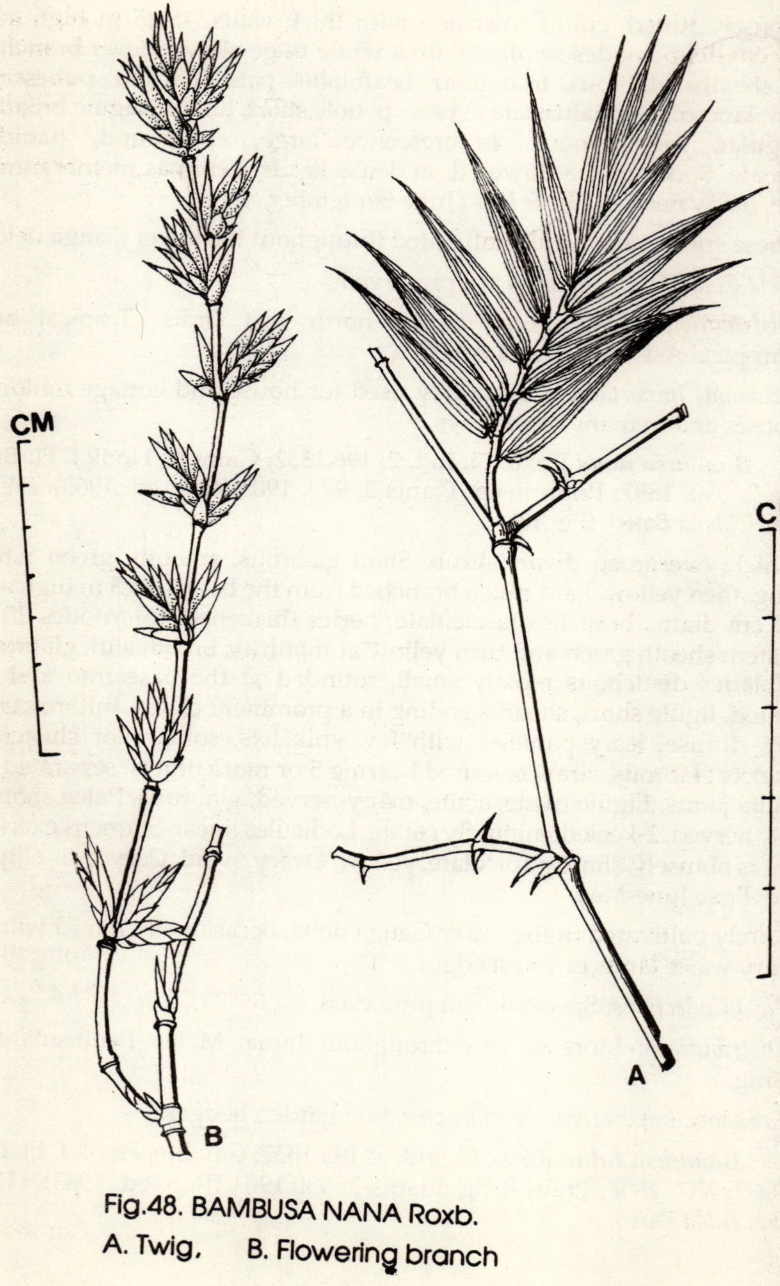

Fig.48. BAMBUSA NANA Roxb.
A. Twig, B. Flowering branch

Stem unarmed, slender, thick-walled, dark green, hollow, smooth; node not swollen. Leafblades usually linear oblong or lanceolate leaf-sheath striate, ending smooth callus. Inflorescence spikes, glossy and sessile. Spikelets many flowered, glumes several. Ovary with short styles. Caryopsis cylindric with obtuse and hairy apex. *Fls. & Frts.:* Occasionlly from July-October.

Cultivated almost throughout the lower Ganga delta.

Field Collections: Specimens not preserved.

Distributions: Abundant in the plains of north-east India. Burma, Sri Lanka and Malaya.

Economic Importances: Culms are commonly used for house and cottage building purposes, making fences, mats, fans, baskets and fish traps, valuable paper pulps and a number of articles of commerce and very common articles in the village lives. Young pulp is also cooked as vegetable.

9.5. ***Bambusa vulgaris*** Schrader, Wendl. Collect 2: 26, t. 47, 1810; Gamble, Hook.f. Fl. Brit. India 7: 391.1897; Prain Beng. Plants 2: 930.1903 (Rep. ed. 1963). *(Vern. Name: Basini Bans).*

Stem unarmed, generally polished; nodes scarcely swollen with ring brown hairs. Leaf-blades hairy-below, mature leaves glabrous, tips twisted, ligule short, hairy, petiole short. Spikelets-clusters, pedicelled.Lodicules unequal densely hairy, nerved. Anthers hairy at apex. Ovary haiy with 3 short stigmas. *Fls. & Frts.:* July August.

Sometimes cultivated in the lower Ganga delta.

Field Collection: Specimen not preserved.

Distributions: Throughout India. Believed to be a native of Java, but commonly cultivated in most Tropical countries.

Economic Importances: Culms are used in building and other domestic purposes. Used for preparing/fabricating fish traps, fishing poles, paper pulps and serveral other purposes in the village areas.

10. *BOTHRIOCHLOA* Kuntze

Key to the Species

1a	Spikes numerous, paniculate, rare-few racemose; glume 1 not pitted	1. *B. intermedia* A. Camus.
1b	Spikes 3-7, digitately sub-fascicled; glume 1 coriaceous, pitted	2. *B. pertusa* A. Camus

10.1. ***Bothriochloa intermedia*** (R. Br.) A.Camus., Ann. Soc. Linn. Lyon, 1930, N.S.76: 164.1931; Bor, grass. Brum., Ceyl., Ind., Pak. 108. 1960. *Andropogon intermedies* R. Br., Prodr. 202. 1810; Hook.f. Fl. Brit. India 7: 175. 1897; Prain Beng. Plants 2: 907. 1903 (Rep. ed. 1963). *A. intermedia* R. Br. var. *genuinus* Hackel, DC. Monogr. Phan. 6: 485. 1889. *Amphilopis intermedia* (R. Br.) Stapf. in Agric. News, W. Ind. 15: 179. 1916. *(Vern. Name: Gandha Gourana).*

Culms-perennial, over 1 m high, tufted, erect, nodes with a tuft of hairs, slightly flattened on one side. Leafblades flat, glaucous, ciliate towards base, sheaths hairy at top, terete, ligule short and truncate. Inflorescence panicles, spikes variable in length, fragile, joints and pedicels, slender and densely ciliate. Spikelets sessile, pale green or purplish. *Fls. & Frts.:* October- March.

Gregariously grow on the moderate soil, garden edges, road-sides and moist waste lands; generally in shady places.

Field Collection: Surjyapur, *K.N. - 289.*

Distributions: Throughout India. Westward to Caucasus, China, Malaya, Pacific and Tropical Africa.

Economic Importance: Young grass is good fodder.

10.2. ***Bothriochloa pertusa*** (L.) A.C. Camus, Ann. Soc. Linn. Lyon (1903) N.S. 76: 176; Bor. Grass. Burm., Ceyl., Ind. & Pak. 109. 1960. *Holcus pertusus* L. Mant. Alt. 301. 1771. *Andropogon pertusus* (L.) Willd. Sp. Pl. 4: 922. 1806; Hook.f. Fl. Brit. India 7: 173. 1897; Prain Beng. Plants 2: 907.1903 (Rep. ed. 1963). *Amphilophis pertusa* Nash ex Stapf., Agric. News W. Ind. 15: 179. 1916. *(Vern. Name: Chandi-ghash).*

Perennial grass. Culms creeping and rooting from the nodes, usually as ascending from a leafy base, slender; nodes sometimes densely bearded, innovations, extra or more often intravaginal. Leafblades dull green, sheath hardly compressed, glabrous or villous, ligule short. Inflorescence panicles. Spikelets sessile. *Fls. & Frts.:* August-March.

Gregariously grow on the railway tracks; creeping and always forming large association.

Field Collection: Garia, *K.N. - 330.*

Distributions: Throughout India. Westward to Sicily, Tropical Asia and Australia.

11. *BRACHIARIA* Griseb.

Key to the Species

1a	Spikelets 2.0 - 2.5 mm long	(2)
1b	Spikelets 1.5-1.8 mm long; lower glume nerveless; spikelets crowded	6. *B. reptans* Gard. & Hubb.
2a (1)	Spikelets turgid, crowded, may be distant	(3)
2b (1)	Spikelets not turgid, crowded or rarely distant	(4)
3a (2)	Aquatic, racemes 8-many; upper lemma obtuse, without mucro; lower glume 1-nerved, nodes long bearded	3. *B. mutica* (Forssk.) Stapf.
3b (2)	Semi-aquatic, racemes 5-many; upper lemma apiculate; lower glume 3-5 nerved, nodes glabrous or pubescent	5. *B. ramosa* (L.) Stapf.
4a (2)	Lower glume more than 5-nerved, pedicel flat; spikelets glabrous, acute	2. *B. milliformis* (Presl.) Chase
4b (2)	Lower glume 3 or rarely faintly 5-nerved; spikelets pubescent acuminate	7. *B. setigera* (Retz.) Hubb.
4c (2)	Racemes 2-4, if more spikelets distant	(5)
5a (4)	Lower glume 5-7 nerved; spikelets 2.5-3.0 mm long	1. *B. distichya* (L.) Stapf
5b (4)	Lower glume 3-5 nerved; spikelets 2.0-2.5 mm long	4. *B. kurzii* (Hook.f.) Campus

11.1. ***Brachiaria distachya*** (L.) Stapf., Prain Flor. Trop. Afr. 9: 565. 1919; Bor, Grass. Burma, Ceyl., Ind. & Pak. 281.1960. *Panicum distichyon* L. Mant. Act. 183. 1771; Hook.f. Fl. Brit. India 7: 37.1896; Prain Beng. Plants 2: 886.1903 (Rep. ed. 1963). *Digitaria distichya* (L.) Pers. Syn. Pl. 1, 85.1805. *(Vern. Name: Nari (Santal)).*

Culms branched, straggling and creeping below. Leaf-blades linear, lanceolate, glabrous; sheaths carinate, striate, open, lax, mouth hairy. Ligule a thickened hairy ring. Inflorescence panicles, racemes subconjugate. Spikelets elliptic obovate short, pedicelled, solitary, sub-sessile pale green. *Fls. & Frts.:* June-October.

Straggling high lands grass or a common weed of the waysides and waste lands; often found in the cultivated fields. It can grow under shade; very difficult to eradicate due to presence of nodal roots.

Field Collection: Media, *K.N. - 1695.*

Distributions: Plains of India from Punjab to Upper Assam. Burma, Sri Lanka, Malaya, Malacca, China and Australia.

Economic Importances: Acts as soil binder on the sand-dunes, a good fodder plant but an obnoxious weed.

11.2. ***Brachiaria milliformis*** (Presl.) A. Chase, Contrib. U.S. Nat. Herb. 22: 35. 1920; Bor Grass. Burma, Ceyl., Ind. & Pak. 283.1960. *Panicum milliforme* J. Presl. ex C. Presl. Rel. Haenk 1: 300. 1830.

Culms decumbent at base and rooting, smooth and glabrous, slender, often branching below. Leafblades lanceolate, acuminate, glabrous, rarely with few hairs, margin scabrid often with a few cilia on the rounded base, sheath striate, glabrous or sparsely hairy, margins, ciliate, close terete, a few hairs on the throat, ligule a ridge of dense white hairs. Inflorescence panicles, made up of 4-6 racemes. Spikelets solitary, rachis flattened, striate and convex on the back, pedicels short, about the length of a spikelets. *Fls. & Frts.:* May-October.

Common grass, growing abundantly near the cultivated grounds, it prefers moderate soil and water; shade loving plant. It is a troublesome weed for its creeping and nodal roots and very difficult to eradicate.

Field Collection: Lashmikantapur, *K.N. - 1779.*

Distributions: South & East India. Sri Lanka, Cocoa Islands and Malaya.

Economic Importance: An excellent fodder grass, but an obnoxious weed.

11.3. ***Brachiaria mutica*** (Forssk.) Stapf., Prain Fl. Trop. Afr.9: 526.1919; Naskar Aq. & Semi-Aq. Plnts 334.1990. *Panicum muticum* Forssk., Fl. Aegypt.-Arab 20.1775. *P. purpurascens* Raddi, Agrost, Bras 47.1823. *P. punctulatum* Arn. ex Steud, Syn.Pl. Glum.1: 62.1854. *(Vern. Name: Nardul).*

Clums branched, robust, geniculate base, ascending, compressed, striate, glabrous, nodes with long white tuft of hairs. Aquatic or semi-aquatic grass, grow profusely on the marshes, every where. Naskar (l.c.) described this taxon after collecting from Belghoria *(K.N. - 1395).*

11.4. ***Brachiaria kurzii*** (Hook.f.) A. Campus, Lecomte.Flor. Gen. De. L, Indo China Val. 7: 438.1922; Bor Grass. Burma, Ceyl., Ind. & Pak. 283.1960.

Panicum kurzii Hook.f. Fl. Brit. India 7: 38.1896; Prain Beng. Plants 2: 886.1903.

Culms decumbent and rooting below, flaccid. Leaf-blades lanceolate, acuminate, margins crisped at base long ciliated, cordate. Sheath with villous margins reaching half the length of internode. Ligule bearded. Inflorescence of few false spikes, very distant, with rarely a short branch bearing 2-3 spikelets, rachis glarbous. *Fls. & Frts.:* May-October.

Shade loving, mesophytic grass, commonly growing in great abundance near the edges of the village roads and even on the cultivated lands.

Field Collection: Kestopur/near Baguihati, *K.N. - 1898.*

Distributions: Abundant in the plains of India. Both Tropical and Temperate.

Economic Importance: Common as fodder grass.

11.5. ***Brachiaria ramosa*** (L.) Stepf, Fl. Trop. Apr. 9: 542.1919; Bor Grass. Burma, Ceyl., Ind. & Pak. 284.1960. *Panicum ramosum* (L.) Mant. Pl. 1: 29.1767; Hook.f. Fl. Brit. India 7: 36.1896; Prain Beng. Plants 2: 886.1903 (Rep. ed. 1963). *P. canescens* Roth ex Roem. & Schult Syst. Veg. 2: 457.1817. *Panicum arvense* Kunth, Rev., Gram 1: 391,t.109. 1831. *Panicum patens* Boj., Hort. Maur 365.1837 (non linn. 1753). *(Vern. Name: Brown top-millet (English)).*

Branched and decumbent, glabrous or pubescent, fascicled, slender grass; culms sub-erect; branches leafy. Leaf-blades linear, lanceolate, flaccid, pubescent, base deeply lobed; sheath loose glabrous. Inflorescence panicles, racemes, rachis scabrid, hairy. Spikelets broadly elliptic or ovoid-elliptic, apiculate, alternate often in pairs. *Fls. & Frts.:* March-November.

Common throughout and in all habitats, both mesophytic and in wetlands; dwarf or large and straggling grass; the size is greatly variable.

Field Collections: Kholapota/Basirhat, *K.N. - 572;* Thuba/Taki, *K.N. - 1041;* Hasnabad, *K.N. - 1957;* Bhagajatin/Jadavpur, *K.N. - 1057.*

Distributions: Plains of India. Tropics of the Old World.

Economic Importance: The young leaves are much liked by cattle.

11.6. ***Brachiaria reptans*** (L.) C.A. Gard. & C.E. Hubb. Hook. Icon. Fl. 3363.1938; Naskar Aq. & Semi-Aq. Plants 335.1990. *Panicum reptans* L. Syst.

Nat. (ed.10) 870. 1759. *Panicum umbrosum* Retz., Obs. Bot. 4: 16. 1786. *P. prostratum* Lamk., Tab. Encycl. Meth. Bot. 4, 730.1797; Hook.f. Fl. Brit. India 7: 33-34. 1896; Prain Beng. Plants 2: 886.1903 (Rep. ed. 1963). *P. procumbens* Nees, Agrost. Bras. 109. 1829. *Brachiaria prostrata* (Lamk.) Griseb. Abhandl. Ges. Wiss. Gott. 7: 263. 1857. *Urochloa reptans* (L.) Stapf. Prain Fl. Trop. Afr. 9: 601. 1920. *(Vern. Name: Chhota-Jalganti).*

Culms 15-50 cm long, creeping below, nodes glabrous; much branched innovation shoots intravaginal. Leaf-blades amplexicaul, hairy, ovate-lanceolate, sheaths ciliate not up to the next node. Common weed on the cultivated fields, roadsides and waste lands and create a lot of trouble for eradication due to its extensively creeping and rooting at nodes.

Naskar (1.c.) described this taxon after collecting from lower Ganga Delta.

11.7. ***Brachiaria setigera*** (Retz.) C.E. Hubb.Hook., Icones Pl. 34, 2 (1938) sub. tab. 3369, 1938; Bor Grass. Burma, Ceyl., Ind. & Pak. 286.1960. *Panicum setigerum* Retz., Obs. Bot. 4: 15.1786; Hook.f. Fl. Brit. India 7: 37.1896; Prain Beng. Plants 2: 886. 1903 (Rep. ed. 1963). *P. affine* Poir, Lamk. Encycl. Meth. Bot. Suppl. 4: 273. 1816. *Urochloa setigera* (Retz.) Stapf. Prain, Fl. Trop. Afr. 9: 598. 1920. *(Vern. Name: Bara-Jalganti).*

Culms slender, prostrate and widely straggling below, lower internodes long, nodes rooting, tomentose. Leaf-blades crispedly wrinkled below sheath, light smooth, often with few hairs, margins ciliate branching intravaginal. Ligule a few hairs. Inflorescence panicles racemes obliquely erect, rachis almost straight, rather flat, wide, hairy at base; pedicel with scattered long hairs. Spikelets biseriate, sub-sessile, lanceolate or ovate, acuminate, usually pubescent paired in lower part of the racemes. *Fls. & Frts.:* March-October.

Rather rare, a mesophytic grass.

Field Collection: Ganga Sagar Mela Premises, *A.K. Mukherjee-s.n* (CAL).

Distributions: Almost throughout India. Sri Lanka and East Africa.

Economic Importance: Leaves are soft and a common fodder grass.

12. *CHLORIS* Sw.

Key to the Species

1a	Spikelets 4-awned	2. *C. montana* Roxb.
1b	Spikelets 3-awned	(2)

2a (1)	Upper glume awned, reduced florets one	3. *C. virgata* Sw.
2b (1)	Upper glume never awned, reduced florets two	1. *C. barbata* Sw.

12.1. ***Chloris barbata*** Sw. Fl. Ind. Occ. 1: 200. 1797; Naskar Aq. & Semi-Aq. Plants 335.1990. *Andropogon barbatus* L. Mant. Pl. Alt. 302.1771 (non Linn.1759). *Chloris inflata* Link, Enum. Hort. Berol. 1: 105.1821.

Perennial tufted grass; culms erect, stout, with creeping branched base. Grow everywhere both in dry and wet situations, easily recognised by its reddish racemes. Naskar (1.c.) described this taxon after collecting from Mallickpore, *(K.N. - 81)*; Haroa *(K.N. - 593)*; Bhangankhali *(K.N. - 208)*. Cattle like this grass only when it is young, but avoid when mature.

12.2. ***Chloris montana*** Roxb. Fl. Ind. 1: 331.1820; Prain Beng. Plants 2: 926.1903 (Rep. ed. 1963). *C. decora* Thw., Enum. Pl. Zeyl. 371.1864 (non Nees 1854). *Chloris montana* var. *glauca* Hook.f. Trim., Handb. Fl. Ceyl. 5: 276. 1900.

Culms erect or geniculately ascending from a creeping base. Leafblades narrow, acuminate, sheath compressed, smooth, mouth of the sheath ciliate, ligule very short. Inflorescence spikes, erect, very pale, rachis glabrous. *Fls. & Frts.:* October-January.

Xerophytic grass, commonly grow on the old brick walls, railway tracks and roadsides.

Field Collection: Bamanghata/Salt Lake, *M.K.Ghosh- 3042* (CAL).

Distributions: Hilly districts of Coromondal, Upper and Lower Gangetic plains, West Bihar and Bengal. Sri Lanka and the Tropics.

12.3. ***Chloris virgata*** Sw., Fl. Ind. Occ. 1: 203.1797; Bor Grass. Burma, Ceyl., Ind. & Pak. 468.1960. *Chloris pubescens* Lagasca, Varied. Ciene 2(4): 143.1805. *Rhabdachloa virgata* (Sw.) P. Beauv., Ess. Agrost. 84: 158. 1812. *Chloris compressa* DC., Cat. Hort. Monsp. 94.1813.

Culms stout, flattened, usually leafy decumbent, lower portion branched. Leafblades broad, flat and acute, upper sheaths more or less inflated. Inflorescence spikes, digitate, enclosed at the lower portion by the spathes. *Fls. & Frts.:* July-November.

Xerophytic grass, common on the old brick walls, railway tracks.

Field Collection: Dakshin Barasat, *K.N. - 1651*.

Distributions: Rajputana, both upper and lower Gangetic plains, South and Central Bihar, South India and West Bengal. Tibet, Burma, widely distributed throughout the tropics of both hemispheres.

Economic Importances: Reputed as good fodder grass.

13. *CHIONACHNE* R. Br.

13.1. ***Chionachne koenigii*** (Spreng.) Thw., Enum. Pl. Zeyl. 357. 1864; Naskar Aq. & Semi-Aq. Plants 336.1990. *Coix koenigii* Spreng., Syst. 1: 239. 1825. *C. barbata* Roxb. Fl. Ind. 3: 569.1832. *C. arundinacea* Koen. ex Willd., Sp. Pl. 4: 203.1805 (non Lamk.1791). *Polytoca barbata* Stapf.; Hook.f. Fl. Brit. India 7: 102. 1896. *(Vern. Name: Gurgar).*

Tall branched grass, nodes bearded. Grow in the paddy fields and along the edges of pools and marshes, sporadic in jungle clearings. Naskar (1.c.) described this taxon after collecting from Bahuru *(K.N. - 1629).*

14. *CHRYSOPOGON* Trin. *nom. cons.*

14.1. ***Chrysopogon aciculatus*** (Retz.) Trin., Fund. Agrost. 188.1820; Bor Grass. Burma, Ceyl., India & Pak. 115.1960. *Andropogon aciculatus* Retz., Obs. Bot. 5: 22.1789; Hook.f. Fl. Brit. India 7: 188.1896; Prain Beng. Plants 2: 907.1903 (Rep. ed. 1963). *Andropogon acicularis* Retż. ex Roem. & Schult, Syst. Veg. 2: 812.1817. *(Vern. Name: Chor-kanta).*

Perennial, rhizomatous herb; culm creeping and branching below, 5-10 cm woody, leafy and extensively creeping below. Leafblades flat, often undulated, margins spinulose, densely tufted, sheath terete, ligule a very narrow ciliate membrane. Inflorescence panicles, erect, oblong, pale or purplish, branches capillary, fragile, not enclosed in sheaths spreading, spikes solitary. Caryopsis linear, laterally compressed, embryo 1/2 the length of the grains. *Fls. & Frts.:* June-November.

Extremely common grass, growing both in dry and wet places of villages, pastures and an obnoxious weed of the cultivated grounds. The caryopsis can easily be transferred from place to place by sticking after the cloths and cattle's tails.

Field Collections: Masjidbati, *K.N. - 837;* Bamanpukur, *K.N. - 1087.*

Distributions: Plains and lower hills of India, from Kumaon to Sikkim in the Himalayas, the Khasia Hills and Southward to the Nicobar Islands. Sri Lanka, Tropical Asia, Australia and Polynesia.

Economic Importance: Used as fodder; grazing animals suffer severely from the attached ripe fruits, which stick to thin hairs.

15. *COIX* L.

15.1. ***Coix lachryma-jobi*** L. Sp. Pl. 972.1753; Naskar Aq. & Semi-Aq. Plants 337. 1990. *Coix lachryma* L. Syst. Nat. (ed.10) 1261. 1759. *Coix agrestis* Lour., Fl. Cochinch. 551.1790. *Coix arundinacea* Lamk., Encycl. Meth. Bot. 3: 422.1791. *(Vern. Name: Gurgor).*

Tall densely tufted annual or perennial grass; common along the water courses and ditch-sides. Naskar (1.c.) described this taxon after collecting from Canning *(K.N. - 129);* Basanti *(K.N. - 1280);* Baduria *(K.N. - 1306)* and Falta (*R.Mallick-216* (CAL)).

16. *CYMBOPOGON* Spreng.

16.1. ***Cymbopogon nardus*** (L.) Rendle, Cat. Welw. Afr. Pl. 2: 155.1899; Bor Grass. Burma, Ceyl., Ind. & Pak. 130. 1960. *Andropogon nardus* L. Sp. Pl. 1046.1753; Prain Beng. Plants 2: 906.1903 (Rep. ed. 1963). *Andropogon amplifiorus* Steud., Syn. Pl. Glum. 1: 388. 1854. *(Vern. Name: Gandhabena).*

Aromatic tufted perennial grass. Culms tall, erect or geniculate at the base. Leaves flat, margin cutting. Inflorescence large panicles; racemes at first erect, afterwards reflexed. Spikes without awns. *Fls. & Frts.:* July-November.

Sometime cultivated and occasionally found on the garden edges.

Field Collection: Dakshin Barasat, *K.N. - 1655.*

Distributions: Throughout the hotter parts of India. S.E. Asia, Africa and Australia.

Economic Importances: This grass smell lemon like and yield the citronella oil of commercial importance.

Notes: R. B. Majunder (1956) has also reported *Cymbopogon citratus* (DC.) Stapf, Kew Bull. 1906: 357.1906; but the present author has made no collection or studied any herbarium or others collection.

17. *CYNODON* Rich. ex Pers. *nom. cons.*

17.1. ***Cynodon dactylon*** (L.) Pers., Syn. Pl. 1: 85.1805; Hook.f. Fl. Brit. India 7: 288. 1895; Prain Beng. Plants 2: 925.1903 (Rep. ed. 1963); Bor Grass Burma, Ceyl., Ind. & Pak. 469.1960. *Panicum dactylon* L. Sp. Pl. 58.1753. *Digitaria dactylon* (L.) Scop., Pl. Carn. (ed.2) 69.1787. *Paspalum dactylon* (L.) Lamk., Tab. Encyc. Meth. Bot. 1: 176. 1791. *(Vern. Name: Durba).*

Perennial grass; culms rhizomatous, prostrate, creeping and rooting from the nodes, forming wide mats on the surface of the soil. Leaves short, deep green, linear to lanceolate, tapering to a fine point. Inflorescence a whorl of 2-6 spikes, radiating from the top of a slender peduncle; spikes covered with unilateral rows of small spikelets. *Fls. & Frts.:* Almost throughout the year.

Very common on the moderate grazzed areas/ground, on the road-sides, on the railway tracks, dry lands and paddy fields, during dry season. It stands moderate trampling and the prostrate stems ensure its survival. It is commonly used as lawn grass. Mesophythic, xerophytic and halophytic—in all these areas it grows profusely.

Field Collections: Canning, *K.N. - 191;* Baghajatin, *K.N. - 1062.*

Distributions: Throughout India. All the warmer countries.

Economic Importances: Good fodder grass; sand binder and lawn-grass. Leaves are very auspicious and is largely used in all the religious festivals by the Hindus. Juice of the leaves and the plants are used as astringent, beneficial in fresh cuts and wounds, to stop bleeding from the nose etc.

18. *CYRTOCOCCUM* Stapf

Key to the Species

1a	Inflorescence lax panicle, pedicel very long, leafblade ovate lanceolate	1. *C. accrescens* (Trin.) Stapf.
1b	Inflorescence contracted panicles, pedicel short, leafblades ovate lanceolate	2. *C. trigonum* (Retz.) Commus

18.1. ***Cyrtococcum accrescens*** (Trin.) Stapf, Hook. Icon Pl. Sub. Tab. 3096.1922; Naskar Aq. & Semi-Aq. Plants 338.1990. *C. patens* var. *lactifolium* (Honda) Chwi, Acta Phytotax. Geobot. 11: 47. 1942. *Panicum accrescens* Trin., Sp. Gram. Ic. 1.t.88,1828.

Perennial tall; culms decumbent, often branching at lower nodes, leafy at the tops, nodes glabrous. Leafblades linear-lanceolate, ovate lanceolate or lanceolate, contracted at the base, tapering to a sharp point, many nerved, smooth, margins scarcely thickened, glabrous and smooth. Ligule a narrow membranous rim. Inflorescence panicles, branches and branchlets capillary, straight or flexuous. Spikelets pedicels of side spikelets, slightly swollen and cup shaped at the tip. *Fls. & Frts.:* June-October.

Occasionally grow on the road-sides, canal sides, cultivated fields; prefer moist areas and shade-loving situations.

Field Collection: Taki/near Thuba bus Stand, *K.N. - 1042.*

Distributions: Almost throughout India. Tropics of S.E. Asia.

Notes: Prain (1903a) did not mention this taxon from Bengal.

18.2. ***Cyrtococcum trigonum*** (Retz.) Commus, Bull. Mus. Nat. Hist. Paris 27: 118.1921; Bor Grass Burma, Cycl., India & Pak. 292.1960. *Panicum trigonum* Retz., Obs. Bot. 3, 9, 1783. *P. difforma* Roth ex Roem. & Schult Syst. Veg. 2: 433.1817.

Culms branching at the base; branches erect and slender, lower nodes with solitary roots. Leaf-blades linear-lanceolate, glabrous or laxly hairy. Sheath glabrous or margins ciliate. Ligule rounded. Inflorescence panicles contracted, upright, lax axis, often flexuous, smooth, branches angular. *Fls. & Frts.:* July-March.

Common on the roadsides, cultivated lands, railway tracks and canal sides also; mesophytic habitat.

Field Collection: Herobhanga Sundarbans, *K.N. - 1863.*

Distributions: Throughout India. South east Asia.

Economic Importance: Good forage plant.

Notes: Prain (1903a) did not report this taxon from Bengal.

19. *DACTYLOCTENIUM* Willd.

19.1. ***Dactyloctenium aegypticum*** (L.) Willd., Enum. Hort. Berol 1029. 1809; Bor Grass Burma, Ceyl., Ind. & Pak. 489.1960. *Cynosurus aegypticus* L. Sp. Pl. (ed.) 72. 1763. *Eleusine aegyptica* (L.) Desf., Fl. Atlant. 1: 85.1798 *(aegyptica). Eleusine pectinata* Moench, Meth. Pl. Suppl. 68. 1802. *Choloris mucronata* Michx., Fl. Bor. Amer.1: 59.1803 *(Vern. Name: Makra).*

Erect or prostrate grass, culm profusely branched, nodes swollen and rooting below. Leaves distichous, acute or acuminate, ciliate, sheaths compressed. Inflorescences digitate, spikes, solitary, light olive grey. Spikelets densely crowded, glabrous, glumes almost pungent, sessile. Grains obovoid-globose, corrugated. *Fls. & Frts.:* April-November.

Mesophytic, growing everywhere on the roadsides or railway tracks.

Field Collection: Magrahat, *K.N.* - 257.

Distributions: Throughout India. Burma, Sri Lanka, Malacca and Tropical region of the Old World.

Economic Importances: Said to be rich in Cyanogenetic glucosides; a good forage plant but an obnoxious weed. The grains are eaten during scarcity but is said to have an unplesant taste and cause internal disorders.

20. *DENDROCALAMUS* Nees

20.1. ***Dendrocalamus giganteus*** Munro, Trans. Linn. Soc. 26: 150.1868; Gamble, Hook.f. Fl. Brit. India 7: 406. 1897; Benthall Trees Calcutta 473.1946. *Bambusa gigantea* Wall., Cal. Bot. Gard. Calc. 79. *(Vern. Name: Bara Bans).*

Stem 20-40 m long, 13-30 cm broad, branched; nodes hairy and internodes short. Leaf-blades cuspidate, twisted hairy below, sheaths striate, hairy at throat. Inflorescence large, panicle, leafy at the top, branchlets curved. Spikelets in lax heads, ovoid, acute, spinescent, puberulous, many flowered. Glumes ovate, mucronate, many, nerved. Ovary with bi-fid styles. Caryopsis oblong, nut free. *Fls. & Frts.:* June-October.

Mostly cultivated.

Field Collections: Specimens not preserved.

Distributions: Eastern India and Bombay. Tennaserim, Malaya/Peninsula, Penang and Indo-Malaya China.

Economic Importances: It is largely used in many household purpose and also used as paper pulp.

21. *DESMOSTACHYA* Stapf

21.1. ***Desmostachya bipinnata*** (L.) Stapf, Iyer. Fl. Cap. 7: 632.1900; Naskar Aq. & Semi-Aq. Plants 339.1990. *Briza bipinnata* L. Syst. Nat.(ed.10) 2: 875.1759. *Uniola bipinnata* L. Sp. Pl. (ed.2) 104. 1762. *Cynosurus durus* Forssk, Fl. Aegypt. Arab. 21.1775. *Eragrostis cynosuroides*, Beauv. Agrost.71: 162.1812; Hook.f. Fl. Brit. India 7: 324.1896. *(Vern. Name: Kush).*

Perennial tall grass; branches from the base. Grow gregariously in all habitat, in dry, semi-dry areas. Naskar (1.c.) described this taxom after collecting from the canal side of Surjyapur *(K.N. - 283)* and Santoshpur *(K.N. - 498).*

22. *DICHANTHIUM* Willemet

Key to the Species

1a	Sessile spikelets green	1. *D. annulatum* (Forssk.) Stapf
1b	Sessile spikelets dark brown	2. *D. caricosum* (L.) Camus

22.1. ***Dichanthium annulatum*** (Forssk.) Stapf, Fl. Trop. Afr. 9: 178.1917; Bor Grass. Burma, Ceyl., Ind. Pak. 133.1960. *Andropogon annulatus* Forssk., Fl. Aeg.-Arab. 173.1775; Hook.f. Fl. Brit. India 7: 196.1896; Prain Beng. Plants 2: 908.1903 (Rep. ed. 1963). *A. scandens* Roxb. Fl. Ind. 1: 262. 1820. *A. comosus* Link., Hort. Berol. 1: 239.1827. *(Vern. Name: Loari).*

Densely tufted, woody at the base and strong wiry-root, geniculately ascending, nodes generally bearded with soft hairs. Leaf-blades glabrous or hairy above, with or without tubercled based hairs. Sheaths terete, subcarinate at the top, striate. Ligule oblong, obtuse. Inflorescence spiciform, panicles solitary, soft nodes bearded, generally terminal, purple, rachis many jointed, hairy with fascicle of long hairs at base, articulation oblique, without appendages. Caryopsis ovate-oblong. Compressed. Pedicelled-spikelets. *Fls. & Frts.:* November- March.

Mesophytic grass of various status.

Field Collections: Sitakundu/Baruipur, *K.N. - 57;* Madarhat, *K.N. - 69.*

Distributions: Throughout the hills and plains of India. Burma, Tropical Africa, China, Australia and Pacific Islands.

Economic Importances: A very excellent fodder grass, but act as the obnoxious weed of high cultivated lands.

22.2. ***Dichanthium caricosum*** (L.) Camus, Bull. Muss. Hist. Nat. Paris 27: 549. 1921; Bor Grass Burma, Ceyl., Ind. & Pak. 134.1960. *Andropogon caricosus* L. Sp. Pl. (ed.2) 1480.1762 (excl. Sym. Rumphi); Hook.f. Fl. Brit. India 7: 196. 1896; Prain Beng. Plants 2: 908.1903 (Rep. ed. 1963). *A. serratus* Retz., Obs. Bot. 5: 2.1789. *Andropogon filiformis* Pers. Syn. Pl. 1: 103.1805. *Heteropogon concinnus* Thw., Enum. Pl. Zeyl. 368. 1864. *(Vern. Name: Detara).*

Perennial, erect or decumbent grass. Leafblades-scaberulous margins, median nerves keeled, sheath somewhat loose, shorter than internodes, ligule very short, truncate. Inflorescence-speciform, racemes usually solitary, silky, rachis or racemes fragile, articulatus oblique, without appendages. Spikelets almost distichous, imbricate, persistent. Sessile spikelets bisexual, obovate-elliptic or obovate oblong, dorsally compressed, brownish, dark, hairy. Lemma entire, narrow, almost transformed into awn, slender, column brownish. Palea hyaline. Ovary glabrous. *Fls. & Frts.:* November-April.

Gregariously grow and very common; xerophytic habit, creeping extensively.

Field Collection: Kestopur, *K.N. - 1899.*

Distributions: Plains and low hills of India. Burma, Sri Lanka, Mauritius and China.

Economic Importance: Relishes by the cattles.

23. *DIGITARIA* Heist. ex Fabr.

Key to the Species

1a	Spikelets 0.25 cm long	1. *D. adscendens* Henrard
1b	Spikelets less than 0.25 cm long	(2)
2a (1)	Rachis capillary, trigonous	2. *D. granularis* Henrard.
2b (1)	Rachis narrowly winged	(3)
3a (2)	Spikelets with slender hairs	3. *D. longiflora* Pers.
3b (2)	Spikelets with clavate hairs	4. *D. sticta* Roth ex Roem.

23.1. ***Digitaria adscendens*** (H.B. & K) Henrard, Blumea 1: 92.1934; Bor Grass. Burma, Ceyl., Ind. & Pak. 298.1960. *Panicum adscendens* H.B. & K., Nov. Gen. & Sp. Pl. 1: 97.1816. *Asperella digitaria* Lamk., Tab. Encyl. Meth. Bot. 1: 167. 1791. *Digitaria chinensis* Horn., Suppl. Hort. Bot. Hafn. 8. 1819. *D. fimbriata* Link, Hort. Berol. 1: 226.1827. *Digitaria marginata* Link var. *fimbriata* (Link) Stapf, Fl. Trop. Africa 9: 440.1919. *Digitaria sanguinalis* (Link) Scop. var. *marginata* Fern., Rhodora 22. 103.1920. *Paspalum sanguinale* Lamk. Ill. 1: 176.1791.

Erect, slender grass, ascending from a creeping base. Leafblades long, filiform, with long spreading hairs, sheaths glabrous, ligule rather membranous. Inflorescence digitate racemes. Spikelets lanceolate, acuminate, subsilky. Caryopsis brownish, ovate lanceolate. *Fls. & Frts.:* April-November.

Gregariously grow on the dry waste lands, roadsides and railway tracks.

Field Collections: Magrahat, *K.N. - 255*; Dakshineswar, *R. Mallick-168.*

Distributions: Throughout India, especially in the hilly tracks.

Economic Importances: Good forage grass.

Notes: Prain (1903a & 1905) did not report this plant from Bengal.

23.2. ***Digitaria granularis*** (Trin.) Henrad. Monogr. (Gen. *Digitaria*) 302.1950; Bor Grass Burma, Ceyl., Ind. & Pak. 301. 1960. *Paspalum granulare* Trin. ex Spreng., New Entdeck. 2: 47.1821. *Milium sanguinale* Roxb. Fl. Ind. 1 (ed.2) 315. 1832. *P. pedicellare* Trin. ex Steud. Nom. (ed.2) 2: 272.1841; Hook.f. Fl. Brit. India 7: 19. 1826. *Digitaria pedicellaris* Prain Beng. Plants 2. 890. 1903 (Rep. ed. 1963). *(Vern. Name: Chiri-Chara).*

Tall, erect, branched red. Leafblades finely acuminate, sheath narrow. Inflorescence panicles, alternate, subterminal, rachis slender, hardly winged, pedicels capillary, trigonous, slightly swollen at tip. Spikelets ovate or elliptic, colour pale red, acute, mucronate, spaced. Glumes finely tomentose. *Fls. & Frts.:* July-January.

Frequently grow on the open space or various types of soil.

Field Collection: Dum Dum, *M.K. Ghosh-3942* (CAL).

Distributions: Throughout the plains and low hills of India. Tropical Asia.

Economic Importance: A fodder grass and act as soil binder.

23.3. ***Digitaria longiflora*** (Retz.) Pers., Syn. Plan. 1: 85.1805; Naskar Aq. & Semi-Aq. Plants 339. 1990. *Paspalum longiflorum* Retz., Obs. Bot. 4: 15.1786. *Panicum longiflorum* Gmel. Linn. Syst. Nat. (ed.13) 2: 158. 1791. *Paspalum brevifolius* Flugge, 1: 150.1810. *(Vern. Name: Kanka-Juriya).*

Creeping below, rarely erect or procumbent below. Naskar (1.c.) described this taxon after collecting from the lower Ganga delta, Baguihati *(K.N. - 1890).*

23.4. ***Digitaria stricta*** Roth ex Roem. & Schult., Syst. Veg. 2: 474. 1817; Bor Grass Burma, Ceyl., Ind. & Pak. 305.1960. *Agrostis pilosa* Retz., Obs. Bot. 6: 22. 1791. *Digitaria puberule* Link. Hort. Berol. 1: 223. 1827. *Paspalum royleanum* Nees ex Thw., Enum. Pl. Zeyl. 358.1864; Hook.f. Fl. Brit. India 7: 18.1896; Prain Beng. Plants 2: 890.1903 (Rep. ed. 1963). *Digitaria royleana* (Nees) Prain Beng. Plants 1182. 1903 (Rep. ed. 1963).

Erect or shortly decumbent below; leaves usually erect. Leafblades linear to lanceolate, flaccid. Sheaths rather loose, striate throat hairy. Ligules a membranous rim. Inflorescence panicles. Spikelets pale-yellowish, hairy. *Fls. & Frts.:* October-March.

Grow in lawn or in the cultivated and high lands.

Field Collection: Birati, *R. Mallick-432* (CAL).

Distributions: Hilly districts of India. Sri Lanka and Tropical Africa.

Economic Importance: A good forage grass and act as soil binder.

24. *DIMERIA* R. Brown

24.1. ***Dimeria orinthopoda*** Trin., Fund. Agrost 167. t.14.1820; Hook.f. Fl. Brit. India 7: 104.1896; Prain Beng. Plants 2: 894.1903 (Rep. ed. 1963); Bor Grass. Burma, Ceyl., Ind. & Pak. 142. 1960. *Dimeria filiformis* (Roxb.) Hochst., Hohenacker, Pl. Ind. Or. no. 231.nomen & ex Miq, Verh. Nederl. Inst. 3,4,25.1851.

Annual, leafy slender grass. Culms branched, bearded at nodes. Leafblades soft, linear, erect, sheaths keeled, glabrous. Inflorescence spikes, sessile, reddish or brown, rachis as broad as spikelets. Spikelet linear, callus dorsally bearded, reddish almost sessile acuminate. *Fls. & Frts.:* June-October.

Frequently found on the waste grass lands and paddy fields.

Field Collection: Sagar Islands, *A.K. Mukherjee- 4633* (CAL).

Distributions: Throughout India. Malay Islands, Japan and Tropical Australia.

Economic Importance: Fodder grass and act as soil binder.

Notes: Prain (1903a) did not report this taxon from Bengal.

25. *DINEBRA* Jacq.

25.1. ***Dinebra retroflexa*** (Vahl) Panzer., Denkschr. Acad. Wiss. Munchen 270,t.12.1814; Bor Grass. Burma, Ceyl., Ind. & Pak. 491.1960. *Cynosurus retroflexus* Vahl, Synb. Bot. 2: 20. 1791. *Dinebra arabica* Jacq., Fragm. 17,t.121,f.1, 1807; Watt Dict. Econ. Prod. 3: 115.1890: Hook.f. Fl. Brit. India 7: 1896; Prain Beng. Plants 2: 928.1903 (Rep. ed. 1963).

Annual leafy grass. Culms tufted, branching from base, erect or geniculately ascending. Leafblades flaccid, finely acuminate, hairy, sheath loose often exceeding the upper node. Inflorescence racemose. Spikelets

flowered, reddish, small, crowded. Racemes opposite alternate or whorled, pale-yellow, shining. Caryopsis ovoid, trigonous, wrinkled, free, pale yellow brown, embryo narrowly oblong. *Fls. & Frts.:* July-December.

On the margin of the cultivated fields, grow profusely.

Field Collection: Barasat, *R. Mallick-263* (CAL).

Distributions: Western part of India. East Africa and Afganisthan.

Notes: Prain (1903a) recorded this taxon from the W. Bihar only.

26. *DIPLACHNE* P. Beauv.

26.1. ***Diplachne fusca*** (L.) P. Beauv. Ess. Agrost 80: 163.1812; Naskar Aq. & Semi-Aq. Plants 340. 1990; Naskar & Guha Bakshi Mangrs. Swamps Sundarbans 145. 1987. *Festuca fusca* L. Sp. Pl. (ed.2) 109.1762. *F. indica* Retz., Obs. Bot. 4: 21. 1786.

Perennial, glabrous, tall tufted stout, erect or geniculately ascending, branched clothed with leafsheaths. Strictly halophytic grass, growing chiefly in the marshes of the Sundarbans area. Naskar (1.c.) and Naskar & Guha Bakshi (1.c.) described this taxon after collecting from Goranbose/Sundarbans *(K.N. - 1868).*

27. *ECHINOCHLOA* P. Beauv.

Key to the Species

1a	Stems slender, decumbent, leaves narrow	1. *E. colonum* (L.) Link
1b	Stems stout, erect, leaves broad	2. *E. crusgalli* (L.) Beauv.

27.1. ***Echinochloa colonum*** (L.) Link, Hort. Berol. 2: 209.1833; Naskar Aq. & Semi-Aq. Plants 341. 1990. *Panicum colonum* L. Syst. Nat. (ed.10) 2: 870. 1759. *Milium colonum* (L.) Moench, Methos. Pl. 202. 1794. *(Vern. Name: Shyama Ghas* or *Shyama Dhan).*

Annual, sub-aquatic grass; growing profusely. Naskar (1.c.) described this taxom after collecting from Godkhali/Basanti *(K.N.-844);* Hasnabad *(K.N. - 1046)* and Bamanpukur *(K.N. - 1127).*

27.2. ***Echinochloa crusgalli*** (L.) Beauv. Ess. Agrost. 53: 161. 1812; Naskar Aq. & Semi-Aq. Plants 342. 1990. *Panicum crusgalli* L. Sp. Pl. 56. 1753. *(Vern. Name: Bara-Shyama).*

Annual aquatic grass; grow usually near the margins of the paddy fields, ponds and roadside nalas. Naskar (1.c.) described this taxon after collecting from Taki *(K.N. - 1020)*; Ghusighata/Kulti *(K.N. - 1114)* and Ghoshpur *(K.N. - 1120)*.

28. *ELEUSINE* Gaertn.

Key to the Species

1a	Culms shorter, grains oblong, wild in the area	2. *E. indica* (L.) Gaertn.
1b	Culms longer, grain globose, cultivated	1. *E. coracana* Gaertn.

28.1. ***Eleusine coracana*** (L.) Gaertn., Fruct. 1: 8,t.1,f.11,1789; Watt Dict. Econ. Prod. 3: 237. 1890; Hook.f. Fl. Brit. India 7: 294.1896; Prain Beng. Plants 2: 927. 1903; Bor Grass. Burma, Ceyl., Ind. & Pak. 492.1960. *Cynosurus coraconus* L. Syst. Nat. (ed.10) 2: 875.1759. *(Vern. Name: Marua)*.

Erect, stout, annual grass. Leaves often overlapping the stem with compressed loose sheaths and ligule of hairs. Spikes 4-7, sub-erect, awnless, 3-6 flowered. Seeds globose, dark-bown. *Fls. & Frts.:* September-November.

Generally cultivated, during rainy season.

Field Collection: Baguihati, *K.N. - 1888.*

Distributions: Cultivated throughout India. Widely cultivated in the Old World.

Economic Importances: Grains low quality food; but used to prepare cakes, porridge and sweetmeats; bear is also extracted from the grains. Straw is good fodder for cattles. A minor cereal plant.

28.2. *Eleusine indica* (L.) Gaertn., Fruct. 1: 8.1789; Watt Dict. Econ. Prod. 3: 241. 1890; Hook.f. Fl. Brit. India 7: 293. 1896; Prain Beng. Plants 2: 927. 1903 (Rep. ed. 1963); Bor Grasses Burma, Ceyl., Ind. & Pak. 493. 1960. *Cynosurus indicus* L. Sp. Pl. 72.1753; *Elusine tristachya* Link Ill. 1: 203, t.203.1791; Hook.f. Fl. Brit. India 7: 263.1896. *(Vern. Name: Mal-ankuri)*.

Leafblades flat, narrow, usually crowded at the base, sheath flattened and ciliate. Inflorescence usually digitate spikes, consists of 2-7 digitate racemes, axis terminating in a spikelets. Spikelets 3-8 flowered, flattened and densely imbricate, pointing forward, glabrous, fruit grain long, pericarp loose and hyaline. *Fls. & Frts.:* July-November.

Mesophytic grass, growing profusely everywhere as obnoxious weed. The culms generally spread horizontally and thus whole plant forms a rosette.

Field Collections: Champahati, *K.N. - 828;* Dhosha- Chandaneswar, *K.N. - 878.*

Distributions: Plains of India. Tropical and Sub-Tropical regions.

Economic Importance: A good forage plant and act as soil binder.

29. *ELYTROPHORUS* P. Beauv.

29.1. ***Elytrophorus spicatus*** (Willd.) A. Camus., Lecomte Fl. Gen. de L' Indo-Chine 7: 547.1923; Bor Grass. Burma, Ceyl., Ind. & Pak. 493.1960. *Dactylis spicata* Willd. Ges. Natur.f.Freuds Berlin, Nene Schift 3: 416.1801. *Elytrophorus articulatus* Beauv. Agrost. 67. 1812; Hook.f. Fl. Brit. India 7: 306. 1896; Prain Beng. Plants 2: 918. 1903 (Rep. ed. 1963).

Culm erect, strict, glabrous, simple or branched. Leafblades narrow, slightly rough, sheath loose, thin, glabrous and smooth, ligule obscure, truncate, glabrous. Inflorescence panicle, bristly upto 1 cm wide, pale green, Spikelets 3-4 fit, in globose sessile clusters, sessile compressed, glumes shortly mucronate persistent, rachis jointed at the base and between the lemmas. Fruit grain obliquely oblong, free, pericarp loose at both ends. *Fls. & Frts.:* June-November.

Rare, aquatic plant/grass; grow in the paddy fields during rainy season. Naskar (1.c.) described this taxon after collecting from Goranbose/Sundarbans *(K.N. - 1883).*

Distributions: Assam, Bengal and Deccan Peninsula. Burma, Sri Lanka Tropical Asia, Africa and Australia.

30. *ERAGROSTIS* Beauv.

Key to the Species

1a	Inflorescence spiciform or dense panicle	(2)
1b	Inflorescence open panicles	(3)
2a (1)	Terrestrial, more or less erect grass; spikelets 2-5 mm long	2. *E. coarctata* Stapf. ex Hook.
2b (1)	Terrestrial, viscous herb with spreading almost horizontal culms	11. *E. viscosa* (Retz.) Trin.

3a (2)	Branches and pedicels both long and delicate	(4)
3b (1)	Branches and pedicels long but rigid	(8)
4a (3)	Branches sub-whorled or whorled	(5)
4b (3)	Branches alternate rarely germinate	(6)
5a (4)	Terrestrial grass; spikelets 2.5.-5.0 mm long, very distant, on small panicle, tough and nodding, tip tinged with red	6. *E. pilosa* (L.) Beauv.
5b (4)	Semi-aquatic grass; spikelets 2-3 mm long delicate, 6-14 flowered	3. *E. diarrhena* (Sch.) Steud.
6a (4)	Spikelets 1-4 mm, delicate, never nodding	8. *E. tenella* (L.) Beauv.ex R. & S.
6b (4)	Spikelets tough and nodding over 4 mm long	(7)
7a (6)	Grains 0.5 mm long, reddish brown; annual grass, sheath smooth	5. *E. nutans* (Retz.) Nees ex Wight.
7b (6)	Grains 0.8 mm long, brown	4. *E. gangetica* (Roxb.) Steud.
8a (3)	Lower glume nerveless; florets more or less loose	9. *E. tennuifolia* Hochst.
8b (3)	Lower glume 1-nerved; florets compressed	(9)
9a (8)	Terrestrial small grass; spikelets 4-6 mm long purple	7. *E. poaeoides* Beauv.
9b (8)	Semi-aquatic grass	(10)
10a (9)	Spikelets 5-15 mm long oblong, whitish or greenish green	1. *E. cilianensis* (All.) Vig.
10b (9)	Spikelets 5-10 mm long, ovate oblong, purple	10. *E. uniloides* Retz. ex Steud.

30.1. *Eragrostis cilianensis* (All.) Vignolo-Lutati, Malpighia 18: 386. 1904; Naskar Aq. & Semi-Aq. Plants 344. 1990. *Poa cilianensis* All., Fl.

Pedem. 2: 246.t, 91,f.2.1785. *Briza eragrostis* L. Sp. Pl. (ed.1) 1: 70. 1753. *B. oblonga* Moench, Meth. Pl. 185. 1795. *Eragrostis major* Host, Gram. Austr. 4: 14, t.24.1809.

Annual grass with bad odour. Grow on the margins of the paddy-fields and obnoxious weed grass. Naskar (l.c.) described this taxon after collecting from Goranbose/Sundarbans *(K.N. - 1876).*

30.2. ***Eragrostis coarctata*** Stapf ex Hook.f. Fl. Brit. India 7: 313.1896; Mujumdar, Bull. bot. Soc. Bengal 10(1 & 2): 28.1956; Bor Grass. Burma. Ceyl., Ind. & Pak. 507. 1960.

Perennial shrub, branched; stock base clothed with short, hard sheaths. Leafbase linear, spreading horizontally, finally acuminate, glabrous. Inflorescence spiciform panicle, purplish, branches-hispid. Spikelets slightly compressed, 5-12 flowered shortly pedicelled. Fruit grain shortly ovoid, reddish. *Fls. & Frts.:* September-May.

Dry land grass, common on the railway tracks, road-sides and high cultivated lands; form tufted mass.

Field Collection: Dakshineswar, *R. Mallick-170* (CAL).

Distributions: Bengal and Upper Gangetic plains; apparently confined to India, Burma and Bangladesh.

30.3. ***Eragrostis diarrhena*** (Schult.) Steud., Syn. Pl. Glum. 1: 266. 1854; Naskar Aq. & Semi-Aq. Plants 345. 1990. *Poa diarrhena* Schult., Syst. Veg. Z. Ment. 3: 616. 1827. *P. diandra* Roxb. Pl. Ind. 1: 337. 1820 (non R. Brown, 1810). *Eragrostis interrupta* sensu Hook.f. var. *diarrhena* Stapf ex Hook.f. Fl. Brit. India 7: 316.1896; Prain Beng. Plants 2: 921.1903 (Rep. ed. 1963).

Round smooth, glabrous and polished grass. Prefer wet but sandy soil. Naskar (1.c.) described this taxon after collecting from the lower Ganga delta; Surjyapur *(K.N. - 285)* and Surjyapur *(K.N. - 287).*

30.4. ***Eragrostis gangetica*** (Roxb.) Steud., Syn. Pl. Glum. 1: 266.1854; Naskar Aq. & Semi-Aq. Plants 346. 1990. *Poa gangetica* Roxb. Fl. Ind. 1: 340.1820. *Eragrostis brownei* Nees, Wight, Cat. 1780.1833.

Tufted, perennial, aquatic grass; sub-erect from the spreading base. Commonly found in moist and waste lands. Naskar (1.c.) described this taxon after collecting from the lower Ganga delta.

It is a good forage grass and act as important soil binder.

30.5. ***Eragrostis nutans*** (Retz.) Nees ex Wight. Cat 1776.1833 & ex Steud. Nom. Bot. (ed.2) 563.1840; Naskar Aq. & Semi-Aq. Plants 347.1990. *Poa* *nutans* Retz., Obs. Bot. 4: 19.1786. *P. elegans* Roxb., Hort. Beng. 82. 1814. *Eragrostis interrupta* Beauv. Agrost. 71.1812. *E. stenophylla* sensu Stapf., Hook.f. Fl. Brit. India 7: 318. 1896.

Annual grass; erect smooth, densely tufted, geniculately ascending, rarely found in the moist and waste lands. Naskar (1.c.) described this taxon after collecting from Lower Bengal; Barasat *(R. Mallick-247).*

30.6. ***Eragrostis pilosa*** (L.) Beauv., Ess. Agrost 71: 162.1812; Watt Dict. Econ. Prod. 3: 255. 1880; Hook.f. Fl. Brit. India 7: 323.1896; Prain Beng. Plants 2: 922. 1903 (Rep. ed. 1963); Bor Grass. Burma, Ceylon, India & Pak. 512. 1960. *Poa pilosa* L. Sp. Pl. 68. 1753. *P. eragrostis* Watt, Fl. Carol. 80.1788 (non *Poa eragrostis* L. 1753). *Eragrostis filiformis* Link. Hort. Berol. 1: 191.1827. *E. linkii* Steud., Syn. Pl. Glum. 1: 273.1854. *(Vern. Name: Tanjiniya).*

Tufted, erect, slender, annual, weed grass. Leafblades basal, short, pointed, strict. Inflorescence panicle, delicate open variable in size, flexuous, pilose in axils, often fascicled, rachilla flexous, tough, more or less exposed. Fruit grains, ellipsoid. *Fls. & Frts.:* June-October.

Mesophytic grass, growing plentiful near the railway tracks; often on the cultivated ground as a weed.

Field Collection: Baranagar, *R. Mallick-84* (CAL).

Distributions: Throughout India. Tropical and warmer regions of the Old World.

30.7. ***Eragrostis poaeoides*** Beauv. Ess. Agrost. 162. 1812; Bor Grass Burma, Ceyl., Ind. & Pak. 512. 1960. *Poa eragrostis* L. Sp. Pl. (ed.1) 68. 1753. *Eragrostis minor* Host., Gram. Austr. 4: 15.1809; Hook.f. Fl. Brit. India 7: 321. 1896; Prain Beng. Plants 2: 922.1903 (Rep. ed. 1963).

Annual grass. Culms dense, slender, erect. Leafblades basal, pointed with glandular margin. Inflorescence panicles, branches and lateral pedicels usually shorter, spreading when ripe. Spikelets linear oblong, dark purplish to olive, grey, glistenning, rachilla tough, longer than pedicels. Fruits grain dorsally, slightly flattened or round. *Fls. & Frts.:* Throughout the year.

Mesophytic grass; grows on the roadsides and railway tracks.

Field Collection: Barasat, *R. Mallick - 225* (CAL).

Distributions: Plains of India from Punjab to Bengal. Westward to south Europe and North Asia.

30.8. ***Eragrostis tenella*** (L.) Beauv. ex Roem. & Schult., Syst. Veg. 2: 576. 1817; Naskar Aq. & Semi-Aq. Plants 348. 1990. *Poa tenella* L. Sp. Pl. 69. 1753. *Eragrostis tenella* var. *pulmosa* (Retz.) Stapf; Hook.f. Fl. Brit. India 7: 315. 1896.

Culms slender rooting at the nodes. Leafblades variable in length. Common on the moist places, along the canal sides, ditches and ponds. Naskar (1.c.) described this taxon after collecing from Gobardanga *(K.N. - 537);* Maslandapur *(K.N. - 543)* and Kholapota *(K.N. - 1260).*

30.9. ***Eragrostis tennuifolia*** Hochst. Fl. 24 (1): 20.1841 & ex Steud. Syn. Pl. Glum. 1: 268. 1854; Hook.f. Fl. Brit. India 7: 322. 1896; Prain Beng. Plants 2: 922.1903 (Rep. ed. 1963); Bor Grass. Burma, Ceyl., Ind. & Pak. 514.1960. *Poa tennuifolia* A. Rich., Tent. Fl. Abyss. 2: 425. 1851. *Eragrostis parviglumis* Hochst. ex Steud., Syn. Pl. Glum.1: 267.1854.

Perennial grass; spreading, slender, branches stiff. Leafblades wide, long filiform usually basal mouth of sheath naked. Inflorescence open panicles, oblong to ovate, rachis glabrous, branches solitary or binate, distant, branches rather distant, spreading filiform and stiff. Spikelets 5-15 flowered, olive grey, florets loose. Fruit grains, ovate-oblong, dorsally grooved, remnant of stamens. *Fls. & Frts.:* June-January.

Usually growing in the drier places.

Field Collection: Madhyampur, *K.N. - 947*(CAL).

Distributions: Nilgiri Hills, Chotanagpur, Assam and Bengal, Burma to Tropical Africa.

30.10. ***Eragrostis uniloides*** Retz. ex Steud., Syn. Pl. Glum.1: 264.1854. Naskar Aq. & Semi-Sq. Plants 349. 1990. *Poa unioloides* Retz., Obs. 5: 19.1789; *Eragrostis amabilis* Wt. & Arn. ex Nees, Hook. & Arn. Bot. Beech. Voy 251.1841. *(Vern. Name: Koni).*

Annual, aquatic grass; loosely tufted, erect or geniculately ascending, nodes glabrous. Naskar (1.c.) described this taxon after collecting from Sitakundu *(K.N. - 52);* Ghoramara Islands *(K.N. - 407);* Majerhat *(K.N. - 1213)* and Habra *(K.N. - 1443).*

30.11. ***Eragrostis viscosa*** (Retz.) Trin., Mem. Acad. Sci. Petersb. (Ser. 6)1: 397.1830; Naskar Aq. & Semi-Aq. Plants 349. 1990. *Poa viscosa* Retz., Obs. 4: 20.1786. *P. glutinosa* Roxb. ex Stapf., Hook.f. Fl. Brit. India 7: 316.1896 (nomen).

Spreading, horizontally forming a tuft mass, sweet scented with minute glands all over the surface. *Fls. & Frts.:* June- January. Naskar (1.c.) described this taxon after collecting from Champahati *(K.N. - 834).*

31. *EREMOPOGON* (Hook.) Stapf

31.1. ***Eremopogon foveolatus*** (Del.) Stapf, Flor. Trop. Afr.9: 188.1917; Bor Grass. Burma, Ceyl., Ind. & Pak. 148.1960. *Andropogon foveolatus* Del., Fl. d' Egypte 16, t. 8,f.2.1812. *Andropogon strictus* Roxb. Fl. Ind. 1: 260. 1820. *A. monostachyus* Spreng., Pugill. 2: 9.1815. *Eremopogon strictus* (Roxb.) A. Camus, Ann. L. Lyon, n.s. 69: 208.1922.

Densely tufted, slender grass. Culms branched above, nodes sparsely bearded. Leafblades narrow, rigid, sheaths shorter than the internodes, basal, silky hairy, ligule short, hyaline. Inflorescence racemes, peduncles capillary, erect. *Fls. & Frts.:* June-November.

A dryland grass, not very common.

Field Collection: Panihati, *K.N. - 1606.*

Distributions: Drier parts of India, from Punjab and Sind to Bengal, Chhotanagpur. Warmer countries.

32. *ERIOCHLOA* H.B. & K.

32.1. ***Eriochloa procera*** (Retz.) C.E. Hub. Kew. Bull. 1930: 256.1930; Naskar Aq. & Semi-Aq. Plants 350.1990. *Agrostis procera* Retz., Obs. Bot. 4: 19.1786.

Spreading at the base, ascending from a creeping and short rhizome. Wet and low lands grass; prefer ditches and edges of the paddy fields. Naskar (1.c.) described this taxon after collecting from Gosaba *(K.N. - 678);* Santoshpur *(K.N. - 740);* Kalinagar/Basirhat *(K.N. - 766)* and Basanti *(K.N. - 840).*

33. *HEMARTHRIA* R. Br.

33.1. ***Hemarthria compressa*** (L.f.) R. Brown, Prodr. 207.1810; Naskar Aq. & Semi-Aq. Plants 351. 1990. *Rottboellia compressa* L.f., Suppl. 114.1781. *R. glabra* Roxb. Fl. Ind. 1: 353. 1820. *(Vern. Name: Pansheru).*

Perennial, erect grass; common in the moist waste places; in the crop fields. A very troublesome weed. Naskar(1.c.) described this taxon after collecting from lower Ganga delta (Madhyamgram, *R.Mallick-408* (CAL)).

34. *HETEROPOGON* Pers.

34.1. ***Heteropogon contortus*** (L.) Beauv. ex Roem. & Schult. Syst. Veg. 2: 836.1817; Bor Grass Burma, Ceyl., Ind. & Pak. 163. 1960. *Andropogon contortus* L.Sp. Pl. 1045.1753; Hook.f. Fl. Brit. India 7: 199.1896; Prain Beng. Plants 2: 908.1903 (Rep. ed. 1963).

Densely tufted culms, simple or multiple branched. Leafblades narrow, rigid, acute, glabrous, rigid sheath compressed. Inflorescence racemes usually single. *Fls. & Frts.:* August-January.

Common on dry places, on the railway tracks and roadsides; extremely variable.

Field Collection: Narendrapur, *K.N. - 1978.*

Distributions: Throughout India. Burma, Ceylon, Mediterranian and Tropical regions.

Economic Importances: Good forage/fodder grass. The awned fruits are said to cause sore in the gums and mouth of cattle.

35. *HORDEUM* L.

35.1. ***Hordeum vulgare*** L. Sp. Pl. 84.1753; Hook.f. Fl. Brit. India 7: 371.1896; Prain Beng. Plants 2: 929.1903 (Rep. ed. 1963); Bor Grass Burma, Ceyl., Ind. & Pak. 677.1960. *(Vern. Name: Jab).*

Cultivated crop grains. Culms about 100 cm or more. Leafblades linear, auricled, sheath loose, glabrous. Inflorescence terminal spike, solitary. *Fls. & Frts.:* January-April.

Cultivated field crops, during winter season.

Field Collection: Subuddhipur/Baruipur, *K.N. - 358.*

Distributions: Cultivated chiefly in North India. Egypt and South West Asia.

Economic Importance: Grain flour is edible in the form of bread and cakes.

36. *HYGRORYZA* Nees.

36.1. ***Hygroryza aristata*** (Retz.) Nees ex Wight & Arn., Edinb. New Phil. Journ. 15: 380.1838; Naskar Aq. & Semi-Aq. Plants 351.1990. *Pharus aristatus* Retz., Obs. Bot.5: 23.1789. *(Vern. Name: Jangli Dal).*

Aquatic, free floating glabrous grass. Naskar (1.c.) described this taxon based on collection from Barasat (*R. Mallick-613* (CAL)).

37. *HYMENACHNE* Beauv.

37.1. ***Hymenachne pseudointerrupta*** C. Muell., Bot. Z. 333.1861; Bor Grass. Burma, Ceyl., Ind. & Pak. 313.1960. *H. amplexicaulis* (Rudge) Nees Agrost. Brass. 275.1829 & Fl. Assam 5: 218.1940. *Panicum myurus* H.B. & K.; Hook.f. Fl. Brit. India 7: 39.1896.

Erect, creeping at the base, rooting from the nodes; Naskar (Aq. & Semi-Aq. Plants p.352.1990) described this taxon after collecting from Lower Ganga delta (Handleketki/Sagar Islands).

38. *IMPERATA* Cyrill.

38.1. ***Imperata cylindrica*** (L.) P. Beauv., Agrost 165.1812; Bor Grass. Burma, Ceyl., Ind. & Pak. 169. 1960. *Lagurus cylindricus* L. Syst. Nat. (ed.10) 878.1759. *Imperata arundinacea* Cyrill., Pl. Rar. Neap. 2: 26.1792; Hook.f. Fl. Brit. India 7: 106.1896; Prain Beng. Plants 2: 892.1903. *(Vern. Name: Ulu).*

Culms round, 3-4 noded, solid, with white erect hairs, rootstocks creeping. Leafblades erect; rigid, usually convolute, midrib stout, sheath rather loose, ligule membranous, ciliate, dorsally silky. Inflorescence-silvery with dark anthers and stigmas, pedicels sub-capillary. Spikelets callus hairy pale or purplish. *Fls. & Frts.:* June-October.

A xerophytic plant, growing profusely on dry lands, everywhere; propagate by underground stolons. A much dreaded weed of the cultivated fields, very difficult to eradicate this weed flora.

Field Collection: Baghajatin/Jadavpur, *K.N. - 1059.*

Distributions: Hotter parts of India; warm and temperate parts of Asia, extending to Australia and East and South Africa.

Economic Importance: Used by the Hindus in their religious purposes and also used for mat-making.

39. *ISCHAEMUM* L.

39.1. ***Ischaemum rugosum*** Salisb. Icon. Strip. Rar. 1.t.1.1791; Hook.f. Fl. Brit. India 7: 127.1896; Naskar Aq. & Semi-Aq. Plants 353.1990. *Colladoa distachya* Cav., Ic. 5: 37.1799. *Ischaemum colladoga* Spreng., Syst. 1: 298. 1925. *(Vern. Name: Morako).*

Branches often pinnate at the base. Grow near the edge of the rise-fields, sub-aquatic grass. Naskar (1.c.) described this taxon after collecting from Dum Dum Jn. Railway tracks *(K.N. - 1181).*

40. *LEERSIA* Soland ex Sw. *nom. cons.*

40.1. ***Leersia hexandra*** Swartz., Prod. Ind. Occ. 21.1788; Naskar & Guha Bakshi Mangr. Swamps Sundarbans 146. 1987. *Pharus citiatus* Retz., Obs. Bot. 5: 23.1789. *Homalacenchrus hexandrus* (Sw.) O. Kuntze., Rev. Gen. 777.1891.

Annual, aquatic grass; growing profusely in the marshes of the Sundarbans area and also in the other parts of the lower Ganga delta. Naskar & Guha Bakshi (1.c.) described this taxon after collecting from Sarbanandapur *(K.N. - 260);* Joygram/near Minakhan *(K.N. - 303);* Kalinagar/near Basirhat *(K.N. - 763)* and Dhopa Gachi *(K.N. - 1154).*

41. *LEPTOCHLOA* Beauv.

Key to the Species

1a	Spikelets 2-4 flowered; upper glume longer than 1st floret	2. *L. panicea* (Retz.) Ohwi
1b	Spikelets 4-6 flowered, upper glume shorter than 1st floret	1. *L. chinensis* (L.) Nees

41.1. ***Leptochloa chinensis*** (L.) Nees, Syll. Ratish 1: 4.1824; Naskar Aq. & Semi-Aq. Plants 355.1990. *Poa chinensis* L. Sp. Pl. 69.1753. *P. malabarica* Retz. Obs. 5: 19.1789 (non Linn. 1753). *Leptochloa capillacea* Beauv., Agr. 71.1812. *(Vern. Name: Bara Pininati).*

Culms stout. Aquatic grass, grow profusely along the margins of jheel and paddy fields. Naskar (1.c.) described this taxon after collecting from Joynagar-Mazilpur *(K.N. - 866).*

41.2. ***Leptochloa panicea*** (Retz.) Ohwi, Bot. Mag. Tokyo 55: 311.1941; Naskar Aq. & Semi-Aq. Plants 355.1990 *Poa panicea* Retz., Obs. 3: 11.1783. *L. filiformis* Roem. & Schult, Syst. 2: 580. 1870. *Aira filiformis* Koem. ex Roxb., Fl. Ind. 1: 328. 1820.

Annual, slender grass; a wet low land grass. Naskar (1.c.) described this taxon based on collection from the lower Ganga delta; Ultadanga *R. Mallick-444* (CAL).

42. *MICROCHLOA* R. Brown

42.1. ***Microchloa indica*** (L.) P. Beauv. Ess. Agrost. Explic. Planch 20. 1812; Bor Grass. Burma, Ceyl., Ind. & Pak. 473. 1960. *Nardus indica* Linn.f. Suppl. 105.1781. *Rottboellia setacea* Roxb. Cor. Pl.2: 17.1798. *Microchloa* *setacea* R. Brown Prodr. 208.1810; Hook.f. Fl. Brit. India 7: 380.1896; Prain Beng. Plants 2: 925.1903 (Rep. ed. 1963).

Culms tufted. Leafblades curved, convolute, filiform ligule ciliate. Spike slender, dorsally rounded, terminal, curved filiform, solitary. Spikelets dorsally rounded, minute, biseriate, unilateral, awnless. *Fls. & Frts.:* June-January.

Common on the grasslands and waste places.

Field Collection: Bamanghata/Salt Lake, *K.N. 2311*.

Distributions: Madras and Central India. Malaya, East and West Africa and South America.

43. *MNESITHEA* Kunth.

43.1. ***Mnesithea laevis*** (Retz.) Kunth; Revis. Gram 1: 154.1829; Prain Beng. Plants 2: 899.1903 (Rep. ed. 1963); Bor Grass. Burma, Ceyl., Ind. & Pak. 197. 1960. *Rottboellia laevis* Retz., Obs. Bot. 3: 11. 1783; Hook.f. Fl. Brit. India 7: 158.1896. *(Vern. Name: Kurki).*

Culms erect, solid, rounded, 4-6 noded, simple or branched above; node few. Leafblades flat, conduplicate in veruation, linear from a narrow base, glarbous or base hair, upper acute, lower obtuse, sheaths terete, ligule very short. Inflorescence shortly exerted, excavate on opposite sides with pedicels of opposite spikelets between the opposite fertile spikelets, internodes terete. Caryopsis narrowly oblong; compressed. *Fls. & Frts.:* June-January.

Mesophytic grass, the long and easily detachable inflorescence are under more prominent than the vegetative portions. Never much abundant.

Field Collection: Agarpara, *G.C. Das-67*(CAL).

Distributions: Throughout India, Burma, Sri Lanka, Malaya Islands, Indo-China, Afghanisthan and Java.

44. *MYRIOSTACHYA* Hook.f.

44.1. ***Myriostachya wightiana*** Hook.f. Fl. Brit. India 7: 327. 1896; Naskar & Guha Bakshi Mangr. Swamps Sundarbans 147. 1987. *Eragrostis wightiana* Benth. & Hook.f. Ic. Pl. t. 1381.1882 *(Vern. Name: Nalai).*

Culms densely tufted, erect from a stout sheathed, sometimes floating with long branching roots. Naskar & Guha Bakshi (1.c.) described this taxon after collecting from the lower Ganga delta.

45. *OPLISMENUS* Beauv.

45.1. ***Oplismenus compositus*** (L.) P. Beauv., Ess. Agrost. 54: 168.1817; Hook.f. Fl. Brit. India 7: 66. 1896; Prain Beng. Plants 2: 883.1903 (Rep. ed. 1963); Bor Grass. Burma, Ceyl., Ind. & Pak. 137.1960. *Panicum compositum* L. Sp. Pl.57.1753.

Culms branched from the base, minutely hairy. Leafblades variable, flaccid to rigid, sheath glarbous. Inflorescence panicle, spiciform racemes, rachis flexuous and terminated by a spikelets. Spikelets finely hairy. Caryopsis oblong, dorsally compressed. *Fls. & Frts.:* July-February.

A mesophytic grass, usually shade loving. Grow in the hedge or near the village road-sides.

Field Collections: Dhap Dhapi, *K.N. - 36;* Gosaba, *K.N. - 671.*

Distributions: Tropical India from Kashmir to Upper Assam, Manipur. Burma and most Tropical regions, except Australia.

46. *ORYZA* L.

Key to the Species

1a	Purely cultivated cereal crop plant	1. *O. sativa* L.
1b	Purely wild grass	2. *O. rufipogon* Griff.

46.1. ***Oryza sativa*** L. Sp. Pl. 333.1753; Naskar Aq. & Semi-Aq. Plants 356.1990. *(Vern. Name: Dhan).*

Naskar (l.c.) described this taxon after collecting from the lower Ganga delta.

46.2. ***Oryza rufipogon*** Griff., Notul 3: 5.1851; Bor Grass. Burma, Ceyl., Ind. & Pak. 605.1960; Naskar Aq. & Semi-Aq. Plants 357.1990. *O. sativa* L. var. *fatua* Prain Beng. Plants 2: 891. 1903 (Rep. ed. 1963). *O. sativa* L. var. *bengalensis* Watt. *Oryza fatua* Koem. ex Trin, Mem. Acad. Sci. Petersb. ser.6(2): 177.1839. *O. nivara* Sharma & Shastry, Ind. J. Gen. & Pl. Breed. 25: 161.1965. *(Vern. Name: Buno Dhan).*

Naskar (l.c.) described this taxon after collecting from Dhakuria *(K.N.-1872).* It is closely related to the cultivated species, *O. sativa* L.

47. *PANICUM* L.

Key to the Species

1a	Plants shrubby	(2)
1b	Plants herbaceous	(3)

2a (1)	Spikelets lanceolate	2. *P. maximum* Jacq.
2b (1)	Spikelets ovate or orbicular	5. *P. psilopodium* Trin.
3a (1)	Terrestrial plant	(4)
3b (1)	Aquatic or semi-aquatic plants	(7)
4a (3)	Leaves plicate; spikelets green	1. *P. fasciculatum* Swartz.
4b (3)	Leaves flat; spikelets purple	(5)
5a (4)	Spikelets 2-4 mm long	(6)
5b (4)	Spikelets 1.4-1.5mm long	8. *P. walens* Mez.
6a (6)	Lower glume 5-nerved; upper 9-nerved	7. *P. trypheron* Schult.
6b (5)	Lower glume 7-nerved, upper 11-nerved	3. *P. miliaceum* L.
7a (3)	Plants floating on water; leaves green and flat; spikelets green	4. *P. paludosum* Roxb.
7b (3)	Plants creeping near the edge of water; stoloniferous; leaves glaucous, convolute, spikelets purplish	6. *P. repens* L.

47.1. ***Panicum fasciculatum*** Swartz., Prodr. Veg. Ind. Occ. 22.1788; Majumder, Bull.bot. Soc. Bengal 10 (1&2): 52. 1956 & Ind. Agr. 6(1&2): 204.1962 (Rep. ed. 1963). *(Vern. Name: Banaspati Ghas).*

Culms very variable. Leaf-blades thin, flaccid, softly hairy, plicate, tip drawn into a fine point. Sheath hispidly ciliate, cilia with tubercle bases, compressed, not reached upto the next node. Ligule obscure. Inflorescence branched, spreading, pubescent and bristly, rachis projected beyond the spikelet into a stiff portion. Spikelets sub-secund, sub-sessile, lanceolate, round. Caryopsis enclosed in glumes. *Fls. & Frts.:* June-December.

Plentiful in the different parts of the lower Ganga delta.

Field Collection: Bamanghata/Salt Lake, *M.K.Ghosh-3076* (CAL).

Distributions: Throughout Bengal, introduced grass, from Tropical America.

Notes: D. Prain (1903a) did not mention this taxon from Bengal.

47.2. ***Panicum maximum*** Jacq., Ic. Pl. Rar. 1.2, t.13,1781- 86; Hook.f. Fl. Brit. India 7: 49.1896; Prain Beng. Plants 2: 888.1903 (Rep. ed. 1963); Bor Grass. Burma, Ceyl., Ind. & Pak. 327.1960. *P. polygamus* Sw., Prod. Veg. Ind. Occ.24.1788 (non Forssk.1775).

Culms branching, branches light green erect. Rhizomes short and stout, nodes usually densely hirsute. Leafblades midrib stout, sheath striate, mouth villous, Inflorescence panicles, decompound, lower branches whorled and hairy pedicels shorter than the spikelets, capillary, glistening. Spikelets sub-hyaline, lanceolate,roundish, nerves anastomosing. *Fls. & Frts.:* June-January.

Rarely cultivated in the garden and as garden escape.

Field Collection: Chandpara, *K.N. - 1673.*

Distributions: Cultivated in many parts of India and in many gardens. A native of Tropical America and Tropical Africa; and now introduced in several warmer countries, including india.

Economic Importance: Great value as forage or fodder and also act as important soil binder–a pasture grass.

47.3. ***Panicum miliaceum*** L. Sp. Pl. 58.1753; Hook.f. Fl. Brit. India 7: 45.1896; Prain Beng. Plants 2: 887.1903 (Rep. ed. 1963); Bor Grass. Burma, Ceyl., Ind. & Pak. 327.1960. *Milium panicum* Miller, Gard. Dict. (ed.7) 2: 1.1759. *(Vern. Name: Cheena).*

Annual tufted, leafy grass. Leafblades acuminate. Sheath lax, deeply grooved, hairy with tubercled based hairy. Ligule hairy. Inflorescence open panicle thyrsiform decarved or noding. Spikelets ovate, cispidately acuminate, turgid, green. Fruit caryopsis almost orbicular, palm convex, white. *Fls. & Frts.:* June-January.

Grow wild on the road-sides as small procumbent grass.

Field Collection: Falta Fort, *M.K. Ghosh- 3536*(CAL.)

Distributions: Cultivated or naturalised throughout the hotter parts of India. Africa and other hotter countries.

47.4. ***Panicum paludosum*** Roxb. Fl. Ind. 1: 310.1820; Naskar Aq. & Semi-Aq. Plants 358. 1990. *P. proliferum auct. non.* Lamk. 1797. *(Vern. Name: Kalasnar).*

Culms erect from a floating base; Naskar (1.c.) described this taxon after collecting from Media/near Gobardanga *(K.N. - 1693).*

47.5. ***Panicum psilopodium*** Trin. Gram. Panic. 217.1826; Naskar Aq. & Semi-Aq. Plants 359. 1960. *P flexuosum* Retz., Obs. Bot. 3: 9.1783.

Culms usually leafy; leafblades linear, margin cartilaginous; rarely grow on the open grassy places; Naskar (1.c.) described this taxon after collecting from Ghoramara Islands *(K.N. - 413)* and Baghajatin, near Jadavpur *(K.N. - 1065).*

47.6. ***Panicum repens*** L. Sp. Pl. 87.1762; Naskar Aq. & Semi-Aq. Plants 359.1990. *(Vern. Name: Baranda).*

Perennial erect herb. Extensively creeping stoloniferous; Naskar (1.c.) described this taxon after collecting from the lower Ganga delta. Baguihati *(K.N. - 1191).*

47.7. ***Panicum trypheron*** Schult. Syst. Veg. 2, Mant. 244. 1824; Hook.f. Fl. Brit. India 7: 47.1896; Prain Beng. Plants 2: 888. 1903 (Rep. ed. 1963); Bor Grass. Burm., Ceyl., Ind. & Pak. 331. 1960. *P. tenella* Roxb. Fl. Ind. 1: 309. 1820. (*non* Lamk.1791). *(Vern. Name: Jupi-Kanta).*

Annual grass; culms tufted, erect, or leafy chiefly at the base, nodes rarely rooting. Leafblades narrow, usually flat, tin, acuminated, glabrous, hairy or hirsute on the both or under-surface only. Sheaths loose. Ligule short fimbriate. Inflorescence panicles erect often as broad as long, branches long, stiff and filiform, divaricate, scaberulous. Spikelets often germinate, pedicels, capillary, green yellow or purplish. Glumes divergent, purplish, strongly nerved, broadly ovate. Caryopsis enclosed in hardened lemna and palea. *Fls. & Frts.:* August-December.

A dry land grass, growing plentifully near railway lines or on dry paddy fields.

Field Collections: Ichhapur, *K.N. - 1822.*

Distributions: Punjab and Gangetic plain and south-ward to Madras, Kumaon, upto 500 ft. China, Borneo, Africa, Tropics.

Economic Importance: Used as forage grass.

47.8. ***Panicum walens*** Mez, Engler Bot. Jahrb. 34: 146.1904; Naskar Aq. & Semi-Aq. Plants 360.1990. *P. austroasiaticum* Ohwiin, Act. Phyt. & Geobot. 2(1): 45. 1942; Bor Grass.Burma, Ceyl., Ind. & Pak. 324. 1960. *P. humile* Nees ex Steud. Syn. Pl. Gram. 1: 84. 1854. *(Vern. Name: Jupi Kanka).*

Annual, dwarf, slender and tufted grass, usually the plant is purplish or dark purple. Culms erect or geniculate below, much branched and often flowering from the base. Leafblades glabrous, flat, base narrow, sheath glabrous or mouth slightly hairy, somewhat loose, often hirsute, striate, ligule a ciliolate rim. Inflorescence-broad, lax flowered, branchlets slender, capillary. Spikelets long, pedicelled, pedicel long, slightly thickened at tips, ovate, glabrous, pale or reddish. *Fls. & Frts.:* June-December.

Grow near the edges of the paddy fields; canal sides.

Field Collections: Habra, *K.N. - 1441.*

Distributions: Punjab, Assam and Bengal. Sri Lanka and Borneo.

48. *PASPALIDIUM* Stapf

Key to the Species

1a	Spikes shorter than the internodes	1. *P. flavidum* (Retz.) Campus
1b	Spikes longer than the internodes	2. *P. punctatum* (Burm.) Campus.

48.1. ***Paspalidium flavidum*** (Retz.) A. Campus, Lecomte, Flor. Gen. de L, Indo-Chine 7: 419. 1922; Bor Grass. Burma, Ceyl., Ind. & Pak. 333.1960. *Panicum flavidum* Retz., Obs. 4: 15.1786; Hook.f. Fl. Brit. India 7: 28. 1896; Prain Beng. Plants 2: 885.1903 (Rep. ed. 1963). *P. brizoides* Ropxb. Fl. Ind. 1: 293. 1820.

Culms compressed, leafy. Leafblades bifarous, linear-lanceolate. Sheath longer than internodes, compressed, mouth hairy. Ligule often a ridge of hairs. Inflorescence racemose, generally of simple spikes, racemes longer than internodes, common on axis triquetrous, flexous. Spikelets dorsally compressed, sessile. Caryopsis more or less enclosed in lemma and palea. *Fls. & Frts.:* June-January.

A mesophytic grass, growing profusely everywhere.

Field Collections: Masjidbati, *K.N. - 847;* Taki, *K.N. - 985.*

Distributions: Plains of India. Tropical Asia and Africa.

Economic Importance: A good forage but an obnoxious weed.

48.2. ***Paspalidium punctatum*** (Burm.f.) A. Campus, Lecomte, Fl. Gen., de L' Indo-Chine 7: 419.1928; Naskar Aq. & Semi-Aq. Plants 361.1990. *Panicum punctatum* Burm.f. Fl. Ind. 26. 1768. *(Vern. Name: Patinar).*

Aquatic grass, base floating, rooting at nodes of lower internodes; Naskar (1.c.) described this taxon after collecting from Canning *(K.N. - 198);* Haroa *(K.N. - 593)* and Barasat (*R. Mallick-262* (CAL)).

49. *PASPALUM* L.

Key to the Species

1a	Rachis as wide as spikelets	(2)
1b	Rachis narrower than the distichous spikelets	2. *P. distichum* L.
2a (1)	Spikelets ciliate with long hairs	1. *P. conjugatum* Berg.
2b (1)	Spikelets glabrous or nearly so	3. *P. scrobiculatum* L.

49.1. ***Paspalum conjugatum*** Berg., Act. Helvet. Phys. Math. 7: 129, Tab.8.1772; Hook.f. Fl. Brit. India 7: 11.1896; Prain Beng. Plants 2: 890. 1903 (Rep. ed. 1963); Bor Grass. Burma, Ceyl., Ind. & Pak. 338. 1960.

Culms extensively creeping and stoloniferous, stolons strong, nodes usually conspicously pilese, rooting at the nodes, often erect. Leafblades linear-lanceolate, covered with scattered hairs, margins ciliate, sheath broad, lax, compressed, ligule pubescent ridge. Inflorescence-2 false spikes, very slender, erect, shortly pedicelled, rachis 3-nerved and convex on back, strongly keeled, pedicels alternate, short nearly as wide as spikelets. spikelets biseriate, deciduous, light yellow, compressed, pubescent. Caryopsis elliptic rotundate, rather flat, scutellum of similar shape. *Fls. & Frts.:* June-September.

A mesophytic grass, grow profusely on low dampy ground and under the tree shades.

Field Collection: Ultadanga, *Mallick-442* (CAL).

Distributions: Throughout the hotter parts of India. Sub-Tropical to Tropical regions of the Old World; a native of Tropical America, now Pantropic.

Economic Importance: Much liked by the domestic cattles.

49.2. ***Paspalum distichum*** L. Syst. Nat. (ed.10) 11: 855. 1759; Naskar & Guha Bakshi Mangr. Swamps Sundarbans 142.1987; Naskar Aq. & Semi-Aq. Plants 362.1990.

Culms creeping and rooting; an aquatic grass; grow in the halophytic and helophytic conditions. Naskar & Guha Bakshi (1.c.) and Naskar (1.c.) described this taxon after collecting from Surjyapur *(K.N. - 296);* Chandaneswar/Dhosa *(K.N. - 611)* and Ghushighata *(K.N. - 1124).*

49.3. ***Paspalum scobiculatum*** L. Mant. 1: 29.1767; Naskar Aq. & Semi-Aq. Plants 362. 1990. *(Vern. Name: Khodo dhan).*

Culms leafy, glabrous, rarely hairy. Grow profusely on the paddy fields. Naskar (1.c.) described this taxon after collecting from Dehimedanamallapur/Mallickpur *(K.N. - 84).*

50. *PENNISETUM* Pers.

Key to the Species

1a	Shrubs; inflorescence yellow, rachis hairy	1. *P. orientale* Rich.
1b	Herbs or shrubs; inflorescence purplish or white; rachis glabrous	(2)
2a (1)	Wild herb; bristle silky hairy below the middle	2. *P. setosum* (Sw.) Rich.
2b (1)	Cultivated shrub; bristles scabrid or slightly ciliate	3. *P. typhoides* Stapf & Hub.

50.1. ***Pennisetum orientale*** Rich., Pers. 1: 72. 1805; Hook.f. Fl. Brit. India 7: 86.1896; Prain Beng. Plants 2: 879. 1903 (Rep. ed. 1963).; Bor Grass. Burma, Ceyl., Ind. Pak. 345.1960.

Shrubs, culms erect, densely tufted, often stout, creeping and branching powerful, woody root-stock; leaf-blades narrowly linear, glabrous, hairy, mid-rib prominent, ligule minute. Inflorescence spikes, rachis stout, sub-erect pedicels of involucels hispid, flexuous, often purplish, very unequal villous. Spikelets lanceolate on the short branches, outer spikelets often imperfect. Anthers apiculate. *Fls. & Frts.:* June-January.

Sometime grow on the village thickets and garden surroundings, acts as soil binder.

Field Collection: Barasat, *R. Mallick-733* (CAL).

Distributions: Western Himalaya from Kashmir to Kumaon, the Punjab and Kankan. Asia minor and North Africa; introduced to the West Indies.

Economic Importance: Forage grass of the domestic cattles.

Notes: Prain (1903a) did not report this taxon from Bengal.

50.2. ***Pennisetum setosum*** (Swartz.) Rich, Pers. Syn. Pl. 1: 72. 1805; Hook.f. Fl. Brit. India 7: 87.1896; Prain Beng. Plants 2: 880.1903 (Rep. ed. 1963).; Bor grass. Burma, Ceyl., Ind. & Pak. 348.1960. *Cenchrus setonus* Swartz., Prodr. Veg. Ind. Occ. 26. 1788. *(Vern. Name: Shuti Ghas).*

Annual grass, culms in loose climps 1-2 m tall, geniculate, erect, often stout and fastigiately branched at the nodes. Leafblades flat, flaccid, glabrous, or hairy. Inflorescence excluding bristles, spikes few or more, usually purplish brown, rachis slender, deeply pitted, involucels close set, peduncle glabrous, bristle laxly but copiously ciliate below the middle. Spikelets soilitary in sessile involucels bristles sparsely ciliate. Caryopsis sub-indurete, smooth and shining. *Fls. & Frts.:* August-November.

A xerophytic plant, grow on the top of the old brick-walls and railway tracks.

Field Collection: Sarisa, *K.N. - 1941;* Barasat, *R. Mallick 505.*

Distributions: Upper Gangetic plains at Kanpur, Bihar, Chhotanagpur and Bengal, introduced into India. Africa and Tropical India.

50.3. ***Pennisetum typhoides*** (Burm.f.) Stapf & Hub., Kew Bull.(1933): 271. 1933; Bor Grass, Burma, Ceyl., Ind. & Pak. 350.1960. *Alopecurus typhoides* Burm.f., Pl. Ind. 27. 1768. *Pennisetum alopecuroides* L., Mant. 322. 1771. *P. typhoideum* Rich., Pers. Syn. 1: 72. 1805; Hook.f. Fl. Brit. India 7: 82.1896; Prain Beng. Plants 2: 879.1903 (Rep. ed. 1963). *(Vern. Name: Bajra).*

Cultivated minor cereal grass, culms erect, tomentose, below the inflorescence glabrous, nodes hairy. Leafblades ligule, a dense fringe of hairs. Inflorescence false spikes cylindric, and tapering towards base and apex, rachis slender to stout, villous, closely beset with the stalked persistent involucre, inner bristles stout with silky hairs. Spikelets in clusters of 2-5 or solitary, pedicelled, glume broad, truncate, hyaline and thinly membranous. Caryopsis ovoid or elliptic. *Fls. & Frts.:* September-November.

Cultivated on the garden but very rare in the lower Ganga delta.

Field Collection: Malatipur/near De Ganga, *K.N. - 1588.*

Distributions: Throughout the hotter parts of India; cultivated or an escape. Africa and South Europe.

Economic Importance: Food grains, minor cereal and also as fodder crop.

51. *PEROTIS* Ait.

51.1. ***Perotis indica*** (L.) O. Kuntze, Rev. Gen. Pl. 2: 787.1891; Naskar Aq. & Semi-Aq. Plants 363.1990. *Anthsanthum indicum* L. Sp. Pl. (ed.1) 28.1753. *Saccharum spicatum* L. Sp. Pl. 54.1753. *Perotis latifolia* Ait., Hort. Kew 1: 85.1789.

Annual, tufted, wiry grass. Culms erect from a geniculate base. Common on dampy, shady places near nalas and ditches. Naskar (1.c.) described this taxon after collecting from Baguihati *(K.N. - 1889).*

52. *PHALARIS* L.

52.1. ***Phalaris minor*** Retz., Obs. Bot. 3: 8.1783; Hook.f. Fl. Brit. India 7: 221.1896; Bor Brass. Burma, Ceyl., Ind. & Pak. 616.1960.

Erect or decumbent annuals. Culms branched, nodes swollen. Leaves glabrous, finely pointed, linear lanceolate. Panicles spike-like, contracted, cylindric in outlines erect. Spikelets strongly laterally compressed, flattened, 1-flowered. Empty glumes boat-shaped, strongly keeled, keel broadly winged. Floral glumes about half the length of the empty glumes, keeled. Grains free. *Fls. & Frts.:* January-April.

Common on the cultivated fields.

Field Collection: Sarberia, *K.N. - 788.*

Distributions: Throughout India. Mediterranean regions.

Notes: Prain (1903a) did not mention this taxon from Bengal.

53. *PHRAGMITES* Trin.

53.1. ***Phragmites karka*** (Retz.) Trin. ex Steud. nom. (ed.2), 2: 324. 1814; Naskar & Guha Bakshi Mangr. Swamps Sundarbans 148.1987; Naskar Aq. & Semi-Aq. Plants 364.1990. *Arundo karka* Retz., Obs. Bot. 4: 21.1786. *A. maxima* Forssk., Fl. Aegypt-Arab. 24.1775. *A. phargmites* L. Sp. Pl. 81. 1753. *Phragmites maxima* (Forssk.) Blatt. & Mc Cann., Bomb. Gr. 202.1935. *(Vern. Name: Nal* or *Dharma).*

Culms erect from a stout, creeping rootstock and with extravaginal young shoot. Plentiful in marshes, throughout the lower Ganga delta. Naskar & Guha Bakshi (1.c) and Naskar (1.c.) described this taxon after collecting from Nurpur *(K.N. - 1940)* and Sagar Islands *(L.K.B.-5730).*

54. *POGONATHERUM* Beauv.

54.1. ***Pogonatherum paniceum*** (Lamk.) Hack., Allg. Bot. Zeit. 12: 178. 1906; Bor Grass. Burma, Ceyl., Ind. & Pak. 202. 1960. *P. saccharoideum* Beauv. Agrost. Expl. des. Pl. 9.t.11, f.7.1812; Hook.f. Fl. Brit. India 7: 141. 1896; Prain Beng. Plants 2: 902.1903 (Rep. ed. 1963). *Saccharum paniceum* Lamk. Encycl. Meth. Bot. 1: 595.1785.

Culms densely tufted, wiry. Leafblades medium, nerves prominent below, margins imperfectly thickened sheath, rather loose, auricled, ligule convex upward, continuous with leaf auricle. Inflorescence racemes, rachis rather stout, angled, spikelets oblong, always 2 flowered callus small with hairs, shorter than the spikelets. Caryopsis obovate, slightly, laterally compressed, *Fls. & Frts.:* December-March.

Only available as cultivated plants in the gardens and rarely as wild.

Field Collections: Falta Fort, *M.K. Ghosh-4035* (CAL).

Distributions: Throughout the hilly parts of India, in dry places from Punjab eastwards to Bhutan and Manipur.

55. *PORTERESIA* Takeoka

55.1. ***Porteresia coarctata*** (Roxb.) Takeoka, Bull. Nat. Sci.-Mus.Tokyo 8: 406. 1965; Naskar & Guha Bakshi Mangr. Swamps Sundarbans 148.1987. *Oryza coarctata* Roxb. Fl. Ind. (ed.2) 2. 206.1832. *O. triticoides* Griff. Nat. 3: 8.1851. *Sclerophyllum coarctatum* (Roxb.) Griff., Nat. 3: 8.1851. *(Vern. Name: Dhani Ghas).* (Fig.49).

Culms standing, erect from a stout creeping rhizomatous root-stock; a halophytic grass and very common on the tidal flats and river slopes in the mangrove forests of the Sundarbans; a pioneer species for the halophytic succession. Naskar & Guha Bakshi (1.c) described this taxon.

56. *PSEUDANTHISTIRA* Griff.

56.1. ***Pseudanthistiria heteroclita*** (Roxb.) Hook.f. Fl. Brit. India 7: 219.1896; Prain Beng. Plants 2: 909. 1903 (Rep. ed. 1963); Bor Grass, Burma, Ceyl., Ind. & Pak. 203.1960. *Anthistiria heterocita* Roxb. Fl. Ind. 1: 253.1820.

Annual grass. Culms geniculate, slender. Leafblades linear, glabrous or somewhat ciliate, with tubercled hairs, sheath much shorter than the internodes, ligule membranous. Inflorescence leafy panicles compound with many shortly peduncled fascicles, equal to the spikes. *Fls. & Frts.:* June-November.

Common on the grasslands.

Field Collection: Satzelia, *K.N. - 2162.*

Distributions: Bengal, Bombay and Madras. In the Tropics.

57. *PSEUDORAPHIS* Griff.

57.1. ***Pseudoraphis brunoniana*** Griff., Notul. Plant Asiat. 3: 29.1851; Naskar Aq. & Semi-Aq. Plants 365.1990. *Chamaeraphis spinescens* (Poir.) R. Brown var. *brunoniana* Hook.f. Fl. Brit. India 7: 62.1896.

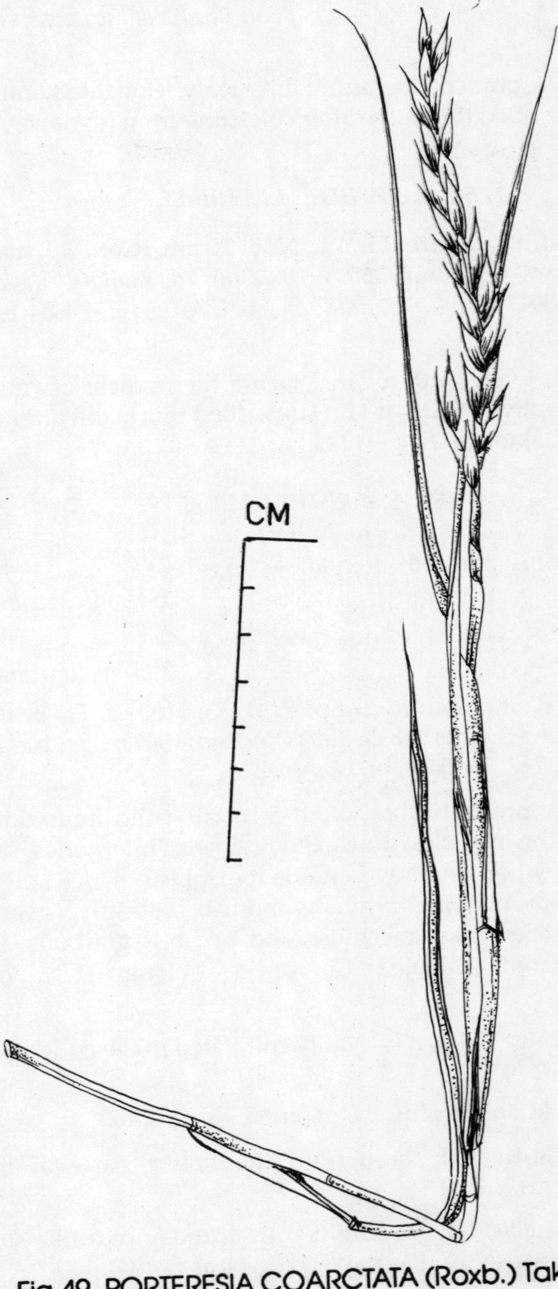

Fig.49. PORTERESIA COARCTATA (Roxb.) Takeoka—with paniclé inflorescence

Floating much branched grass; culms leafy, elongated, rather stout, Naskar (1.c.) described this taxon after collecting from Joynagar/Mazilpur *(K.N. - 1619).*

58. *ROTTBOELLIA* Linn.f.

58.1. ***Rottboellia exaltata*** Linn.f. Nov. Gram. Gen. 40, tab. 1,1779; Naskar Aq. & Semi-Aq. Plants 365. 1990. *Monisuris exaltata* O. Kuntze Rev. Gen. Pl. 2: 779. 1891. *Stegosia exaltata* Mash, N. Amer. Fl. 17: 84.1909. *(Vern. Name: Bara Swati).*

Culms robust, terete with a canal facing the branches; wet low land grass, often mesophytic. Naskar (1.c.) described this taxon after collecting from Atghora/Sitakundu *(K.N. - 1243.).*

59. *SACCHARUM* L.

Key to the Species

1a	Commonly planted; stem juicy, 4-5 m tall	1. *S. officinarum* L.
1b	Generally wild, stem not much juicy and tall	2. *S. spontaneum* L.

59.1. ***Saccharum officinarum*** L. Sp. Pl. 1753; Hook.f. Fl. Brit. India 7: 118.1896; Prain Beng. Plants 2: 895. 1903 (Rep. ed. 1963); Bor Grass. Burma, Ceyl., Ind. & Pak. 212. 1960. *(Vern. Name: Akh).*

Cultivated, perennial shrubs. About 5 m tall, solid internodes, erect, glarbous below the penicles, thick, juicy, lower internodes short and swollen and with whorl and adventitious roots. Leaf-blades long, margin scabrid, with prominent whitish midrib, sheath overlapping. Inflorescence-panicles silky, racemose and pedicels glarbous. Spikelets awnless, 1-flowered, binate. Caryopsis oblong. *Fls. & Frts.:* October-January.

In this lower Ganga delta, it is mainly cultivated in the paddyfields and low gardens.

Field Collection: Malatipur/near De Ganga, *K.N. - 1587.*

Distributions: Cultivated throughout the hotter parts of India. A cosmopolitan species.

Economic Importances: Sugars obtained from the juice of the stem; and dried juice less stem also used as paper pulp or as fire-wood.

59.2. ***Saccharum spontaneum*** L. Mant. 2: 183.1771: Naskar Aq. & Semi-Aq. Plants 366. 1990. *(Vern. Name: Kash).*

Culms very variable, straight or straggling, silky below the panicle, densely waxy below the nodes. Along the bank of rivers, ponds, ditches. Naskar (1.c.) described this taxon from the lower Ganga delta after collecting from Naraharipur/Sagar Islands *(K.N. - 462).*

60. *SACCIOLEPIS* Mash.

Key to the Species

1a	Spikelets gibbous; glume I half as long as III	1. *S. indica* (L.) Chase
1b	Spikelets lanceolate; glume I very minute	2. *S. interrupta* (Willd.) Stapf.

60.1. ***Sacciolepis indica*** (L.) A. Chase., Proc. Biol. Soc. Wash. 21: 8. 1908; Naskar Aq. & Semi-Aq. Plants 367. 1990. *Aira indica* L. Sp. Pl. 63. 1753. *A. spicata* L. Sp. Pl. 63.1753. *Panicum indicum* L. Mant. Pl. 2: 184. 1771 (non *P. indica* Mill. 1768).

Culms slender, spreading, branched, striate, compressed and with leafy base. An aquatic grass, grow very often in the paddy fields. Naskar (1.c.) described this taxon after collecting from Naihati *(K.N. - 1595)* and Barasat *R. Mallick-260* (CAL).

60.2. ***Sacciolepis interrupta*** (Willd.) Stapf, Prain Fl. Trop. Afr. 9: 757. 1920; Naskar Aq. & Semi-Aq. Plants 367. 1990. *Panicum interruptum* Willd., Sp. Pl. 1: 341. 1788. *(Vern. Name: Nardul).*

Erect unbranched, perennial, quite glabrous grass. Common on the dampy grasslands. Naskar (1.c.) described this taxon after collecting from Panihati *(K.N. - 1605).*

61. *SCHIZACHYRIUM* Stapf

61.1. ***Schizachyrium exile*** (Hochst.) Stapf, Prain Fl. Trop. Afr. 9: 191. 1917; Bor Grass. Burma, Ceyl., Ind. & Pak. 216.1960. *Andropogon exilis* Hochst., Fl. 27: 241.1844; Hook.f. Fl. Brit. India 7: 166.1896; Prain Beng. Plants 2: 906.1903 (Rep. ed. 1963).

Slender and erect grass. Culms fastigately branched above and on drying the whole grass become reddish. Leaf-blades very narrow, acuminate, spathe erect, narrowly lanceolate, acute. Inflorescence racemes, silky villous, slender, pedicels short. Spikelets 3, 2-pedicelled and one sessile spikelets much smaller, pedicel hairy, hairs long. Caryopsis slender, linear. *Fls. & Frts.:* July-October.

On the roadside thickets.

Field Collection: Diamond Harbour, *K.N. - 1950.*

Distributions: Bengal, Chotanagpur and Madhya Pradesh. Tropical Africa.

62. *SETARIA* Beauv.

Key to the Species

1a	Bristles smooth or with barbs erect or spreading	(2)
1b	Bristles with berbs deflexed	4. *S. verticillata* (L.) Beauv.
2a (1)	Panicles spiciform cylindric	(3)
2b (1)	Panicles more or less lobed or with elongate branches	3. *S. tomentosa* Kunth.
3a (2)	Barbs of bristles strong, spreading or erect	1. *S. glauca* (l.) Beauv.
3b (2)	Barbs of bristles 0 or minute	2. *S. italica* (L.) Beauv.

62.1. *Setaria glauca* (L.) Beauv., Ess. Agrost. 51: 169.1812; Watt Dict. Econ. Prod. 6(2): 546. 1892; Hook.f. Fl. Brit. India 7: 78. 1896; Prain Beng. Plants 2: 880.1903 (Rep. ed. 1963); Bor Grass. Burma, Ceyl., Ind. & Pak. 360.1960. *Panicum glaucum* L. Sp. Pl.56.1753. *(Vern. Name: Pinginatchi).*

Culms loosely tufted, erect or ascending, branches at base, compressed. Leafblades linear, glabrous or sparsely hairy, acute, flat, sheath thin. Ligule reduced to a ciliate rim. Inflorescence dense, evenly cylindric, false spike-erect, dense, continuous, linear spiciform, densely flowered, usually yellow at maturity, rarely purplish or green, rarely interrupted, sometimes reduced to an ovoid head with few spikelets. Spikelets broadly oblong to elliptic, turgid when mature. Caryopsis in black plano-convex, strongly rugose. *Fls. & Frts.:* June-November.

A mesophytic grass; leafblades are very soft and liked by the domestic cattles.

Field Collections: Masjidbati, *K.N. - 839;* Taki, *K.N. - 1029.*

Distributions: Throughout India. Sri Lanka and most of the Tropical Countries of the World.

Economic Importance: Important forage grass and act as soil binder.

Notes: S. glauca Beauv. var *macrocarpa* Hook.f. was also collected from Khardah, R. Mallick-210 (CAL); but the present author could not study the specimen.

62.2. ***Setaria italica*** (L.) Beauv., Ess. Agrost. 51: 170,178, 1812; Watt Dict. Econ. Prod. 6(2): 548.1892; Hook.f. Fl. Brit. India 7: 78.1896; Prain Beng. Plants 2: 880.1903 (Rep. ed. 1963); Bor Grass. Burma, Ceyl., Ind. & Pak. 362. 1960. *Panicum italicum* L. Sp. Pl.56.1753. *(Vern. Name: Kangu).*

Cultivated, branched, tufted, rooting towards the base, often robust. Leafblades usually long, broad, margins retrosely hispid, sheath smooth, mouth bearded. Inflorescence yellow or purplish, spiciform panicle, variable in diameter, dense, inclined and nodding, simple cylindric or lobed or compressed, bristles few. Spikes broadly oblong or sub-globose, nearly smooth. Caryopsis smooth, shining at maturity, falling away from the rest of spikelets, coloured. *Fls. & Frts.:* April- November.

On the waste places, as garden escape.

Field Collection: Nurpur, *K.N. - 1939.*

Distributions: Usually cultivated throughout India. Most warm, temperate and tropical countries.

62.3. ***Setaria tomentosa*** Kunth, Rev. Gram. 1: 47.1829; Naskar Aq. & Semi-Aq. Plants 368.1990. *Panicum tomentosum* Roxb. Fl. Ind. 1: 303: 1820. *Setaria intermedia* Roem. & Schult. Syst. 2: 489. 1817. *Panicum intermedium* Roth, Nov. Pl. Sp. 47.1821. *Setaria intermedia* sensu Hook.f. Fl. Brit. India 7: 79. 1896. *(Vern. Name: Dora biara).*

Culms geniculately sending, glabrous; commonly found in the moist waste places. Naskar (1.c.) described this taxon after collecting from Narendrapur *(K.N. - 1977).*

62.4. ***Setaria verticillata*** (L.) Beauv., Ess. Agrost. 51: 178.1812; Watt Dict. Econ. Prod. 6(2): 549. 1892; Hook.f. Fl. Brit. India 7: 80. 1896; Prain Beng. Plants 2: 881. 1903 (Rep. ed. 1963); Bor Grass. Burma, Ceyl., Ind. & Pak. 60.1960. *Panicum verticillatum* L. Sp. Pl. (ed.2): 82. 1762. *(Vern. Name: Dora Byara).*

Culms loosely tufted, nodes mostly exerted, compressed and often deeply channeled on the side facing the peduncle, slender, rough below the inflorescence. Leafblades loosely pilose with fine tubercle, based hairs sheaths thin herbaceous. Ligule very short, truncate, ciliate. Inflorescence panicles, erect, but not stiff, sub-pyramidal to sub-spiciform. Spikelets pale green, broadly ovate, compressed on one side, green. Caryopsis translucent, broadly elliptic, finely rugose. *Fls. & Frts.:* Almost throughout the year.

A low shrubby grass, mesophytic and sub-erect. Can be easily identified by its sticky inflorescence.

Field Collection: Gosaba, *K.N. - 662.*

Distributions: Throughout India. Sri Lanka, Temperate and Tropical regions.

Economic Importance: Forage grass and act as soil binder.

63. *SORGHUM* Moench. *nom. cons.*

Key to the Species

1a	Rachis of spike fragile; spikelets softy silky	1. *S. halepense* Pers.
1b	Rachis of spike tenacious	2. *S. vulgare* Pers.

63.1. ***Sorghum halepense*** (L.) Pers., Synops. Pl. 1: 101.1805; Watt Dict. Econ. Prod. 6(3): 280. 1893; Bor Grass. Burma, Ceyl., Ind. & Pak. 222.1960. *Holcus halepensis* L. Sp. Pl. 1: 1047. 1753. *Andropogon halepensis* Brot., Fl. Lus. 1: 89. 1804; Hook.f. Fl. Brit. India 7: 182. 1896; Prain Beng. Plants 2: 907.1903 (Rep. ed. 1963). *(Vern. Name: Kala-Mucha).*

Culms tall, perennials, stolons present with creeping roots, erect leafy, simple or branched, nodes pubescent. Leafblades broad, narrowed, rounded, sub-amplexicaul base, glabrous, smooth, midrib broad, margins scabridly serrulate. Inflorescence decompound panicles, peduncles not enclosed in sheaths. *Fls. & Frts.:* July-December.

Only seen as culivated grass.

Field Collection: Goranbose, *K.N. - 1875.*

Distributions: Throughout India. Sri Lanka, Burma and most warmer countries; a native of Tropical America, introduced in India at an early time.

Economic Importance: Excellent forage grass.

63.2. ***Sorghum vulgare*** Pers., Syn. Pl. 1: 101. 1805; Watt Dict. Econ. Prod. 6(3): 289. 1893; Majumder, Bull. bot. Soc. Bengal 10(1&2): 90. 1956. *Holcus* *sorghum* L. Sp. Pl. 1047.1753. *Andropogon sorghum* Brot. Fl. Lusit. 1: 88.1804; Hook.f. Fl. Brit. India 7: 183.1896; Prain Beng. Plants 2: 907.1903 (Rep. ed. 1963). *(Vern. Name: Dedhan).*

Annual, without any creeping stolons; roots fibrous. Leafblades wide. Inflorescence thyrsiform loose panicle with whorl or erect branches. Spikelets usually red. *Fls. & Frts.:* June- November.

Occasionally cultivated for the cereals or grains.

Field Collections: Specimens not preserved.

Distributions: Cultivated in the Northern and Western India. A native of Tropical Africa and Asia.

Economic Importance: Forage grass and grains are also used as forage for the domestic cattles.

64. *SPOROBOLUS* R. Br.

Key to the Species

1a	Panicle very narrow, its branches short	(2)
1b	Panicle contracted	3. *S. tremulus* (Willd.) Kunth.
2a (1)	Spikelets 1.0-1.50 cm long	1. *S. diander* (Retz.) Beauv.
2b (1)	Spikelets 0.02 cm long only	2. *S. indicus* R. Brown

64.1. ***Sporobolus diander*** (Retz.) Beauv., Ess. Agrost. 26: 147.1812; Naskar & Guha Bakshi Mangr. Swamps Sundarbans 150.1987. *Agrostis diander* Retz., Obs. Bot. 5: 191. 1789. *(Vern. Name: Bena-Joni).*

Culms slender. Leafblades mostly basal, crowded, narrow entire and often convolute, leaf-sheath smooth, somewhat loose, ligule a few hairs. Inflorescence pyramidal, rachis slender, spreading, almost at right angle from the axis. Spikelets spindle shaped, 1-flowered. Fruit grains pericarp, loose. *Fls. & Frts.:* June-November.

Xerophytic grass, but grow also in the low land, very abundant.

Field Collections: Beliachandi/Gocharan, *K.N. - 262;* Surjyapur, *K.N. - 289;* Thuba/Taki, *K.N. - 1039* and Dhopagachi, *K.N. - 1156.*

Distributions: Throughout India. Burma, Asia and Australia.

Economic Importance: Often grazed by the domestic cattles.

64.2. ***Sporobolus indicus*** (aucct, *non.* Linn.) R. Brown, Prodr. 170.1810; Hook.f. Fl. Brit. India 7: 247. 1896; Prain Beng. Plants 2: 914.1903 (Rep. ed. 1963); Bor Grass. Burma, Ceyl., Ind. & Pak. 630.1960.

Perennial grass. Leafblades convolute or flat, mostly basal, crowded, papery, ligule a few hairy, sheath compressed, smooth, margins with soft hairs, blade with minute stiff hairs. Inflorescence often interrupted, rachis, stout, erect, panicles very narrow, branches-short appressed, covered to the base with crowded spikelets. Fruit grain broadly obovoid, loose within glume and palea, pericarp or its epidermis loose, brown often with isolated reddish. *Fls. & Frts.:* June-November.

Xerophytic grass.

Field Collection: Sashan/Baruipur, *K.N. - 1174.*

Distributions: Throughout India. Burma, Sri Lanka, and all warmer countries.

Economic Importance: Young leaves are forage grass for domestic cattles.

Notes: Prain (1903a) did not mention this taxon from Bengal proper.

64.3. ***Sporobolus tremulus:*** (Willd.) Kunth, Rev. Gram. 1: 67.1829; Naskar Aq. & Semi-Aq. Plants 369. 1990. *Agrostis tremula* Willd. Sp. Pl. 1: 372. 1798.

Perennial grass; a wet low land grass, much creeping. Naskar(1.c.) described this taxon after collecting from Masmari/Sundarbans, *K.N. - 1865.*

65. *THYSANOLENA* Nees

65.1. ***Thysanolaena maxima*** O. Kuntze, Rev. Gen. Pl. 2: 794.1891; Bor Grass. Burma, Ceyl., Ind. & Pak. 650.1960. *Agrostis maxima* Roxb. Fl. Ind. 1: 319. 1820. *Thysanolaena agrostis* Nees, Edins. N. Phil. Journ. 18: 180. 1835; Hook.f. Fl. Brit. India 7: 61. 1896; Prain Beng. Plants 2: 882.1903 (Rep. ed. 1963).

Culms round, hard, glabrous. Leafblades, coriaceous, broadly-lanceolate, amplexicaul, glaucous, margin scabrid, sheaths hard, shining, ligule cartilaginous, truncate. Inflorescence large, soft, panicles. Spikelets flowered, ovoid, lanceolate, pedicellate, rachilla shortly produced as a short stipe. Caryopsis free. *Fls. & Frts.:* November-March.

Cultivated in the gardens for its graceful foliage and much branched inflorescence.

Field Collections: Mollakhali/Sundarbans, *K.N. - 1854.*

Distributions: Throughout India. New Guinea & South East Asia. An ornamental garden grass.

66. *TRAGUS* Hall. *nom.cons.*

66.1. ***Tragus biflorus*** (Roxb.) Schult., Mant. 2: 205.1824; Bor Grass. Burma, Ceyl., Ind. & Pak. 682. 1960. *Bappado biflorus* Roxb. Fl. Ind. 284. 1820. *Tragus tacemosus* Hook.f. Fl. Brit. India 7: 97. 1896; Prain Beng. Plants 2: 893. 1903 (Rep. ed. 1963).

Tufted, prostrate, ascending or erect grass, simple or branched, leafy below. Leaves linear lanceolate, rigid, with pectnately ciliate margins. Racemes spikes-like, solitary, terminal. Grains linear oblong, pale brown. *Fls. & Frts.:* July-November.

Commonly found in tufts on gravelly soils.

Field Collection: Shyamnagar, *K.N. - 1817.*

Distributions: Throughout the plains of India; S.E. Asia and East Africa. A common forage grass.

67. *TRITICUM* L.

67.1. ***Triticum aestivum*** L. Sp. Pl. 85. 1753; Bor Grass. Burma, Ceyl., Ind. & pak. 679. 1960. *T. mybernum* L. Sp. Pl. 86.1753. *T. sativum* Lamk. Fl. Fr. 3: 625. 1778. *T. vulgare* Vill., Hist.Pl. Dauph.2: 153. 1787; Hook.f. Fl. Brit. India 7: 367.1896; Prain Beng. Plants 2: 929. 1903 (Rep. ed. 1963). *(Vern. Name: Gam).*

Annual, cultivated crop (cereal) plant. Leafblades linear, auricled sheath smooth, ligule short. Inflorescence erect spikes curved at maturity, rachis hairy on margins, spikelets placed flat-wise to rachis. Spikelets solitary, laterally compressed, 3-5 flowered, uppermost sterile. *Fls. & Frts.:* December-March.

Occasionally cultivated in the lower Ganga delta.

Field Collection: Barasat, *R. Mallick-607* (CAL).

Distributions: Cultivated in many parts of North India and the Deccan Peninsula, especially in N. W. Pantropic species; a native of S.W. Asia and Egypt.

Economic Importances: Grain flour contain high strach, which is the staple food throughout the world. Also used as fodder for cattles and poultry birds.

68. *UROCHLOA* Beauv.

68.1. ***Urochloa panicoides*** Beauv., Ess. Agrost. 52. t.11, fig.1.1832. *Panicum javanicum* Poir., Lamk. Encycl. Meth. Bot. Suppl. 4: 279. 1816; Hook.f. Fl. Brit. India 7: 35. 1896; Prain Beng. Plants 2: 886.1903 (Rep. ed. 1963). *P. urochloa Desb.*, Opusc. 82. 1831. *(Vern. Name: Falganti).*

Annual, decumbent or rooting below, nodes pubescent. Culms decumbent and rooting below, nodes pubescent to subvillous, branching. Leafblades finely hirsute with tubercle based hairs, branches intravaginal, margins crisped and undulate, sheath loose and laxly hairy. Inflorescence spikes, 4-7, more or less speciform, racemes moderately dense, rachis flat, pedicels solitary. Spikelets lanceolate, slightly depressed abaxially, convexed on back, rachis star-shaped, acuminate, hispidly hairy, pedicillate, pedicel with long cilia. Caryopsis long compressed. *Fls. & Frts.:* April-October.

Mesophytic grass, grow usually near the village roads, under shade of the big trees.

Field Collection: Ichhapur, *K.N. - 1823.*

Distributions: Plains of India from Kashmir to Bengal. Sri Lanka and Tropics in general.

Economic Importance: Much liked by the domestic cattles.

69. *VETIVERIA* Lamk. Lisane

69.1. ***Vetiveria zizanioiodes*** (L.) Nash, Small Fl. South East U.S. 67. 1903; Bor Grass. Burma, Ceyl., Ind. & Pak. 258.1960. *Phalaris zianioides* L. Mant. Pl. 183. 1771. *Andropogon squarrossus* Linn.f. Suppl. Pl. 433. 1781; Hook.f. Fl. Brit. India 7: 186.1896; Prain Beng. Plants 2: 907.1903 (Rep. ed. 1963). *(Vern. Name: Bena* or *Khas-Khas).*

Culms from an aromatic rhizome, covered by sheath, stout, roots spongy, aromatic, sparingly branched. Leafblades sub-bifarious narrow, acute, erect, keeled, glabrous, margins scabrid sheaths equitant, glabrous, compressed. Inflorescence panicles, conical, erect, filiform, flexuous, branches simple, strict, spikes many, pedicels glabrous. Spikelets awnless and muricate. Caryopsis oblong, top slightly oblique. *Fls. & Frts.:* July-November.

Adaptable to any habitat, usally grow in full vigour in the low lands, & cultivated lands.

Field Collections: Salipur, *K.N. - 86;* Madhyampur/Taki, *K.N. - 945.*

Distributions: Throughout the plains and lower hills of India. Burma, Sri Lanka, Java and Tropical Africa.

Economic Importance: A good forage/fodder grass; but very troublesome weed in the agriculture fields.

70. *ZĖA* L.

70.1. ***Zea mays*** L. Sp. Pl. 971. 1753; Hook.f. Fl. Brit. India 7: 102. 1896; Prain Beng. Plants 2: 911. 1903 (Rep. ed. 1963); Bor Grass. Burma, Ceyl., Ind. & Pak. 270. 1960. *Mays zea* Gaertn., Fruct. 1: 6,t.1. 1789. *Zey segetalis* Salisb., Prodr. Strip. 28.1796. *(Vern. Name: Bhutta).*

Cultivated annual tall shrubby grass. Culms solid, the lower nodes with whorl of adventitious roots. Leaf-blades flat and broad, distichous, sheaths overlapping. Inflorescence-terminal inflorescence is in male and large, spikelets, racemes are long, numerous, spreading. Axillary inflorescence in female or pistillate, the spikelets are arranged in many vertical rows on a thickened inarticulate axis, which at young condition is spongy and at maturity becomes woody, spathes many and large and foliaceous. Caryopsis grain at maturity greatly exceeding the glumes, large, sub-globose or flattened. *Fls. & Frts.:* September-January.

Culitvated throughout the lower Ganga delta, particularly in the gardens and high lands.

Field Collections: Specimens not preserved.

Distributions: Cultivated almost throughout India. A native of America, but now cultivated in all warmer countries.

Economic Importances: Grains are edible after burning or frying; also used as forage and fodder plants; fibrous stem may also used as paper pulp.

71. *ZOISIA* Willd.

71.1. ***Zoisia matrella*** (L.) Merr., Philippine. Jour. Sci. Bot. 7: 230; Naskar & Guha Bakshi Mangr. Swamps Sundarbans 150.1987; Naskar Aq. & Semi-Aq. Plants 369. 1990. *Agrostis matrella* L. Mant. Pl. 2: 185.1771.

Small branched, rigid glabrous grass. Halophytic and abundant in low land and forming tuft masses. A weed of rice fields. Naskar & Guha Bakshi (1.c.) and Naskar (1.c.) described this taxon after collecting from the lower Ganga delta. Bamanghata, *K.N. - 1737.*

Remarks: Aeluropus lagopoides (1.) Trin. ex Thw. Enum. Pl. Zeyl. 374.1864; Bor Gross. Burma, Ceyl., Ind. & Pak. 380. 1960; Naskar & Guha Bakshi, Jour. Econ. Tax. Bot. 7(1): 200.1985; Naskar & Guha Bakshi Mangr. Swamps Sundarbans 144.1990. *(Vern. Name: Nona Durba).*

A common halophytic grass on the sandy sea-shore or river flats; Naskar & Guha Bakshi (1.c.) described this taxon after collecting from the Sagar Islands *(K.N. - 424).* It is reported for the first time by Naskar & Guha Bakshi (1985), from the lower Ganga delta.

SELECTED REFERENCES

Plant Wealth of the Lower Ganga Delta

Agharkar, S.P. & A.K. Ghosh (1931). The Composition of the Bengal Flora. *Proc. Indian Sci. Cong*. 278.

Aiton, W. (1789). *Hortus Kewensis* (ed.1) 3 Vols. London.

Anderson, G. (1862). *Catalogue of Plants Indigenous in Neighbourhood of Calcutta with Direction for Examinations and Preservations of Plants.* Calcutta.

Anonymous (1896). *List of the Principal Crops Grown in Bengal*. Calcutta (ed.2)

Anonymous (1948-1976). *The Wealth of India.* (A Dictionary of Indian, Raw Materials). C.S.I.R., New Delhi, Vols. I-XII.

Anonymous (1976). *Report of the National Commission on Agriculture,1976*. Govt. of India, New Delhi. Part I-XV.

Backer, C.A. (1951). Pontederiaceae. *Fl. Males. ser.I*, 4(3): 255-261.

Backer, C.A. & R.C. Bakhuizen van der Brink (1963-1968). *The Flora of Java. Vol. I-III.*

Basak, R.K. (1973). Th Bibliography on the Flora and Vegetation of Bengal with an Introductary Note. *Bull. bot. Surv. India 15* (1 & 2): 22-38.

Basak, R.K. & D.N. Guha Bakshi (1977). Floristic Studies on the State of West Bengal: Present Status and Future Strategies. *Bull. bot. Surv. India 19* (1-4): 42-55.

Bennet, S.S.R. (1968). Taxonomic Studies on the Flora of Howrah District (West Bengal). Ph. D. Thesis, C.U. (Published-1980).

Benthall, A.P. (1946). *Trees of Calcutta and its Neighbourhood*, Calcutta.

Benthum, G. & J.D. HOOKER (1862-1883). *Genera Plantarum ad exemplaria imprimis in herbariis Kewensi bus servata definita*. 3 vols. London.

Blasco,F. (1975). The Mangroves of India. Inst. Fr. Pondichery. *Trav. Sec. Sci. Tech.14:* 1-175.

Bor, N.L. (1960). *The Grasses of Burma, Ceylon, India & Pakisthan* (excluding Bambuceae). London. (Rep. ed. 1973).

Bruhl, P. (1908). Recent Plant Immigrants. *Jour. & Proc. Asiatic Soc. Bengal* (N.S.) 4: 603-656.

Chatterjee, D.(1939). Studies on the Endemic Flora of India and Burma. *J. Asiat. Soc. Bengal* 5: 19-67.

Chopra, I.C., K.H. Handa & L.D. Kapur (1958). *Indigenous Drugs of India*.. Calcutta, (2nd ed.).

Chopra, I.C. & B.S. Verma (1969). *Supplement to Glossary of Indian Medicinal Plants*, New Delhi.

Cooke, T.(1901-1908). *The Flora of Presidency of Bombay*. 2 Vols. London.

Das, S.R. (1968). *The Flora of Nadia District*. Thesis for Ph.D. (Sc.), Kalyani University.

Den, Hartog, C. (1957). Hydrocharitaceae. *Fl. Males. ser. I.* 5(4): 381-413.

Duthie, J.P. (1903-1929). *The Flora of Upper Gangetic Plains and the Adjacent Sivalik and Sub-Himalayan Tracts*. 3 Vols. Calcutta.

Dutta, N.M. & A.K.Ganguly (1967). *Contribution to the Flora of Calcutta and Neighbourhood*. I. Commonly Occurring Herbs and Shrubs, Calcutta.

Engler, A. (1900-1937). *Das Pflanzenreich Fgni Vegetabilis Conspectus*, etc. Leipzig. Hofte. 1-102.

Engler, A. & K. Prantl (1887-1915). *Die Naturlichen Pflanzenfamilien*, etc. Leipzig. 23. Vols. ed 2,8, incomplete, 1942.

Forsskal, P. (1775). *Flora of Ageyptica-Arabica*. Copenhazen.

Gamble, J.S. & C.E.C. Fisher (1915-1936). *Flora of the Presidency of Madras*. London. Vols. I-III.

Gohsh, A.K.(1941). Submerged Forests in Calcutta. *Sci. & Cult. 6*: 669-670.

Guha Bakshi, D.N. (1984). *Flora of Murshidabad*, West Bangal. Scientific Publisher, Jodhpur.

Guha Bakshi, D.N. & Subir Sen (1969). A botanical exploration of Sagar Island in 24-Parganas. *Bull. bot. Soc. Bengal*. *23*: 31-34.

Guha Bakshi, D.N. & K.R. Naskar (1991). Mangrove Afforestation in the Sundarbans. *In, Plant Science and Man*: Problems and Prospects. Proceeding of the International Botanical Conference, 10-12 January, 1991, Dhaka. pp. 75-83

Hains, H.H. (1921-1925). *Botany of Bihar and Orissa*. 6 parts, London.

Hills, A.P. (1952). *Economic Botany*, New York.

Hooker, J.D. & T. Thompson (1855). Flora Indica. Vol. I. London.

Hooker, J.D. (1872-1897). *The Flora of British India*. Vol. I-VII. London.

Kammathy, R.V. & R.S. Rao (1961). Notes on Indian Commelinaceae III. Cytological Observations. *Bull. bot. Surv. India 3-4*: 393-394.

Kammathy, R.V. & R.S. Rao (1964). Notes on Indian Commelinaceae: Cytotaxonomic Observations. IV. *Bull. bot. Surv. India 6*(1): 1-6.

Kenneth, M. Backer (1973). A comparison of Angiosperm Classification System. *Taxon* 22 (1) 19-50. Feb. 1973.

Kern, J.H. (1956). Notes on the Malaysian and Some South East Asian Cyperaceae III. *Blumea 8*: 110-169.

Kern, J.H. (1955). Notes on the Malaysian and Some South East Asian Cyperaceae IV. *Reinwardtia 4*: 89-97.

Kern, J.H. (1961). Cyperaceae of Thailand (Exc.*Carex*). *Reinwardtia 6*(1): 25-83.

Kirtikar, K.R. & B.D. Basu (1918). *Indian Medicinal Plants*, (2nd ed) Vols. I-IV. Allahabad.

Kunth, C.S. (1833-1850). *Enumeratio Plantarum*. Stuttgart & Tubingen. Vols. I-V.

Kuntze, C.E.O. (1891-1898). *Revisio Generum Plantarum Vascularium etc*. Vols. I-III. Leipzig.

Linnaeus, C. (1753), *Species Planterum*, Vols. I-II. Stockholm.

Linnaeus, C.(1767-1771). *Mantissa Plantarum. Generum Editionis Viet Specierum Editionis*. II.2 parts, Stockholm.

Linnaeus, C. von fil. (1781). *Supplementum Plantarum Systematis Vegetabilium Editionis Decimae* etc. Braunschweig.

Link, J.H.F. (1821-1822). *Enumeratio Plantarum Hortiregii Botanici Berolinensis Altera*. 2 parts, Berlin.

Mahapatra, A.K. & R.K. Chakraverty (1977). A Survey of the Aquatic Flora of the Sagar Island and Their Economic Aspects with Reference to Integrated Rural Development. *J. Sci. Club*. 31(2): 66-83.

Maheshwari, P. & K. Singh (1965). *Dictionary of Economic Plants of India*. New Delhi.

Maiti, G.G. & D.N. Guha Bakshi (1981). Invation of Exotic Weeds in West Bengal Since 1903: Dicotyledons and Monocotyledons. *J. Econ. Tax. Bot*. *2*: 1-21.

Majumdar, R.B. (1956). Studies on the Grasses of 24-Parganas, West Bengal. *Bull. bot. Soc. Bengal 10*: 1-114.

Majumdar, R.B. (1962). Weed Flora of the District 24-Parganas, W.B. *Ind. Agr.* 6: 89-213. (Rep. ed.).

Majumder, N.C. (1965). Aquatic and Semi-Aquatic Flora of Calcutta and Adjacent Localities. *Bull.bot.Soc. Bengal.19:* 10-17.

Mallick. K.C. (1966). A contribution to the Flora of Purulia District, W.B. *Bull. bot. Surv. India 8:* 45-59.

Mazi, S. (1981). *Flora of the Midnapore District,* West Bengal. Ph.D. Thesis, Calcutta University (Unpub.).

Mc Cann, C. (1942). Observation on Indian Duckweeds, Lemnaceae. *J. Bombay nat. Hist. Soc. 43:* 148-162.

Mitra, J.N. (1952). Review and Revision of Commelinaceae of Eastern India. *Bull. bot. Soc. Bengal 6:* 63-70.

Moore, H.P.Jr. (1957). *Musa* and *Ensete*. The Cultivated Bananas. *Baileya 5:* 167-194.

Morgan, J.P. & W.G. Mc Intire (1959). Oquaternary Geology of the Bengal Basin, East Pakistan and Burma. *Bull. Geol. Soc. Am.* 70:319-342.

Mukerjee, S.K. (1940). A revision of Labiatae of the Indian Empire. *Rec. bot. Surv. India 14:* 1-228.

Mukerjee, S.K. (1945). A revision of Indo-Burmese Species of *Lindernia* All. *Journal. Indian bot. Soc.* 24 : 127- 134.

Mukherjee, S.K.(1959). Central National Herbarium. *Bull. bot. Surv. India 1:* 30-32.

Mukherjee, S.N.(1956). *A Brief Agricultural Geography of West Bengal.* Calcutta.

Mukherjee, P.K. (1974). *Flora of Burdwan District.* PRS Thesis, Calcutta University (unpublished).

Narayana, D.V. (1934). Aquatic Weeds in Deccan Irrigational Canals. *Journal Eco. 27:* 66.

Naskar, K.R. (1981). *Floristic Studies of the District 24-Parganas in West-Bengal (India) with Special Reference to the Mangrove Vegetation of Sundarbans.* (Ph. D. Thesis, Calcutta University, unpublished).

Naskar, K.R., D.N. Guha Bakshi & G.G. Maity (1979). On the occurrence of *Moghania macrophylla* (Willd.) O. Kuntze—from the Lower Ganga Plain. *Bull. bot. Soc. Bengal 33:* 39-43.

Naskar, K.R. & D.N. Guha Bakshi (1981). Some of the Important Medicinal Plants of Apocynaceae from West Bengal. *Bull. bot. Soc. Bengal 35:* 7-14.

Naskar, K.R. & D.N. Guha Bakshi (1982). Sundarbans–the World Famous Mangrove Forest of the District 24-Parganas in West-Bengal (India). *J. Econ. Tax. Botany 3:* 883-918.

Naskar, K.R. (1983). Halophytes and Their Unique Adaptation on the Sundarbans' Mangrove Swamps in India. *J. Indian Soc. Coastal agric Res. 1*(2): 91-105.

Naskar, K.R. & D.N. Guha Bakshi (1983a). A brief review on some less familiar plants of the Sundarbans. *J.Econ. Tax. Botany* 4(3): 699-711.

Naskar, K.R. & D.N. Guha Bakshi (1983b). An Eco-Taxonomical Studies of the Typical Halophytic Flora of the Sundarbans in the District 24-Parganas, West Bengal with Special Reference to their Socio-economic Impact. *J. Econ. Tax. Botany* 4(3): 713-746.

Naskar, K.R. & N.M. Chakraborty (1984). Studies on the Economic Fauna from the Sundarbans Delta in West Bengal. *J. Indian Soc. Coastal agric. Res.* 2(1): 17-25.

Naskar, K.R. & D.N. Guha Bakshi (1984). Identification and Extraction Techinques of Crude-drugs (Phyto-origin) from the Wild Sources, Commonly Utilized in Ethno-Pharmacology. *Proc. Nat. Sem. Appl. Bio. Tech. Med. & Arom. Plants (Bot. Lab. Pharm. Anatomy, Calcutta University).* pp. 437-447.

Naskar, K.R., S.K. Saha and P.K. Saha (1985). Compost of Water-hyacinth for the Field of Agriculture and Aquaculture. *Environ. & Ecol.* 3(2): 247-253.

Naskar, K.R., G.N. Chattopadhyay & A. Ghosh (1985). A notes on *Ruppia maritima* L., a submerged aquatic weed from the Brackishwater Fisheries of the Sundarbans. *J. Indian Soc. Coastal agric. Res.* 3 (1): 69-70.

Naskar, K.R. (1985). Population Pressure on the Tidal Mangrove Forests in the Sundarbans – its Causes and Problems. *Monograph* 4: 159-161. *Nat. Atlas & Them. Map. Organisation,* Calcutta.

Naskar, K.R. (1985). A Short History and the Present Trends of Brackishwater Fish Culture in the Paddy Fields of the Kulti- Minakhan Areas of the Sundarbans in West Bengal. *J. Indian Soc. Coastal agric Res.* 3 (2): 115-126.

Naskar, K.R. & D.N. Guha Bakshi (1985). *Parthenium hysterophorus* L. an Exotic Poisonous Weed in India – A Review. *J. Econ. Tax. Botany* 7 (3) 741-748.

Naskar, K.R., A. Banerjee, N.M. Chakraborty & A. Ghosh (1985). Yield of *Wolffia arrhiza* (L.) Horkel ex Wimmer from Cement Cisterns with Different Sewage Concentration and its Efficacy as Carp Feed. *Aquaculture* (Neitherland) 51: 211-216.

Naskar, K.R. & S.C. Santra (1986). Studies on *Enteromorpha tubulosa* in Brackish water Mixed Sewage fed Fisheries from Sundarbans, W.B. *Sci. & Cult.* 52: 210.

Naskar, K.R. (1986). Economic Importance of the Dominant Mangrove Family Rhizophoraceae From the Sundarbans Delta of Twenty Four Paraganas, W.B. *Journ. Econ. Tax. Botany* 8(2): 359-372.

Naskar, K.R. (1986). Role of *Avicennia* L. Plantation on the Brackishwater Fisheries with Special Reference to Their Taxonomy and Ecological Notes in the Tidal Mangrove Forests of the Sundarbans. *J. Indian Soc. Coastal agric. Res.* 4 (2) 139-142.

Naskar, K.R. & D.N. Guha Bakshi (1986). On the Verge of Extinction of some Important Mangrove Species From the Sundarbans Delta in West Bengal. *Journ. Econ. Tax. Botany* 8(2): 431- 437.

Naskar, K.R., P.K. Saha, Rita Mazumder & D.N. Guha Bakshi (1986). Compost Value of Water Hyacinth (*Eichhornia crassipes* (Mart) Solms.) in the Field of Agriculture & Aquaculture. *Bull. bot. Soc. Bengal* 40 (1&2): 11-17.

Naskar, K.R. (1987). Mangrove Ecology of the Sundarbans Delta in West Bengal and its Role on the Brackishwater Fisheries. *Journ-Aqua.* 1(1): 1-9.

Naskar, K.R. & D.N. Guha Bakshi (1987). *Mangrove Swamps of The Sundarbans: An Ecological Perspectives.* Naya Prokash, Calcutta-700 006, India pp. 1-264.

Naskar, K.R. & D.N. Guha Bakshi (1987). Different Phyto-Ecological zones in the 24-Parganas District of West Bengal with special Reference to Its Land Utilization Patterns. *J. Indian Soc. Coastal agric. Res.* 5(1): 183-187.

Naskar, K.R. & N.M. Chakraborty (1987). Studies of the Family- Lemnaceae From the Lower Ganga Delta. *Bull. bot. Soc. Bengal. 41*: 5-12.

Naskar, K.R. & P.K. Saha (1987). Studies of Duck-Weed Culture Utilizing Domestic Organic Wastes. *Bull. bot. Soc. Bengal 41:* 35-43.

Naskar, K.R. (1987). Recent Trends of Aquatic Weeds Management Through Proper Utilizations. *Plant Science Research*. Department of Botany, Visva Bharati, Santiniketan, India. pp. 188-193.

Naskar, K.R. (1987). Mass Production of Duckweeds for Carp Culture. *Proc. USAID Training Programme on 'Freshwater Aquaculture'*, CIFA, Bhubaneswar, pp. 56-65.

Naskar, K.R., S.K. Saha & A.K. Datta (1988). Utilization of Domestic Wastes for Rice Production. *Environ. & Ecol. 6* (1): 125-128.

Naskar, K.R. & S.K. Saha (1988). Utilization of Domestic Wastes For Integrated Farming in the Tropics. *J. Aqua. Trop. 3*(1): 37-41.

Naskar, K.R. & Santunu Biswas (1988). Hydrocharitaceae from the Lower Ganga Delta and Its Role in Aquaculture & Agriculture, *Environ.* & Ecol. 6(3): 620-626.

Naskar, K.R. (1988). Economic Potentialities of the Tidal Mangrove Forests of the Sundarbans in India, *J. Indian Soc. Coastal agric Res. 6*(2) 140-157.

Naskar, K.R. & D.N. Guha Bakshi (1989). Distributions of Mangroves in India and Abroad and Their Role in The Estuarine Ecosystems. *Monograph 9*: 22-23. *(National Atlas & Thematic Mapping Organisation*, Calcutta).

Naskar, K.R.(1989). Studies of Macrovegetation in Beels. *Bulletin* No. *63 (CICFRI)*, July, 1989, P.32-P.56.

Naskar, K.R. & S.C. Santra (1985). A notes on *Enteromorpha tubulosa* in Brackish-mixed-Sewage-Fed Fisheries from the Sundarbans in West Bengal. *Indian Soc. Coastal agric Res.* 5 (2): 471-472.

Naskar, K.R. & D.N. Guha Bakshi (1989). Stages of Land Formation and Comparative Studies of Different Microecological Zones of Sundarbans, West Bengal (India). *Fragm. Florist. Geo.Bot.* 34(314): 341.354.

Naskar, K.R. & A.K. Ghosh (1989). Status of the Tidal and Inter-Tidal Mangrove Forests of The Sundarbans In India. Proceeding of the International Seminar on 'Coastal Environment in Asian Region Potentialities and Problems', December-8,1989, Calcutta. pp.23-29.

Naskar, K.R. & N.N. Mazumdar (1990). Some Fish Poisonous Plants From The Sundarbans Mangrove Ecosystems. *In Contribution to the 'Fisheries of Inland Open Water Systems in India, Part- I.* pp. 184-188.

Naskar, K.R. (1990). *Aquatic & Semi-Aquatic Plants of The Lower Ganga Delta.* Daya Publishing House, Delhi-110006, pp.1-408

Naskar, K.R. & D.N. Guha Bakshi (1990). Present Needs to the Silvicultural Practices in the Tidal Mangrove Forests of The Sundarbans in West Bengal, India. *In Recent Research In Ecology, Environment & Pollution;* Today & Tomorrow Publishers, New Delhi. *4: 105-118.*

Naskar, K.R. (1991). Status of the Sundarbans Mangrove Forests of India And Their Ecological Stresses. *Hydrobiologia* (In Press).

Nicolson, D.H. & J. Saldhana (1976). *Flora of Hassan District*, Karnataka, India.

Oldham, R.D.(1893). *A Manual of the Geology of India*. London.

O' Malley, L.S.S. (1914). *Bengal District Gazetteers, 24- Parganas*, pp.1-268.

Paria, N.D.(1978). A Contribution to the Flora of Ballygunge Science College Campus. *Bull. bot. Soc. Bengal 31*: 62-73.

Paria, N.D. & S.K. Sahoo (1979). A Contribution to the Herbaceous Flora of The District of the 24-Parganas. *Bull. bot. Soc. Bengal 33*: 119-129.

Prain, D. (1903a). *Bengal Plants*. 2 Vols. (Revised ed. 1963), BSI, Howrah.

Prain, D. (1903b). Flora of Sundribuns. *Rec. bot. Surv. India* 2: 231-370.

Prain, D.(1905) The Vegetation of the District of Hugli, Hourah and 24-Pergunnahs. *Rec. bot. Surv. India* 3: 143-329.

Rai, H.N. Choudhuri & G.N. Tribedi (1976). On the Occurrence of some Medicinal Plants in 24 Parganas, West Bengal. *Bull. bot. Surv. India 18* (1-4): 161-165.

Raizada, M.B. (1950). Review of some name changes in Indian Grasses. *Ind. For. 76*: 523-525.

Raizada, M.B.(1958). Name Changes of Common Indian Plants. *Ind. For. 84*: 438-538.

Raizada, M.B. (1959). Name Changes of Common Indiae Plants. *Ind. For. 85.*: 473-509.

Roth, A.W. (1821). *Nova Plantarum species Presertim Indiae Orientalis Ex Collections Doct, B. Heynei cum descriptionibus et Observationibus*. Halbersadt.

Roxburgh, W. (1814). *Hortus Bengalensis or a Catalogue of the Plants Growing in the Honourable East India Company's Botanic Garden* at Calcutta. Serampore.

Roxburgh, W. (1820-1832). Flora Indica: or description of Indian Plants by late W. Roxburgh (ed. by W. Carey and N. Wallich); Vol.1, 1820; Vol.2,1824 and Vol. 3. 1832 (ed.C.B. Clarke).

Sen, S. (1972). *Floristic Studies of the Hooghly District* (West Bengal). (Thesis for D. Phil (Sc.), Calcutta University, Unpublished).

Sikdar, J.(1980). *Studies of the Maldah District Flora of West Bengal*. (Ph.D. Thesis, Calcutta University, Unpublished).

Sinclair, J.(1955) The Flora of Cox's Bazar, East Pakistan. *Bull. bot. Soc. Bengal 9*(1): 84-116.

Steenis, C.G.G.J.van(1948-1960). *Flora Malesiana*. Djakarta & Groningen.

Steenis, C.G.G.J.van (1948). Aponogetonaceae. *Fl. Males.* ser. *I,4:* 1-12.

Voigt. J.O. (1845). *Hortus Suburbanus Calcuttensis*, Calcutta,

Wallich, N. (1828-1849). *A numerical list of dried specimens of Plants in the East India Company's Museum*. London (usually referred to as Wallich Catalogue).

Wallich, N.(1840-1853). *Icones Plantarum Indiae Orientalis or Figures* Of Indian Plants. 6 Vols. Madras.

Ward, H.B. & G.C. Whipple (1959). *Freshwater Biology*,(ed.2). W.T. Edmondson, John Wiky & Sons, inc. N.Y. London.

Watt, G. (1889-1896). *The Dictionary of The Economic Products of India*. 6 Vols. Calcutta.

Wight, R. (1838-1853). *Icones Plantarum Indiae Orientalis*, Madras, 6 Vols.

Willdenow, C.L. (1797-1830). *Coroli a Linne Species Plantarum editio Quarta etc*. ed.5. Vols. 6. Berlin.

Willis, J.O. (1966). *A Dictionary of Flowering Plants and Ferns*. 7th. ed. Revised by H.K. Airy Shaw.

Wilson, F.(1950). *Biological Control of Weeds,* New Biology Penguin Books *8*: 51-74.

Wolverton, B.C.(1976). *Water Hyacinth for Upgrading Sewage Lagoons to Meet Advanced Waste Water Treatment Standards*. Part-III, Nasa Technical Memorandum T.M.-X-72730, Washington,DC.

INDEX TO THE BOTANICAL NAMES OF THE LOWER GANGA PLANTS

(Valid Families and Genera names are in capitals; valid Species are italics)

(Species marked with Asterisk '*' are mangroves/halophytes)

INDEX TO VERNACULAR NAMES (BENGALI)

INDEX TO THE ECONOMIC IMPORTANCES OF THE LOWER GANGA DELTA PLATNS

Economic Importances *Dealt in the Pages*